Springer Proceedings in Physics 58

Springer Proceedings in Physics

Managing Editor: H. K. V. Lotsch

44 *Optical Fiber Sensors*
Editors: H. J. Arditty, J. P. Dakin, and R. Th. Kersten

45 *Computer Simulation Studies in Condensed Matter Physics II: New Directions*
Editors: D. P. Landau, K. K. Mon, and H.-B. Schüttler

46 *Cellular Automata and Modeling of Complex Physical Systems*
Editors: P. Manneville, N. Boccara, G. Y. Vichniac, and R. Bidaux

47 *Number Theory and Physics*
Editors: J.-M. Luck, P. Moussa, and M. Waldschmidt

48 *Many-Atom Interactions in Solids*
Editors: R .M. Nieminen, M. J. Puska, and M. J. Manninen

49 *Ultrafast Phenomena in Spectroscopy*
Editors: E. Klose and B. Wilhelmi

50 *Magnetic Properties of Low-Dimensional Systems II: New Developments*
Editors: L. M. Falicov, F. Mejia-Lira, and J. L. Morán-López

51 *The Physics and Chemistry of Organic Superconductors*
Editors: G. Saito and S. Kagoshima

52 *Dynamics and Patterns in Complex Fluids: New Aspects of the Physics-Chemistry Interface*
Editors: A. Onuki and K. Kawasaki

53 *Computer Simulation Studies in Condensed Matter Physics III*
Editors: D. P. Landau, K. K. Mon, and H.-B. Schüttler

54 *Polycrystalline Semiconductors II*
Editors: J. H. Werner and H. P. Strunk

55 *Nonlinear Dynamics and Quantum Phenomena in Optical Systems*
Editors: R. Vilaseca and R. Corbalán

56 *Amorphous and Crystalline Silicon Carbide III*
Editors: G. L. Harris, M. G. Spencer, and C. Y.-W. Yang

57 *Evolutionary Trends in the Physical Sciences*
Editors: M. Suzuki and R. Kubo

58 *New Trends in Nuclear Collective Dynamics*
Editors: Y. Abe, H. Horiuchi, and K. Matsuyanagi

59 *Exotic Atoms in Condensed Matter*
Editors: G. Benedek and H. Schneuwly

60 *The Physics and Chemistry of Oxide Superconductors*
Editors: Y. Iye and H. Yasuoka

61 *Surface X-Ray and Neutron Scattering*
Editors: H. Zabel and I. K. Robinson

62 *Surface Science: Lectures on Basic Concepts and Applications*
Editors: F. A. Ponce and M. Cardona

63 *Coherent Raman Spectroscopy: Recent Advances*
Editors: G. Marowsky and V. V. Smirnov

Volumes 1–43 are listed on the back inside cover

Y. Abe H. Horiuchi
K. Matsuyanagi (Eds.)

New Trends in Nuclear Collective Dynamics

Proceedings of the Nuclear Physics Part
of the Fifth Nishinomiya-Yukawa
Memorial Symposium, Nishinomiya, Japan,
October 25 and 26, 1990

With 90 Figures

Springer-Verlag
Berlin Heidelberg New York
London Paris Tokyo
Hong Kong Barcelona
Budapest

Professor Dr. Yasuhisa Abe
Yukawa Institute for Theoretical Physics,
Kyoto University, Kyoto 606, Japan

Professor Dr. Hisashi Horiuchi
Professor Dr. Kenichi Matsuyanagi
Department of Physics,
Kyoto University, Kyoto 606, Japan

ISBN-13:978-3-642-76381-6 e-ISBN-13:978-3-642-76379-3
DOI: 10.1007/978-3-642-76379-3

Softcover reprint of the hardcover 1st edition 1992

Typesetting: Camera ready by authors

54/3140-543210 – Printed on acid-free paper

Preface

This volume contains written versions of the lectures presented at the fifth Nishinomiya-Yukawa Memorial Symposium on "New Trends in Nuclear Collective Dynamics" held on October 25 and 26, 1990 in Nishinomiya.

Major subjects discussed are

Order and Chaos in Finite Quantum Systems,
Dissipation in Heavy-Ion Collisions,
Damping Mechanism of Collective Motions in Warm Nuclei,
Time-Dependent Mean-Field Theory with Collision Terms,
Nuclear Fissions and Multi-Dimensional Tunnelings,
Large-Scale Collective Motions,

in which significant progress has been made in recent years. Special emphasis is placed on promoting research toward understanding collective and statistical aspects of nuclear dynamics in a unified way. We are very pleased to have had outstanding invited lecturers, who have all played important roles in this progress. Together with the workshop subsequently held on October 29–31, 1990, at the Yukawa Institute for Theoretical Physics (YITP), this symposium attracted many young nuclear physicists in Japan, and provided them with an opportunity to hold discussions with the invited lecturers in an informal atmosphere.

We thank all the invited lecturers for giving their inspiring lectures and for the trouble they took, to ensure that their lecture notes were excellent reviews for *New Trends in Nuclear Collective Dynamics*. We believe that this volume will be very useful, especially to young nuclear physicists; many open problems of fundamental significance are suggested here as well as possible future directions of nuclear physics. Three papers by Professor Weidenmüller are included in the form of reprints, since their contents constituted parts of his lecture. We thank Elsevier Science Publishers for permission to reproduce the papers. Professor Berry has two contributions: a guide to recent publications, which is a substitute for the main part of his lecture note, and an original paper, which describes his recent work briefly discussed in his lecture. We deeply regret that Professor Bohigas' lecture note was not received in time for inclusion.

The organizing committee of the symposium consisted of:

Yasuhisa Abe (YITP, Kyoto University)
Masayasu Ishihara (University of Tokyo and RIKEN)
Hisashi Horiuchi (Kyoto University)
Masayuki Matsuo (YITP, Kyoto University)
Kenichi Matsuyanagi (Kyoto University, Chairman)
Hidetoshi Nishioka (Konan University)
Noboru Takigawa (Tohoku University)

To celebrate the quinquennial, the fifth Nishinomiya-Yukawa Symposium was organized such that it consisted of two parts in parallel: nuclear physics and high-energy physics. The title of the latter symposium was "Present and Future of High-Energy Physics" and its proceedings will be published as a separate volume in the series "Springer Proceedings in Physics".

The symposium was sponsored by Nishinomiya City, where Professor Hideki Yukawa was living when he wrote his paper on the meson theory in 1935, and was supported by the Yukawa Institute for Theoretical Physics, Kyoto University, and by the Physical Society of Japan. We express our sincere gratitude to Mr. Yoneji Yagi, mayor of Nishinomiya City, for his generous support. We thank the staff members of the Nishinomiya Municipal Comprehensive Education Center for their excellent work supporting the symposium. Finally, we thank Ms. Naomi Tezuka and Mrs. Kiyoe Yokota for their help with the editorial work.

Kyoto,
February 1991

Yasuhisa Abe
Hisashi Horiuchi
Kenichi Matsuyanagi

Contents

Collective Versus Statistical Aspects of Nuclear Motion
By H.A. Weidenmüller 1

Tests of Fundamental Symmetries in Compound-Nucleus Scattering
By H.A. Weidenmüller (With 4 Figures) 11

Chaos in Nuclei
By H.A. Weidenmüller (With 4 Figures) 21

Ericson Fluctuations Versus Conductance Fluctuations:
Similarities and Differences
By H.A. Weidenmüller (With 4 Figures) 31

Hot Nuclei – Landau Theory, Thermal Fluctuations
and Dissipation
By Y. Alhassid (With 19 Figures) 41

Dissipation and Thermal Fluctuations in Heavy-Ion Collisions
By P. Fröbrich (With 10 Figures) 93

Multi-Dimensional Tunneling and Nuclear Fission
By A. Iwamoto (With 4 Figures) 125

Phase-Space Dynamics of Heavy-Ion Reactions
By W. Cassing, S. Hannuschke, U. Mosel, K. Niita,
and G. Wolf (With 17 Figures) 143

Quantum Chaology: Our Knowledge and Ignorance
By M.V. Berry 177

True Quantum Chaos? An Instructive Example
By M.V. Berry 183

Toward the Fundamental Theory of Nuclear Matter Physics:
The Microscopic Theory of Nuclear Collective Dynamics
By F. Sakata, T. Marumori, Y. Hashimoto, H. Tsukuma, Y. Yamamoto,
J. Terasaki, Y. Iwasawa, and H. Itabashi (With 12 Figures) 187

Mean Field Dynamics and Low-Lying Collective Excitations
By M. Matsuo (With 10 Figures) 219

Rotational Motion in Warm Atomic Nuclei
By T. Døssing and B. Herskind (With 6 Figures) 239

Collectivity and Chaoticity in Nuclear Dynamics
By V.G. Zelevinsky 255

Index of Contributors 297

Collective Versus Statistical Aspects of Nuclear Motion*

H.A. Weidenmüller

Max-Planck-Institut für Kernphysik,
W-6900 Heidelberg, Fed. Rep. of Germany

*Dedicated to Professor H.J. Mang on the occasion of his 60th birthday

The interplay between collective and statistical features in nuclear spectroscopy and nuclear reactions is discussed. Collectivity and stochasticity appear as the only two alternative forms of nuclear motion. They must be considered as equally important and they relate to fundamental aspects of classical Hamiltonian dynamics: regularity versus chaos.

1. Introduction: Aim

In this first contribution to the 1990 Yukawa Memorial Symposium on Nuclear Physics, I intend to give a survey of current themes at the interface between collective dynamics and statistical theory. The present survey is supplemented by reprints of three contributions I wrote in recent months on related topics. In particular, I wish to show

(i) that collectivity and statistics are the two fundamental, mutually complementary aspects of nuclear motion. Here, collectivity stands for all effects relating to single-particle motion, to collective excitations of RPA type, and to large-amplitude collective motion as encountered in superdeformed nuclei or in heavy-ion reactions. It relates to ordered or regular motion. Statistics, on the other hand, stands for the typical behaviour of the vast majority of non-collective levels: local fluctuation properties of eigenvalues and eigenvectors of the GOE type (where GOE is the Gaussian orthogonal ensemble of random matrices) relating to chaotic motion.

The point of view that collectivity and statistics are equally relevant for the understanding of nuclear motion is of fairly recent origin, and signals a change in attitude and thinking of the nuclear physics community.

(ii) that collectivity and statistics are manifest in nuclei in many different ways, making their joint study an enormously rich field, both experimentally and theoretically.
This statement bears testimony to the fact that low-energy nuclear physics is a field full of challenging problems.

(iii) that many features of and problems in nuclei have closely similar counterparts in other areas of physics: atomic and molecular physics, physics of clusters, problems of wave propagation in irregularly shaped cavities, speckle interferometry in astrophysics, conductance properties of disordered solids, to name but a few.

This indicates that we must not be parochial. We can learn much about our own problems from progress in other fields, and should strive to make our results interesting and known to people working in other areas so that we can contribute to their work.

Springer Proceedings in Physics, Vol. 58 New Trends in Nuclear Collective Dynamics
Eds: Y. Abe, H. Horiuchi, K. Matsuyanagi

(iv) that the theory of collectivity and statistics is incomplete in many ways, and poses difficult and interesting questions, some related to fundamental problems of present-day theoretical physics.

Taking a step back from an innocent-looking, data-oriented problem one may find that it turns into a deep problem which is of general significance. The abstraction required in defining a theoretical problem brings out its generic features and unifies physics.

To accomplish these aims, I touch upon a number of select phenomena and problems, without going into the details of any one of them. Such details may either be found in the references, or the supplementary reprints following the present paper.

The manner in which collectivity and statistics shape nuclear spectra and nuclear reactions is the topic of sections 2 and 3, respectively. The ensuing technical problems in modelling the stochastic dynamics are addressed in section 4. Ways of attaining an understanding from first principles of the nuclear dynamics form the content of section 5. Section 6 contains a brief summary.

2. Spectra: Collectivity, Chaos, and Level Statistics

Starting point of our discussion is a conjecture originally formulated by Bohigas, Giannoni and Schmit[1] and meanwhile well supported by a variety of numerical and analytical studies. It says:

Consider a closed quantum system S with time-reversal symmetry. Assume that in the classical limit ($h \to 0$), S is fully chaotic, with all parts of phase space equally accessible to any chaotic trajectory. Then, in the semiclassical regime the local fluctuations of eigenvalues and eigenvectors of S that have fixed symmetry coincide with the fluctuation properties of the same quantities for the Gaussian Orthogonal Ensemble (the "GOE").

Lack of space does not allow me to discuss either the justification for, or the open questions connected with this hypothesis[2]. In the investigation of experimental or of computer-generated spectra, one often uses the converse of this hypothesis:
An experimental or computer-generated spectrum with spectral fluctuations of the GOE type indicates chaotic dynamics.

I recall that the analysis of nuclear data[3] obtained by the study of neutron s-wave resonances in the 1970 s by Rainwaters et al., and by high-resolution proton spectroscopy in the late 1970 s by the TUNL group, showed clear evidence for GOE type fluctuations of levels and partial width amplitudes. Therefore, the statistical nuclear properties established at excitation energies of several MeV support the idea of chaotic nuclear dynamics.

At the same time it goes without saying that we have very strong evidence for collective motion in nuclei, both in the ground-state domain and at excitation energies of several MeV. Since collective models typically are simple, integrable models, this evidence supports the idea that the nuclear dynamics is at least partly regular. Examples are

(i) the single-particle states, rotational bands and vibrational states in the ground-state domain;

(ii) the maxima in the s-wave neutron strength function at neutron threshold, the giant resonances for electromagnetic and Gamov-Teller transitions, and the superdeformed bands discovered recently;

(iii) the formation of quasimolecules in heavy-ion reactions and the dynamics of deeply inelastic heavy-ion reactions with their characteristic dissipation of angular momentum and energy, suggesting a weak coupling between collective and stochastic degrees of freedom.

We conclude that all this evidence - to which we could easily add more - shows that collectivity and chaos are the dominant features of nuclear motion: A comprehensive picture of nuclear physics is obtained by adding to the comparatively few and select collective modes of nuclear excitation, the multitude of others which are stochastic, and by studying the interplay between these two types of behaviour.

Much of the current research into chaotic dynamics, both within and outside the realm of nuclear physics, is related to the provisos used in the formulation of the conjecture by Bohigas, Giannoni and Schmit, given above. For instance, the restriction to closed systems has been lifted in recent years, and the investigation of open systems, both classically and quantum-mechanically, has founded the new field "chaotic scattering", of relevance to Sect. 3 of the present paper. The restriction to time-reversal symmetric systems can be lifted. In particular, it is of interest to study chaotic systems in which this or another fundamental symmetry like parity is weakly broken. Lack of space prevents me from dealing with this interesting topic here; it forms the subject of one of the accompanying papers. - The restriction to fully chaotic systems has recently been lifted, and the dynamics of partly chaotic systems has been found to offer fascinating aspects. This issue is taken up in section 5, and in the talk by Bohigas given at this meeting.

In the remainder of the present section, I wish to dwell on current research related to the provisos that "all parts of phase are equally accessible", that we study fluctuation properties "in the semiclassical regime", and that we consider "eigenvalues and eigenvectors of states with fixed symmetry". Finally, I will also dwell on the universality of GOE type fluctuations implied by the conjecture.

The assumption that all parts of phase space be equally accessible is not always met. A counterexample is provided by isospin symmetry breaking in light nuclei. A case in point is ^{26}Al, where a complete set of 100 levels of positive parity is known between the ground state and about 8 MeV excitation energy, of which 75 levels have isospin $T = 0$, the rest has $T = 1$. The fluctuation properties of these levels have been examined by G. E. Mitchell et al.[4], and the question arose: how does the (weak) isospin symmetry breaking caused by the Coulomb force influence the spectral statistics? Analysis of the data yielded a Δ_3 statistic with values between that of two superposed GOE's (typical for the superposition of two non-interacting random spectra) and that of a single GOE (typical for complete isospin symmetry breaking).

A theoretical model[5] for this problem uses a Hamiltonian matrix of dimension $N_0+N_1 = 100$, with $N_0 = 75$ and $N_1 = 25$. The two diagonal blocks of dimension N_0 and N_1 model the $T = 0$ and $T = 1$ states, respectively, and consist of two uncorrelated GOE matrices, with variances chosen in such a way as to reproduce the mean level spacings d_0 and d_1. The matrix elements in the off-diagonal blocks are taken to be uncorrelated

Gaussian random variables (in as much as they are not identified by the requirements of Hermitecity and reality) with common variance v^2. Changing v^2 from zero to a maximum value, we change the theoretical model from that of two disjoint GOE's to that of a single one. Fitting v^2 numerically to the data for ^{26}Al yields[8] $v \simeq 20$ keV and a spreading width $\Gamma^{\downarrow} = 2\pi v^2/d_0 \simeq 35$ keV , in keeping with previous work[6] on isospin symmetry breaking in nuclei.

Up to this point I have described a standard statistical-model analysis of a nuclear physics problem, seemingly not related very closely to the general problems addressed above. A moment's reflection shows, however, that the problem is of general interest, for the following reasons. (i) With $v \simeq 20$ keV and $d_0 \simeq 60$ keV, we have $v < d_0$. This shows that a small symmetry-breaking interaction may cause a major change in the fluctuation pattern and indicates the great sensitivity of stochastic tests to a symmetry violation, as outlined in more detail in one of the accompanying papers[7]. (ii) In the limit $N_0, N_1 \rightarrow \infty$ and $h \rightarrow 0$ (classical limit), not all parts of phase space are equally accessible: The mixing time between the two parts of phase space corresponding to $T = 0$ and $T = 1$ depends on the mixing parameter v. This connects to the problem of Hamiltonian transport in section 5. (I do not pay attention here to the fact that discrete symmetries lose their meaning in the context of classical theory.) (iii) The analytical problem posed by the mixing of two GOE's (which likewise arises in the theory of precompound reactions involving two classes of exciton states) is very difficult and yet unsolved, even in the limit $N_0, N_1 \rightarrow \infty$. It is a very interesting problem, however, because its solution would pave the way towards an analytical understanding of the localization problem in quasi one-dimensional mesoscopic devices, or for electromagnetic waves in a one-dimensional array of irregularly shaped cavities, with nearest neighbours linked by thin pipes. To see this connection, consider two irregularly shaped cavities connected by a thin pipe. In case the pipe is absent, the spectrum of standing electromagnetic waves in each cavity has GOE type fluctuations, because of the irregular shape of the cavity. Linking the two cavities by a thin pipe accomplishes the same effect as the introduction of the mixing parameter v^2 in the model of two coupled GOE's described above. Linking many, rather than two, cavities with each other changes this problem into one of the forms of the localization problem for quasi one-dimensional systems, and poses the question: are the eigenfunctions spread out more or less uniformly over all cavities, or are they each localized over a finite number of cavities, with exponential tails beyond this region (localization)?

The second of the above-mentioned provisos states that one considers fluctuation properties of eigenvalues and eigenfunctions "in the semiclassical regime". This in fact is a tautology: to establish the statistics of levels and wavefunctions, it is necessary to investigate a very large number, and this implies going into a semiclassical regime. I have nevertheless used this formulation because much of the recent work on nuclear eigenvalue fluctuations has concentrated on the ground-state domain. A brief review of this work, and further references may be found in one of the accompanying papers[8].

The restriction to "eigenvalues and eigenvectors of states of fixed symmetry" is too rigid for nuclear physics applications where collectivity mainly manifests itself as a strong enhancement of the transition strength between levels carrying different quantum numbers. This point is further discussed in ref.[7].

The universality of the GOE type fluctuations implied by the conjecture is a most remarkable feature because it applies to any Hamiltonian irrespective of any dynamical details as long as the dynamics is chaotic. This makes it possible to model the Hamiltonian (or parts of it) in terms of GOE matrices. (For details, see the examples given in section 3, the isospin-mixing problem discussed above, and the talk by O.Bohigas at this conference). The merit of such a procedure and its practical consequences are considerable: Finding the eigenvalues and eigenfunctions of a classically chaotic Hamiltonian is a very difficult and demanding task because the instability of the classical trajectories is paralleled by a substantial instability of the quantum problem under small perturbations like numerical rounding errors. Replacing this task by the need to solve a stochastic matrix problem which is known to give the same fluctuation properties amounts to a great simplification. Therefore, random-matrix modelling, which is very useful for the description of disordered solids, finds a meaningful place also in the theory of classically chaotic systems. Such modelling, however, requires the calculation of ensemble averages; technical tools developed for this task are briefly mentioned in section 4.

3. Reactions: The Universality of Fluctuations

Stochastically fluctuating observables are met in many areas of physics, even when the system studied is isolated (not coupled to a heat bath). Such fluctuations have one of two causes: the dynamics of the system is chaotic, or the system is disordered. In the latter case, electrons moving through a crystal encounter dislocations and/or impurities by which they are scattered classically; the precise location of these scattering centres is unknown; the scattering process therefore is simulated by a random potential. A similar situation arises in the passage of laser light through a medium with an irregularly varying index of refraction.

In the case of scattering or reaction processes, the stochastic fluctuations are manifest in very similar features of the intensities. Examples are:

(i) The functional dependence of the total CN cross-section for neutron scattering on medium-weight and heavy nuclei versus energy in the domain of non-overlapping resonances, i.e. at neutron energies up to $\sim$ 100 keV above threshold. Here, the mean level spacing d is large compared to the average width Γ of the resonances, and isolated peaks in the cross-section mark the positions of the s-wave resonances[9]. Although the experimental cross-sections are deterministic and reproducible, the fluctuations caused by the resonances are, from the theoretical point of view, unpredictable and random: as described at the end of section 3, a theoretical effort at reproducing the cross-section would require both a knowledge of the nuclear Hamiltonian and an accuracy of the numerical calculation better than the average level spacing $d \approx$ 10 eV. Such a goal is unattainable. Moreover, it is not worth attaining because of the known fact that the resonances show GOE fluctuation properties which implies that individually they carry no information content.

(ii) Ericson fluctuations[10] of nuclear reaction cross-sections in the domain of overlapping CN resonances ($\Gamma > d$). Many of the statements under (i) remain valid except that now peaks (valleys) in the cross-section are not caused by the presence (absence) of isolated resonances but rather by the constructive (destructive) interference of many ($\approx \Gamma/d$) resonances with random amplitudes.

(iii) Non-overlapping resonances seen in the reflection of microwaves on a billiard in the shape of a stadium[11]. A microwave is fed via a cable into a cavity shaped like a quarter of a stadium. The reflected intensity is measured as a function of microwave frequency. The spacings between the resonances show fluctuations of GOE type, consistent with the fact that the classical motion in the stadium is chaotic.

(iv) Overlapping resonances seen in the intensity of resonance fluorescense light in molecules[12]. Organic molecules are excited into a singlet level by laser light. Changing a strong external magnetic field, one shifts the $m = \pm 1$ components of triplet states past the singlet state. At each anticrossing, the resonance fluorescense intensity of the singlet state shows a resonance-like minimum. When the density of triplet states is sufficiently high, the resonances overlap. The fluctuation patterns are consistent with GOE predictions both for overlapping and for some non-overlapping resonance cases, thereby signalling that the dynamics in these molecules is chaotic.

(v) Speckle patterns seen in the transmission of light through media with an irregularly varying index of refraction[13]. This is a case of overlapping resonances in an impurity scattering problem; it is of considerable interest in astronomy.

(vi) Conductance fluctuations in mesoscopic systems[14]. Wires or semiconductor devices are cooled down below 1 K so that typical length scales, amounting to μm, are larger than the inelastic mean free path for electron-phonon scattering. The transmission of electrons through the sample is affected only by elastic impurity scattering. Conductance fluctuations are seen by changing the gate voltage or an external magnetic field and correspond to overlapping resonances. The phenomenon is of interest in applied physics and because of its wave-mechanical aspects. It obviously connects to the localization problem.

In any of these phenomena (with the exception of (iii)), it is not possible and would not be of interest to reproduce exactly from a theoretical calculation the experimental observations. Rather, interest focuses

(a) on the average intensity (averaged over many mean spacings d between resonances) and its dependence on external parameters like length or magnetic field strength;

(b) on the variance of the intensity as a measure of the magnitude of the fluctuations and its dependence on external parameters; in particular one wishes to understand why the intensity is not self-averaging in the semi-macroscopic devices described under (v) and (vi) and whether the transition to localization can be observed and understood;

(c) on higher moments of the intensity or the entire probability distribution for the intensity, especially near the localization transition;

(d) on the intensity autocorrelation function as a measure of the width of the fluctuations and the dependence of this width on external parameters;

(e) on the relation between these quantal features and classical concepts like delay time and the Ljapunov coefficient.

Examples (i) to (vi) and points (a) to (e) define a field of intense studies involving many areas of physics. In nuclear physics, one is also interested in the interplay between such statistical processes and collective features. Examples are inelastic neutron scattering in the CN domain to a final 2^+ collective state; deexcitation of the giant dipole resonance in a proton-induced pre-equilibrium reaction; the coupling between collective and stochastic degrees of freedom manifest in the appearance of transport coefficients in the theory of heavy-ion reactions.

In all the cases (i) - (vi) listed above, the intensity is given by the square of a wave-mechanical amplitude S_{ab}, the element of the scattering matrix connecting suitably defined channels a and b, or by a sum over such squares. A theoretical description might therefore be based on a suitable model for S_{ab} which takes account of the stochastic features of the system caused by chaos or disorder, and which describes the resonances properly. This is possible by using the unitary N-level formula

$$S_{ab} = \delta_{ab} - 2i\pi \sum_{\mu,\nu=1}^{N} W_{a\mu} \left[D^{-1}\right]_{\mu\nu} W_{\nu b} \,. \tag{3.1}$$

Here, $W_{a\mu}$ is the amplitude coupling channel a with level μ, and D^{-1} is the propagator matrix given by its inverse

$$D_{\mu\nu}(E) = E\delta_{\mu\nu} - H_{\mu\nu} + i\pi \sum_{c} W_{\mu c} W_{c\nu} \,, \tag{3.2}$$

with E the energy and $H_{\mu\nu}$ the Hamiltonian in the N-dimensional subspace of quasi-bound states. Expression (3.1) is manifestly unitary. For $N = 1$, it coincides with the Breit-Wigner single-level formula where H_{11} is the resonance energy and $2\pi \sum_c W_{1c} W_{c1}$ is the resonance width.

To endow $S_{ab}(E)$ with stochastic features, it is convenient and completely in line with the arguments of section 2 to consider $H_{\mu\nu}$ as a random matrix. The precise random-matrix modelling of $H_{\mu\nu}$ depends on the system considered. For cases (i), (ii) and (iii), $H_{\mu\nu}$ is taken as a single GOE. In the case (iv), special account has to be taken of the singlet state as a doorway state to the GOE-type triplet states. In cases (v) and (vi), the length of the system must be taken into account by modelling $H_{\mu\nu}$ as a block matrix, each diagonal block containing a GOE matrix, with Gaussian distributed coupling matrix elements connecting neighbouring blocks. Each block represents the Hamiltonian over a piece of the disordered system of length l where l is the elastic mean free path. These examples show that finding a random-matrix model for $H_{\mu\nu}$ which is suitable for the problem under consideration is usually easy.

The calculation of average intensities, intensity variances and all other quantities mentioned under (a) to (e) from eqs. (3.1) , (3.2) and the assumed stochastic model for $H_{\mu\nu}$ is a difficult theoretical task which has not been accomplished completely in all cases. Partial answers do exist, however; in almost all cases they have been obtained in terms of an asymptotic expansion in N^{-1} where $N \gg 1$ is the number of resonances in the system. Limitations of space do not allow me to describe these answers here. In addition to the references given in conjunction with examples (i) to (vi), I mention for further reading ref.[15] for cases (i) to (iii), and ref.[16] for cases (v) and (vi). Moreover, much work relating to cases (v) and (vi) has been done by Altshuler et al.[17], Imry et al.[18] and Lee et al.[19]; these authors have not taken the point of view that the physical process is described as a scattering problem. A comparison between Ericson fluctuations (case (ii)) and conductance fluctuations (case (vi)) emphasizing physical similarities and differences is given in an accompanying paper[20].

4. Tools

The phenomena listed under (i) to (vi) in section 3 have led us to the formulation of a stochastic scattering matrix. In all cases, it has the form of eqs. (3.1) and (3.2) . The Hamiltonian is modelled as a random matrix of a form which is suitable for the specific situation.

Intensities are squares of amplitudes. To calculate average intensities, second moments of intensities, or the intensity autocorrelation function, we must calculate averages over products of up to four S-matrix elements each of which depends in a complex way on a large number ($N \to \infty$) of random variables. In the present section, I mention ways of coping with this problem. Basically, three procedures are available: (i) Monte-Carlo simulation; (ii) a disorder perturbation expansion; (iii) the method of the generating functional.

In Monte-Carlo simulations, one calculates the average over a finite set of S-matrices which is meant to represent the ensemble. The set is generated numerically from random drawings of the random elements in the Hamiltonian. This method is straightforward for the average intensity, becomes involved for the second moment of the intensity, and almost prohibitively time-consuming for the autocorrelation function. It is not suitable for investigating the parameter-dependence of average quantities. Often it is helpful, however, in establishing trends, checking analytical approximations, and in detailed comparison with experiment.

The disorder perturbation expansion uses a Born-series expansion of the scattering matrix in terms of the random part of the Hamiltonian. Averages of products of elements of S are calculated by making use of the Gaussian distribution of the random variables: by calculating all pairwise contractions of random variables in the series. This leads to a diagrammatic representation of the algorithm. Results are obtained by partial resummation of the resulting series. The method hinges on the existence of a small parameter.

The method of the generating functional expresses S_{ab} as the logarithmic derivative (with respect to an auxiliary variable J_{ab}) of a generating functional $Z(J_{ab})$, evaluated at the point $J_{ab} \equiv 0$. This causes the random elements in the Hamiltonian to appear in the exponent of the integrand in the expression for $Z(J_{ab})$. Therefore, calculating the average of Z or of products of Z, is trivial. But it is not trivial to calculate the average over $\ln Z(J_{ab})$, the entity defining S_{ab} by differentiation.

Two ways exist to overcome this difficulty. The replica trick uses the identity

$$\ln Z = \lim_{n \to \infty} \left[\frac{1}{n} (Z^n - 1) \right]$$

and determines the average of $\ln Z$ from the average of Z^n calculated for positive integer values of n. The replica trick is obviously not a rigorous procedure; the analysis of simple cases (where the exact solution is known) suggests that its domain of validiy is similar to that of the impurity diagram technique: The existence of a small parameter is required.

The method of anticommuting integration variables (Grassmann integration) normalizes $Z(0)$ to unity by introducing integrals over anticommuting variables. This is the only method known which leads to a complete analytical solution of the CN problem. For $N \to \infty$, the average of $Z(J_{ab})$ is mapped onto a non-linear sigma model. For disordered systems of infinite extension, one obtains in this way a field-theoretical model. Given a small parameter, asymptotic solutions can be constructed by perturbation expansion[15,16].

5. Theoretical Foundations

To obtain a deeper understanding of the interplay between collective and stochastic motion in nuclei, beyond the phenomenological modelling described above, it is perhaps useful to be guided by some recent developments in the Hamiltonian theory of partially chaotic systems, worked out for two degrees of freedom systems, and their quantum analogues[21]. This topic has been covered at the workshop following the Nishinomiya symposium by O. Bohigas; I have referred to it in one of the accompanying papers[8]. I believe that it is most interesting to see that the chaotic domains of phase space possess internal structure, and that this structure is also quantum-mechanically visible where it suggests a random-matrix modelling of the type mentioned in section 3 for various physical systems. It seems to me that in this way, a solid and deep understanding of random-matrix modelling of physical systems can be attained, at least for sufficiently simple examples but perhaps also in general.

And where does collectivity come in? Most theoretical efforts have aimed at deriving nuclear collective motion from the TDHF approximation. The link between these ideas and the ideas of Hamiltonian transport referred to above is still missing, and provides a challenge to nuclear and many-body theory.

6. Conclusions

I hope to have fulfilled my promise given in the introduction. The combination of collective and stochastic aspects of the dynamics which we face in nuclear physics makes this science into a particularly rich field, with many unsolved and challenging problems.

References

1. O. Bohigas, M.J. Giannoni and C. Schmit: Phys. Rev. Lett. **52**, 1 (1984)
2. Cf. the review by M.J. Giannoni and O. Bohigas in Lecture Notes in Phys. **209** (1984) and the talk by M. Berry at this Conference

3. U. Haq, A. Pandey and O. Bohigas: Phys. Rev. Lett. **48**, 1086 (1982)
4. G.E. Mitchell, E.G. Bilpuch, P.M. Endt and J.F. Shriner: Phys. Rev. Lett. **61**, 1473 (1988)
5. T.Guhr and H.A. Weidenmüller: Ann. Phys. **199**, 412 (1990)
6. H.L. Harney, A. Richter and H.A. Weidenmüller: Rev. Mod. Phys. **58**, 607 (1986)
7. H.A. Weidenmüller, Nucl. Phys. A522, 293c (1991)
8. H.A. Weidenmüller, Nucl. Phys. A520, 509c (1990)
9. A. Bohr & B.R. Mottelson: Nuclear Structure I, Benjamin, Reading (1969)
10. T.O. Ericson and T. Mayer-Kuckuck: Ann. Rev. Nucl. Sci. **16**, 183 (1966)
11. H.J. Stöckmann and J. Stein: Phys. Rev. Lett. **64**, 2215 (1990)
12. L. Leviandier, M. Lombardi, R. Jost and J.P. Pique: Phys. Rev. Lett. **56**, 2449 (1986)
13. I. Freund, M. Rosenbluh, S. Feng: Phys. Rev. Lett. **61**, 2328 (1988)
14. S. Washburn and R.A. Webb: Adv. Phys. **35**, 375 (1986)
15. J.J.M. Verbaarschot, H.A. Weidenmüller and M.R. Zirnbauer: Phys. Rep. **129**, 367 (1985)
16. S. Iida, H.A. Weidenmüller and J. Zuk: Ann. Phys. (N.Y.) **200**, 219 (1990)
17. B. Altshuler and B.I. Shklovski: Sov. Phys. JETP (Engl. Transl.) **64**, 127 (1986)
18. J. Imry: Europhys. Lett. **1**, 249 (1986)
19. P. Lee, A.D. Stone and H. Fukuyama: Phys. Rev. B **35**, 1039 (1987)
20. H.A. Weidenmüller: Nucl. Phys. A **518**, 1 (1990)
21. O. Bohigas , S. Tomsovic and D. Ullmo: Phys. Rev. Lett. **64**, 1479 (1990) and **65**, 5 (1990)

Tests of Fundamental Symmetries in Compound-Nucleus Scattering *

H.A. Weidenmüller

Max-Planck-Institut für Kernphysik,
W-6900 Heidelberg, Fed. Rep. of Germany

In compound-nucleus reactions, the violation of parity and/or time-reversal symmetry is enhanced by several orders of magnitude. I present a brief review of facts and causes, and discuss recent progress in experiment and theory.

1. INTRODUCTION

In the last decade, it has been established that compound-nucleus (CN) reactions provide an extremely sensitive testing ground for the violation of a fundamental symmetry like parity. Experimentally, a very large enhancement of parity violation was found in the first half of the 1980's by the Dubna group[1]. This work was based on a suggestion by Sushkov and Flambaum[2] and later in its essentials confirmed by experiments at IAE[3], at KEK[4] and at Los Alamos[5]. The violation of parity was tested by a measurement of the asymmetry parameter

$$\varepsilon = \frac{N_+ - N_-}{N_+ + N_-} , \tag{1}$$

where $N_\pm$ denotes the number of neutrons with helicity $\pm$ transmitted by an unpolarized target. In the case of a ^{139}La target and neutrons with energy $E_n = 0.74$ eV populating a known p-wave resonance in the CN ^{140}La, ε was found[4,5] to be about 10 per cent. We recall that the strength of the weak parity-violating interaction is about 10^{-7} times that of the strong interaction and observe that a value $\varepsilon \cong 0.1$ signals an enhancement of about 5 orders of magnitude, indicating the enormous possibilities offered by CN reactions to test fundamental symmetries [parity (P), time reversal-symmetry (T), and the combination (PT) of both]. In addition to large enhancement factors, CN reactions also offer the advantage that data on symmetry breaking directly lead to a determination of (or a bound on) the relative strength α of the symmetry-breaking part of the effective nucleon-nucleon interaction in the shell model.

What is the interest in such tests? The answer depends on the point of view taken, and on the symmetry under consideration. From the point of view of many-body physics, the CN is strongly chaotic. It is of interest to study the action of a symmetry-breaking force in such a system. From the point of view of symmetry violation, we must look at P, T and PT separately. We denote by α_P, α_T, α_{PT} the strength (relative to the strong interaction) of the effective nucleon-nucleon interaction in the shell model that violates P, T, and PT, respectively. Common to all three is the fact that the CN reaction, being stochastic, calls

* Nuclear Physics A522 (1991) 293c-304c © 1991 – Elsevier Science Publishers B.V. (North-Holland)

for a statistical evaluation of the data. This fact obviates the need to know wavefunctions of individual nuclear levels and allows for a fairly direct determination of the α's. [Contrast this statement with the theoretical subtleties encountered in the analysis of data on parity violation from transitions between select pairs of low-lying nuclear states!].

In the case of P violation, the most recent data on ^{239}U yield[6] $\alpha_P = 4 \cdot 10^{-7}$, in rough agreement with expectations based on the relative strength of the weak interaction. Interest in P violation in CN reactions focusses on stochastic aspects and on the question whether symmetry breaking in a chaotic system is sufficiently well understood to yield a reliable value for α_P. This requires further experimental work at different energies and in other nuclei to check theoretical expectations. It also requires further theoretical work on the reduction procedure leading from the data to α_P. Finally, it is important to establish the relation between the symmetry-breaking bare nucleon-nucleon interaction, and its effective counterpart in the shell-model. A good understanding of P violation is vital if tests of T and PT are to provide reliable information.

In the case of T violation, theory predicts that CN reactions can reduce the upper bound on α_T set by the present best data on the electric dipole moment of the neutron by two orders of magnitude. This assumes the same sensitivity for the T violation experiment (transmission of polarized neutrons by an aligned target) as for the P experiment on ^{239}U. And in the case of the PT combination, there is hope that in future an experiment involving the transmission of polarized neutrons by a polarized target, may yield a positive signal. In view of the very considerable efforts undertaken in atomic and elementary-particle physics to study fundamental symmetries, it is my contention that the nuclear-physics community would be well advised to avail itself of the unique possibilities offered by CN reactions to pursue these problems as energetically as possible.

After this broad overview, I devote the main body of the present paper to a presentation of the theoretical issues encountered in tests of fundamental symmetries on CN reactions: It is necessary to attain a quantitative understanding of the enhancement effects, to identify the observables showing greatest sensitivity (including an identification of optimum energy and mass number ranges); to study the relationship between data and the strength parameters α_P, α_T, and α_{PT}, and the relation of the latter with the symmetry-breaking bare nucleon-nucleon force. I will focus attention on the domain of weakly overlapping CN resonances (average width Γ very small compared to average spacing d), for the following reasons. Theoretical work[7,8] on T violation has shown that this is a domain with strong enhancement; the arguments apply also to P violation. The necessary experimental energy resolution $\Delta E \lesssim \Gamma$ is available here. And in this domain, a set of statistically significant data[6] on parity violation exists. I will mainly address P and T for which all the extant work has been done. In both cases, the asymmetry parameter ε in eq. (1) is directly proportional[9] to the difference $\sigma_+ - \sigma_-$ of the two cross-sections for neutrons with $\pm$ polarization. In the case of P, measured on p-wave resonances, we have[9] $\sigma_+ - \sigma_- \sim \beta_P = \mathrm{Re}[S_J(0\frac{1}{2}, 1\frac{1}{2})]$ where $S_J(lj, l'j')$ is the S-matrix element for CN scattering with total spin J, angular momentum $l'(l)$ and total projectile spin $j'(j)$ in entrance and exit channels, respectively. For T violation, I consider the case where the CN is populated and depopulated by p-wave neutrons. Then[9], $\sigma_+ - \sigma_- \sim \beta_T = \mathrm{Im}[S_J(1\frac{1}{2}, 1\frac{3}{2}) - S_J(1\frac{3}{2}, 1\frac{1}{2})]$.

In the organization of this paper and in the material presented, I draw heavily on a forthcoming paper[9] by V. E. Bunakov, E. D. Davis, and myself. I am grateful to both my coauthors for all I have learned from them. I have also profited from conversations with H. L. Harney and A. Müller on parity violation. And many discussions with G. E. Mitchell during the last few years have been very helpful to me in widening my perspective and identifying relevant problems. I am grateful to the authors of refs. 6, 14 for informing me of their results prior to publication.

In section 2, I recall the model of two isolated resonances introduced by Bunakov and Gudkov[10,11,12]. This model shows the origin of the big enhancement. It also shows the need for a statistical treatment. In section 3, I recall the work by E. D. Davis and others[7,8,13,14] on ensemble-averaging. This work avoids the shortcomings of section 2. But in the domain of isolated resonances, it fails to take account of an experimental bias: Symmetry violation will only be tested where it is expected to be biggest, i.e. on identified isolated resonances. A statistical approach which includes this bias is presented in section 4. Section 5 deals with the reduction problem: How are α_P and α_T obtained from data on symmetry violation? This problem has been addressed by French et al.[15] with the methods of statistical spectroscopy. Little work seems to have been done recently on the connection between the symmetry-breaking bare nucleon-nucleon interaction and its effective counterpart in the shell model. This is why this issue receives no further mentioning in the text although I consider it an important open problem. Section 6 contains the conclusions.

2. MODEL OF TWO ISOLATED RESONANCES[10–12]

This model was designed to identify the conditions for optimum enhancement of symmetry violation. For P, we consider an s-wave and a p-wave resonance, with equal total widths Γ and a spacing $D \gg \Gamma$. Choosing the neutron energy equal to the energy of the p-wave resonance, we find that β_P (see section 1) is proportional to

$$\beta_P \sim \gamma_n^{s1/2} \gamma_n^{p1/2} \cdot \frac{1}{\Gamma} \cdot \frac{v^P}{D} . \tag{2}$$

In the case of T, we consider two p-wave resonances under the same assumptions and find β_T to be proportional to

$$\beta_T \sim (\gamma_{n0}^{p1/2} \gamma_{n1}^{p3/2} - \gamma_{n1}^{p1/2} \gamma_{n0}^{p3/2}) \cdot \frac{1}{\Gamma} \cdot \frac{v^T}{D} . \tag{3}$$

The $\gamma_n^{s1/2}$ etc. are the partial width amplitudes in obvious notation, and v^P and v^T are the symmetry-breaking matrix elements which mix the two resonances. The expressions (2), (3) contain the factors v^P/D and v^T/D, respectively. In CN reactions, D is typically a few eV. The smallness of this denominator is a cause of substantial enhancement ("dynamic enhancement"). Comparing the expressions (2), (3) with their counterparts evaluated at a neutron energy taken between the two resonances, we find another enhancement factor $(D/\Gamma)^2 \gg 1$ ("resonance enhancement"). It reflects the fact that the long lifetime $\hbar/\Gamma$ of the CN enhances symmetry breaking. And for P, the occurrence of the product $\gamma^s\gamma^p$ gives another enhancement factor γ^s/γ^p if we compare the symmetry-breaking and the symmetry-conserving CN reaction on the same p-wave resonance. [We recall that near neutron threshold, $|\gamma^p| \ll |\gamma^s|$].

The various enhancement factors explictly contained in this model give a good semiquantitative understanding of the Dubna results. However, for quantitative purposes, the model fails, because of the following reasons. (i) The CN is a chaotic system. The matrix elements v^P and v^T are random variables. Beyond a signal that symmetry is broken, a measurement of v^P or v^T on an isolated, well-chosen pair of resonances can therefore not yield quantitative information an α_P or α_T. Only the r.m.s. values of v^P and v^T, taken from many data points, carry significant information. (ii) The assumption that only a pair of resonances need be considered, is untenable. Symmetry violation will mix in *all* resonances in the vicinity of a given one. The parameters of these resonances (positions and partial width amplitudes) may or may not be known.

3. STATISTICAL APPROACH: ENSEMBLE AVERAGING

These difficulties are avoided in the statistical approach[7,8,13,14,16]. The elements of the nuclear S-matrix are modelled from the outset as random variables. This is accomplished by assuming that the projection of the Hamiltonian onto the space of CN levels is an element of an ensemble of random matrices. For P violation, the matrices have 2×2 block structure. The 2 matrices in the diagonal blocks refer to the positive and negative parity states, respectively; each is modelled as a member of the Gaussian orthogonal ensemble (GOE). The 2 matrices in the off-diagonal blocks are related by Hermitecity and T; the elements of one are assumed to be uncorrelated Gaussian random variables with zero mean and common variance σ_P. It is the aim of this approach to determine σ_P from the data. To this end the variance of β_P is calculated to lowest nonvanishing order in σ_P. For T violation, the Hamiltonian matrix in the space of CN levels is modelled as the sum of a GOE matrix and a small admixture of a matrix of GUE type. The GUE is the T-violating Gaussian unitary ensemble. One proceeds as in the case of P violation, aiming at an upper bound on σ_T, the variance of the symmetry-breaking part of the Hamiltonian.

It has been possible to calculate the necessary ensemble averages analytically and for all values of Γ/d. The connection with the data is made by assuming ergodicity (equality of ensemble average and energy average). It is this work that points to the domain $\Gamma \ll d$ as an area of big resonance enhancement.

For the case of P violation, Figure 1 shows calculated values of $\overline{\epsilon^2}^{1/2}$ for various nuclei versus neutron energy. A value $\alpha_P = 10^{-7}$, the associated estimate of $\overline{(v^P)^2}$ (cf. section 5) and standard optical-model parameters were used. We observe that $\overline{\epsilon^2}^{1/2}$ is much enhanced over the value of α_P, which shows that enhancement survives ensemble-averaging. We note that $\overline{(\epsilon^2)}^{1/2}$ increases with neutron energy up to a few 100 eV; this increase is due to increasing p-wave penetration. The subsequent decrease is due to decreasing values of d/Γ. We also note that the sensitivity is not at all the same for different nuclei; this fact is mainly due to differences in level density. A similar calculation for T violation in p-wave scattering is shown in figure 2. The parameter S is given by $\overline{\epsilon^2}^{1/2}/(\alpha_T n)$ where n is the number of target nuclei per unit area. The steep increase is due to p-wave penetration; the decrease is caused by the experimental resolution of the LANSCE facility which is taken into account in the calculation.

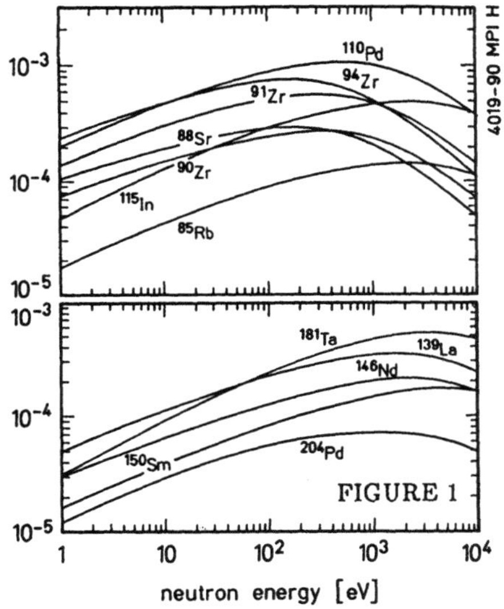

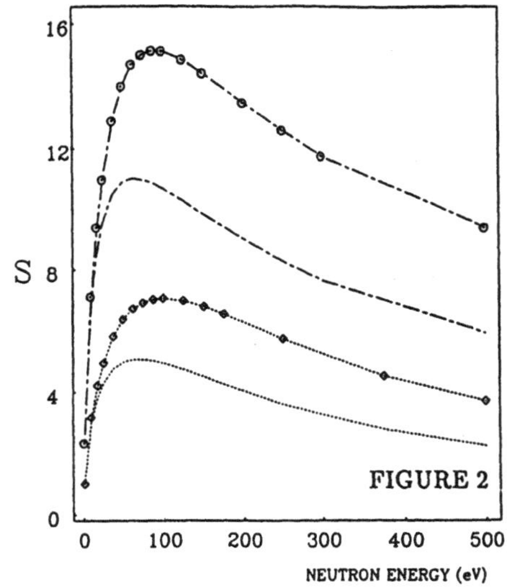

FIGURE 1

Expected size of the r.m.s. transmission asymmetry $\overline{\epsilon^2}^{1/2}$ for P violation. Taken from ref. 14.

FIGURE 2

Sensitivity of the LANCSE facility for T violation. Taken from the first entry in ref. 16.

Calculations of this type, vital for the analysis of data obtained in the domain $\Gamma \gtrsim d$ (example: tests of T-violation through detailed balance[13,8,17]) and important in defining areas of maximum resonance enhancement, fail to take into account the experimental bias mentioned in the Introduction. This is clear from figure 1: The curve for ^{139}La is nowhere near the value 0.1 found experimentally at the 0.74 eV p-wave resonance. The average over energy reduces the strong enhancement found within the resonances by roughly a factor $(\Gamma/d)^2$. Hence it is necessary to retain the statistical approach, but to pay attention to the experimental bias on well-identified resonances.

4. BIASED ENSEMBLES

I use two examples to demonstrate the approach taken in ref. 9, one for P violation and one for T violation.

In the case of P, we consider a p-wave neutron resonance with energy E_p. The energy of the incident neutron is chosen to be E_p. This resonance mixes with N s-wave resonances with known energies E_{sj} and known partial width amplidudes $\gamma_{nj}^{s1/2}$, $j = 1, \ldots, N$. The mixing matrix element between the p-wave resonance and the jth s-wave resonance is denoted by v_j^P. Assuming the widths of all resonances to be the same (Γ) and $\Gamma \ll d$, the quantity β_P of section 1 is proportional to

$$\frac{\gamma_n^{p1/2}}{\Gamma} \cdot \sum_{j=1}^{N} \frac{\gamma_{nj}^{s1/2}}{(E_{sj} - E_p)} \cdot v_j^P = \sum_{j=1}^{N} A_j v_j^P \ . \tag{4}$$

This is a sum of the N uncorrelated Gaussian random variables v_j^P with known coefficients A_j (all v_j^P have the same probability density). Therefore, the expression (4) is itself a Gaussian random variable with mean value zero and a second moment given by $(\sum_j A_j^2)\overline{(v^P)^2}$. The sum $\sum_j A_j^2$ obviously depends on the p-wave resonance under consideration. This shows that the observable (4) is not ergodic: The average of (4) evaluated over many p-wave resonances is not equal to the ensemble average. We consider instead of (4) the ergodic observable $\delta_P = (\sum_{j=1}^N A_j v_j^P)/(\sum_{j=1}^N A_j^2)^{1/2}$. Obviously, δ_P is Gaussian with variance $\overline{(v^P)^2}$. We conclude that a series of measurements of δ_P on a set of p-wave resonances yields both $\overline{(v^P)^2}$ and, via the known distribution of δ_P and the maximum likelihood method, a confidence interval.

Parallel to the work of ref. 9, this approach has been taken in ref. 6. Table 1 shows the energies of the 17 p-wave resonances in ^{239}U which were investigated, and the measured values of δ_P together with the errors. These values show a stochastic pattern. A maximum likelihood analysis which takes into account that the spins of the p-wave resonances are unknown, yields the value[6]

$$\overline{(v^P)^2}^{1/2} = 0.58 {}^{+0.50}_{-0.25} \text{meV} . \quad (5)$$

The limits corresponds to a 68 % confidence interval. In section 5, I return to the analysis of this result.

In the case of T, the observable β_T, simplified with assumptions as used for (4), is proportional to

TABLE 1

p-wave neutron resonances in ^{239}U (in eV) and associated values of the ergodic variable δ_P (in meV). Taken from ref. 6.

E_j	δ_P
10.2	−0.06 ±0.03
11.3	0.13 ±0.07
45.2	−0.4 ±0.6
57.9	0.8 ±0.4
63.5	0.7 ±0.1
83.7	1.4 ±0.6
89.2	−0.5 ±0.2
93.1	−0.01 ±1.0
98.0	−0.4 ±0.2
125.0	1.1 ±0.9
152.4	−0.4 ±1.4
158.9	−0.4 ±1.7
173.1	1.4 ±1.0
242.7	−1.3 ±1.3
253.9	−0.5 ±2.0
263.9	−0.05 ±1.7
282.4	0.8 ±2.7

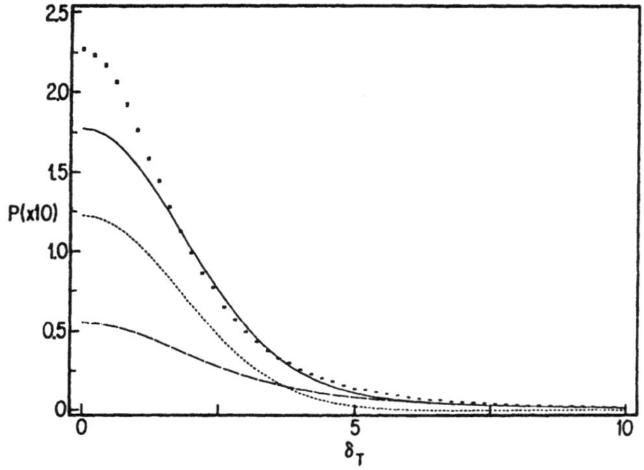

FIGURE 3

The probability density $\mathcal{P}(\delta_T) = \mathcal{P}(-\delta_T)$ for the ergodic observable δ_T corresponding to expression (5) in two approximations. Dots: Monte-Carlo simulation; Solid line: Analytical approximation. Taken from ref. 9.

$$\Gamma^{-1} \sum_{j=1}^{N} (\gamma_{n0}^{p1/2} \gamma_{nj}^{p3/2} - \gamma_{nj}^{p1/2} \gamma_{n0}^{p3/2}) \frac{(E_j - E_0) v_j^T}{(E_0 - E_j)^2 + \Gamma^2/4} . \tag{6}$$

The parameters E_0, $\gamma_{n0}^{p1/2}$, $\gamma_{n0}^{p3/2}$ of the p-wave resonance under study are assumed to be known but the parameters E_j, $\gamma_{nj}^{p1/2}$, $\gamma_{nj}^{p3/2}$ of the N admixed p-wave resonances are usually not known and must themselves be considered as random variables, the γ_{nj}'s having a Gaussian distribution and the eigenvalues E_0, E_j having a probability distribution given by the GOE. Therefore, the probability density $\mathcal{P}(\delta_T)$ for the ergodic variable δ_T proportional to (6) must be worked out from this input and from the fact that the v_j^T are uncorrelated Gaussian random variables with identical probability densities. If data on T violation are to be used, via the maximum likelihood method, to work out an upper bound on $\overline{(v^T)^2}$ then $\mathcal{P}(\delta_T)$ must be known. Figure 3 shows the result of an approximate analytical evaluation of $\mathcal{P}(\delta_T)$ — expected to be very good for large $|\delta_T|$ — and the result of a Monte Carlo simulation — expected to be very good for small $|\delta_T|$. The agreement between both curves is sufficient to apply with confidence the maximum likelihood method. For the case that the parameters of the p-wave resonance labelled zero in (6) are likewise unknown, a similar analysis is possible[9].

An experiment on T violation that has the same sensitivity as the P experiment (which yields the result (5)) could lower the existing upper bound on α_T by two orders of magnitude. The argument is this[18]. The present experimental upper bound on the electric dipole moment of the neutron is $D \leq 10^{-25}$ e cm. Taking $2 \cdot 10^{-14}$ e cm as a natural scale for this quantity and recalling that a nonzero dipole moment requires both P and T to be broken, we use $\alpha_P = 2 \cdot 10^{-7}$ and $\alpha_T = \gamma\, 2 \cdot 10^{-7}$ and obtain the theoretical estimate $D \cong 2 \cdot 10^{-14}\, \alpha_P \alpha_T$ e cm $\cong \gamma\, 2 \cdot 10^{-27}$ e cm. Comparing this with the experimental upper bound, we find $\gamma \leq 100$. On the other hand, the parity experiment on ^{239}U has a sensitivity which permits to test α_P on the scale of 10^{-7}. An experiment on T violation would test α_T on the same scale, i.e. on the scale $\gamma \cong 1$. Unfortunately, there is one weak point in the argument: I have tacitly assumed that the strength of the fundamental T violating interaction and α_T are equal. In view of the forthcoming test of T by the TRIANGLE group, it is important that this assertion be theoretically examined.

The two examples displayed in this section should lend credibility to the claim that the theory of biased ensembles has been developed to the point that data on symmetry breaking in isolated resonances can reliably be converted into values for (or bounds on) the mean square matrix elements of the symmetry-breaking interaction.

5. THE SPREADING WIDTH

How can one reduce a result like the one in eq. (5) to a value for α_P, the relative strength of the parity-breaking effective nucleon-nucleon interaction in the shell model? French et al.[15] have proposed the following procedure.

Take $\overline{(v^P)^2}$ and the actual mean level density ϱ_s of the s-wave resonances to calculate the spreading width for the parity-violating interaction, $\Gamma_p^{\downarrow} = 2\pi\overline{(v^P)^2}\varrho_s$. In the case of ^{239}U, this yields[6] $\Gamma_p^{\downarrow} = 1.0 \cdot 10^{-7}$ eV. The ratio of $\Gamma_p^{\downarrow}$ to the spreading width $\Gamma_s^{\downarrow}$ of the strong interaction gives directly α_P^2. Qualitatively speaking, the rationale behind this procedure is this. In contrast to the quantity $\overline{(v^P)^2}$, the spreading width $\Gamma_p^{\downarrow}$ (or any other spreading width for another symmetry violation) is nearly independent[15] of excitation energy and mass number. The complexities of the many-body system therefore do not affect either $\Gamma_p^{\downarrow}$ or $\Gamma_s^{\downarrow}$, and their ratio directly yields α_P^2. French et al.[15] have advocated a value $\Gamma_s^{\downarrow} = 2\pi \cdot 10^5$ eV for the spreading width of the strong interaction; with this value, one finds[6] $\alpha_P = 4 \cdot 10^{-7}$. This value is in rough agreement with expectations based on the relative strength of the weak interaction.

While certainly correct qualitatively, this procedure is probably not quantitative, for the following reason. Spreading widths for isospin symmetry breaking in CN reaction have been determined for a large number of nuclei and excitation energies. The latest compilation is shown in figure 4. We observe that the spreading widths, although confined to an energy interval between about 10 keV and 100 keV, do have a substantial scatter. This raises the question whether the constancy of $\Gamma_p^{\downarrow}$ assumed in the procedure described above is quantitatively correct, and calls for a theoretical investigation. Moreover, applying the procedure described above to the present case and using for the isospin spreading width an approximate mean value of 30 keV, I find for the above-mentioned ratio of spreading widths the value 1/20. This is considerably bigger than what a naive estimate ($\cong 1/137$) for the relative strength of the electromagnetic interaction would give, suggesting that the value of $\Gamma_p^{\downarrow}$ should be increased by a factor around 5, eventually yielding $\alpha_P \cong 2 \cdot 10^{-7}$. This last value is probably more in line with weak interaction theory than the value $4 \cdot 10^{-7}$ reported above.

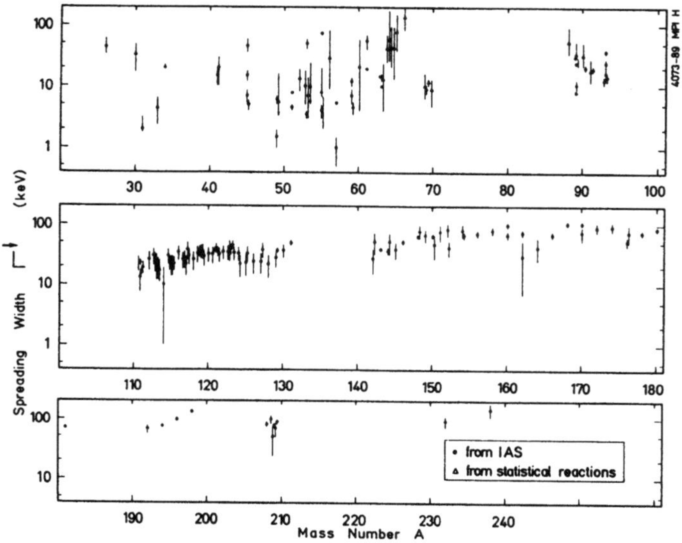

FIGURE 4

Spreading widths for isospin symmetry breaking in CN reactions. Taken from refs. 19, 20.

CONCLUSIONS

Experiments on P violation in CN reactions show large enhancement factors. Present theory accounts for the enhancements and predicts similar effects also for T violation, and probably also for PT violation. This makes the CN a very attractive field for the study of violation of fundamental symmetries.

The CN is a chaotic quantum system. Symmetry breaking in such a system is of interest in its own right. The theoretical approach to symmetry breaking must, however, be based on statistical concepts. Individual symmetry-breaking matrix elements carry no information beyond the statement that symmetry is broken. Only mean values of squares of matrix elements yield significant information.

The theoretical approach based on ensemble-averaging does take account of these facts. In the domain of overlapping resonances, it is the only viable approach. In the domain of isolated resonances, it fails to account for the enhancement connected with experiments on individual resonances.

To overcome this shortcoming, it is necessary to develop a theory of biased ensembles. Recent developments are such that the probability density for the ergodic observable can be determined with an accuracy that suffices to apply with confidence the maximum likelihood method to data.

The TRIPLE experiment on ^{239}U yields a parameter-free value of $\overline{(v^P)^2}$ and, via the spreading width $\Gamma_p^{\downarrow}$ and the comparison with the spreading width for the strong interaction, a value $\alpha_P = 4 \cdot 10^{-7}$ which is in keeping with theoretical expectations. There remain, however, uncertainties about the reduction procedure which require further theoretical work.

Experiments on other nuclei and/or at other excitation energies are needed to clarify this issue. The connection between α_P, α_T and the relative strength of the symmetry-breaking bare nucleon-nucleon interaction requires more theoretical work.

Given the same sensitivity and resolution as for the P experiment on ^{239}U, the five-fold correlation experiment on T violation may decrease the best upper bound on α_T derived from experiments on the dipole moment of the neutron by two orders of magnitude. This is therefore a very important experiment.

Finally, there is a chance that as time passes, sophistication and resolution of the experimental work will be refined to the point that an experiment on PT violation might show a positive signal. In view of the paucity of information available on the PT violating interaction, this seems like a very worthwhile goal, and a most interesting possibility.

Taken in all, the study of symmetry breaking in CN reactions seems to have a bright future in the 1990's.

REFERENCES

1) V. P. Alfimenkov et al., Nucl. Phys. A 398 (1983) 93 and Sov. Phys. Usp. 27 (1984) 797.
2) O. P. Sushkov and V. V. Flambaum, JETP Lett. 32 (1980) 352.
3) S. A. Biryukov et al., Sov. J. Nucl. Phys. 45 (1987) 937.
4) Y. Masuda et al., Nucl. Phys. A 504 (1989) 269.
5) C. D. Bowman, J. D. Bowman and V. W. Yuan, Phys. Rev. C 39 (1989) 1721.
6) J. D. Bowman et al., submitted to Phys. Rev. Lett..
7) E. D. Davis, Phys. Lett. 226 B (1989) 197.
8) E. D. Davis and U. Hartmann, submitted to Phys. Rev. C.
9) V. E. Bunakov, E. D. Davis and H. A Weidenmüller, submitted to Phys. Rev. C.
10) V. E. Bunakov and V. P. Gudkov, Z. Phys. A 303 (1981) 285; Nucl. Phys. A 401 (1983) 93; and in "Tests of Time-Reversal Invariance in Neutron Physics", N. Robertson, C. Gould and J. Bowman, editors, World Scientific, Singapore (1987) p. 175.
11) V. E. Bunakov, Phys. Lett. 60 (1988) 2250 and Phys. Rev. Lett. 60 (1988) 2250.
12) V. E. Bunakov and H. A. Weidenmüller, Phys. Rev. C 39 (1989) 70.
13) D. Boosé, H. L. Harney and H. A. Weidenmüller, Phys. Rev. Lett. 56 (1988) 2012 and Z. Phys. A 325 (1986) 363.
14) A. Müller, E. D. Davis and H. L. Harney, submitted to Phys. Rev. Lett..
15) J. B. French, A. Pandey and J. Smith, as in the last entry of ref. 10 p. 80.
16) E. D. Davis, Workshop on Spin and Symmetries, edited by D. Ramsay and W. T. H. van Oers, TRIUMF preprint TRI-89-5 (1989) p. 105 and Fundamental Symmetries and Nuclear Structure; J. N. Ginocchio and S. P. Rosen, editors, World Scientific, Singapore (1988) p. 52
17) E. Blanke et al., Phys. Rev. Lett. 51 (1983) 355.
18) Yu. Khriplovich, private communication.
19) H. L. Harney, A. Richter and H. A. Weidenmüller, Rev. Mod. Phys. 58 (1986) 607.
20) H. L. Harney, private communication and Z. Phys. A, in press.

Chaos in Nuclei*

H.A. Weidenmüller

Max-Planck-Institut für Kernphysik,
W-6900 Heidelberg, Fed. Rep. of Germany

I review recent progress relating to chaotic dynamics in the areas of statistical spectroscopy, global indicators of chaotic motion, and Hamiltonian transport.

1. INTRODUCTION

During the last decade, the interest in chaotic motion and non–linear dynamics has risen sharply among physicists, chemists, and biologists, not to speak of more distant fields like economic theory. It affects the study of both conservative and dissipative systems. In nuclear physics, where we deal with a conservative Hamiltonian, this development has given a fresh impetus to statistical nuclear spectroscopy[1], to the statistical theory of nuclear reactions[2], and to dissipation in large–amplitude collective nuclear motion[3]. In my opinion, present–day nuclear physics has two frontiers: The study of subnuclear degrees of freedom, and the investigation of the dynamics of nuclei in terms of the interplay between regular and chaotic motion. In this talk, I attempt to review some recent developments, using as my starting point the papers cited above. In doing so, I wish to stress the generic aspects at the expense of addressing purely nuclear physics applications.

In section 2, I review recent progress in statistical spectroscopy and its connection with chaotic motion. Section 3 addresses a related problem: Is it possible to obtain information on level statistics directly from some piece of data without going through the tedious analysis of very many levels? In section 4, I turn to some recent work in statistical nuclear reaction theory and Hamiltonian transport. Section 5 contains the conclusions.

2. STATISTICAL SPECTROSCOPY

After earlier work by several authors, especially Berry and Zaslavsky, the study of the distribution of the eigenvalues of the Sinai billiard led Bohigas, Giannoni and Schmit[4] to the following conjecture: The fluctuation properties of generic quantum systems with (without) time–reversal symmetry which in the classical limit are fully chaotic, coincide with those of the Gaussian Orthogonal (Unitary) Ensemble, i.e. GOE (GUE). Studies of level spacing distributions and partial width distributions in nuclei are based on this conjecture and usually identify chaos in nuclei via the agreement with GOE statistics.

Although a complete analytical theory of the implications of fully chaotic classical motion on the spectral statistics of the associated quantum system is still lacking, various investigations, notably by Berry, have shed some light on the domain of validity of the

*Nuclear Physics A520 (1990) 509c-519c © 1990 – Elsevier Science Publishers B.V. (North-Holland)

conjecture cited above. It has become apparent that the conjecture is generically valid in the semiclassical limit ($\hbar \to 0$) provided that all regions of phase space are equally accessible to (almost) any trajectory. Counterexample: The motion of a single particle in two irregularly shaped cavities which are connected by a thin pipe. Here, a trajectory starting in cavity 1 will generically fill the phase space associated with this cavity long before it enters into cavity 2; the analogous quantum problem is expected to be simulated by two weakly coupled GOE's. More generally, a quantum problem showing localization is expected to display Poissonian level statistics, although the dynamics is fully chaotic in the classical limit.

An issue not addressed in the above-mentioned conjecture concerns the possible importance of quantum tunneling[5]. Consider a particle with mass m and momentum $\vec{p}$ in two dimensions, subject to elastic reflection at the surface of the rectangle $0 \leq x \leq a$ and $0 \leq y \leq b$, and governed by the Hamiltonian $H = (2m)^{-1}\vec{p}^{\,2} + \delta(x - y)$. Classically, the repulsive delta-function potential completely separates the phase space into two disconnected domains. The level statistics of the quantum problem should accordingly correspond to the superposition of two independent distributions and should, therefore, differ from the GOE. (There is no repulsion of levels belonging to different domains of the phase space). In actual fact, the two regions of phase space are quantum-mechanically coupled because the wave functions tunnel through the delta-function barrier. This causes mixing and level repulsion, and pushes the spectral statistics towards GOE behaviour. An example encompassing 289 levels for $m = 30$, $a = 1$ and $b = \sqrt{3}$ is shown in figure 1.

Much attention has recently been payed to "scars" in stationary eigenfunctions of quantum systems which are fully chaotic in the classical limit[6]. Such scars are related to unstable periodic orbits of the classical problem; orbits of this type are known to be important in the semiclassical limit; with an eye on collective nuclear dynamics, they have also been studied in simple model systems[7].

With the above-mentioned provisos in mind, let me now turn to tests of GOE level statistics in nuclei. Interest in such tests is due to the following reasons. (i) Most conservative systems studied analytically and numerically have two degrees of freedom. It is of interest

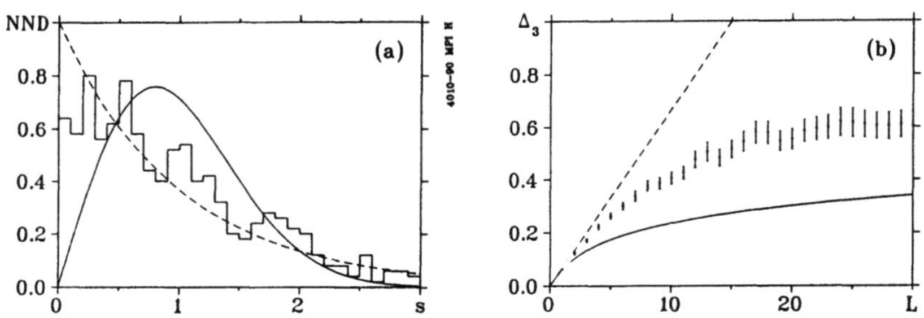

FIGURE 1

Nearest-neighbour spacing (upper part) and Δ_3-statistic (lower part) for the 2 dimensional problem described in the text (taken from ref. 5). Abscissa: $x = s/d$ with s = level spacing, d = mean level spacing (upper part) and length of integration interval L (in units of d) (lower part).

to compare results obtained in these studies with the behaviour of a genuine many-body system like the nucleus. (ii) Independent particle motion, rotational bands, giant resonances and large-amplitude collective motion suggest that the dynamics of nuclei is at least partly regular. Yet there is also evidence for GOE statistics. This poses the question: Are nuclei partially chaotic? Does the degree of chaotic motion (measured, for instance, by the fraction of phase space filled by chaotic trajectories) depend on excitation energy? Is there an order-to-chaos transition that separates regular motion at low energy from chaotic motion at higher energy? (iii) What is the influence of the exclusion principle on the chaotic dynamics? I recall that in a system with d degrees of freedom, the average level spacing is proportional to $\hbar^d$. Hence, the limit $\hbar \to 0$ becomes more effective with increasing d. But in the vincinity of the Fermi energy, the exclusion principle "freezes" most degrees of freedom. How does this fact influence spectral fluctuation patterns? (iv) Whenever the fluctuation patterns of eigenvalues and of matrix elements involving eigenfunctions are GOE-like, the system is highly instable: A small perturbation of the Hamiltonian, with matrix elements typically of the order of the mean level spacing, will quantitatively change eigenvalues and eigenfunctions. It follows that in such a situation, the information content carried by *individual* levels and by the associated eigenfunctions is very small, so that there is little interest in the spectroscopic information contained in *individual* levels. (Average quantities like the mean level spacing or the various strength functions do, however, carry information). Aside from establishing at which excitation energy in nuclei the transition to GOE behaviour occurs, statistical spectroscopy thus also defines the maximum energy up to which the detailed spectroscopic investigation of *individual* levels is physically meaningful.

Following the analysis of Haq, Pandey and Bohigas[8] of 1726 level spacings obtained from the study of isolated neutron and proton resonances, and the investigation of the statistics of partial width amplitudes obtained via inelastic proton scattering by Shriner et al.[9] (both of which showed complete agreement with GOE predictions), G. E. Mitchell et al.[10] investigated the positive parity states in ^{26}Al. This nucleus is exceptional in that a complete set of levels is supposed to be known from the ground state up to proton threshold. The analysis showed largly agreement with GOE statistics. This result was surprising, for two reasons. First, no indication of regular motion was found even in the ground-state domain where independent-particle motion might have been expected to dominate. Second, the levels in ^{26}Al consist of two classes with isospin $T = 0$ and $T = 1$; the data showed strong isospin mixing in the sense that a separate analysis of the levels in either class yielded the same fluctuation measures as the entire ensemble. The second result has since been understood in a schematic model[11] in which two separate GOE's (each simulating one class of states with fixed isospin) are weakly coupled; it is found that such weak coupling produces local GOE fluctuations whenever the coupling matrix elements are comparable in magnitude with the mean level spacing. (As a corrollary it follows that any weak perturbation will generically drive a system towards GOE statistics which shows that deviations from GOE statistics are strongly indicative of regular motion). The value of the isospin spreading width, $\Gamma^{\downarrow} \cong 20$keV, determined from this analysis is consistent with the systematics established earlier[12].

As to the first result, there is growing evidence that it is not an isolated finding. Indeed, extensive (n, γ) measurements, mostly by von Egidy et al.[13] have yielded novel information

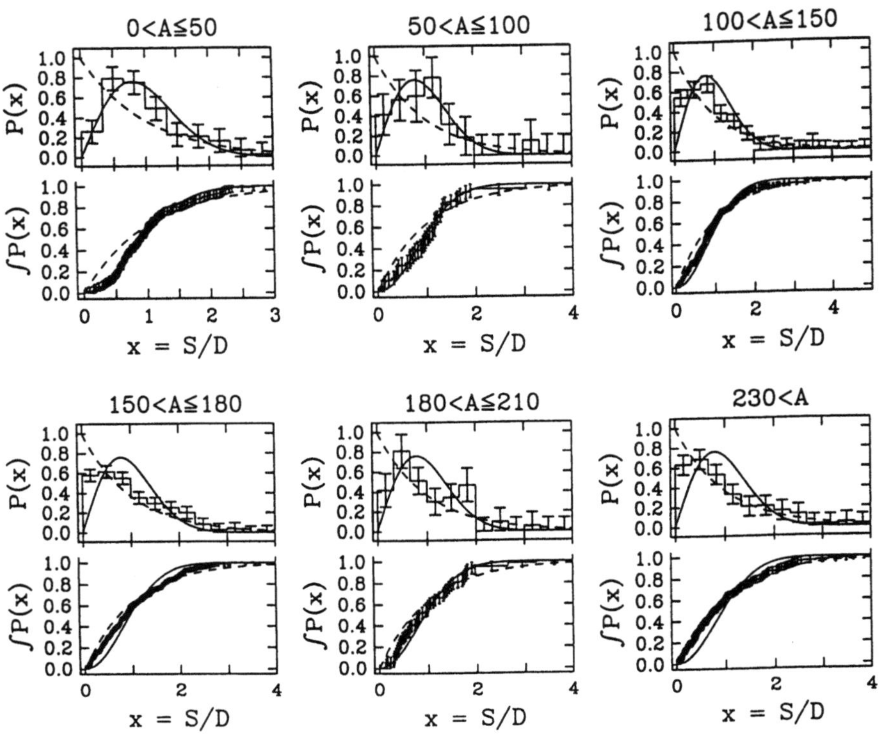

FIGURE 2

Nearest-neighbour spacing distribution $P(x)$ and its integral versus x (as explained below figure 1) for groups of nuclei with mass number A as indicated. (Taken from ref. 15).

on the spectra of many nuclei; this information has been analysed with regard to spectral fluctuations[14,15]. Typical results for the nearest-neighbour spacing distribution $P(x)$, and for the statistically more accurately determined quantity $\int_0^x P(y)dy$, are shown versus $x = s/d$ in figure 2.

The results for $A \leq 50$ (124 levels in all) support the ^{26}Al findings in that they are close to GOE statistics, while the data for $A > 230$ (157 level in all), on the contrary, are close to Poisson. Breaking the data set up into finer details (even–even versus odd and odd–odd nuclei, spherical versus deformed nuclei, 2^+ and 4^+ states versus the rest) shows[15] further characteristic trends, largely in keeping with earlier results[14]. I believe that these trends are very interesting, although it is premature to speculate about them, and that further efforts to substantiate and to analyse them are much worthwhile. In view of the remarks made above, it is particularly intriguing to find evidence for Poisson statistics signalling strong regularity. Hopefully such and other studies (see section 3) can lead to partial answers to the problems listed above under (i) to (iv), and to an understanding of the order-to-chaos transition which is perhaps characteristic of nuclear dynamics.

The order-to-chaos transition has also been studied theoretically in the context of various nuclear models. Early work on shell-model matrices[16] and on chaos in the Lipkin model[17] was followed by interesting studies of the influence of rotational motion[18] and of the influence of two coalescing potential wells[19] (typical for heavy–ion reactions) on single–particle dynamics. The order–to–chaos transition can also be observed in the interacting Boson–Fermion model far away from parameter values that signal a dynamical symmetry of the Hamiltonian[20]. While such studies have an interest of their own, it is presently difficult to place them into the wider context of the nuclear many-body problem. In my opinion, further work is necessary, addressing the above–mentioned problems (i) to (iv).

3. GLOBAL INDICATORS OF CHAOTIC DYNAMICS

The experimental work which provides the basis of statistical nuclear spectroscopy relies on a painstaking analysis of a large number of resolved individual nuclear levels and/or resonances. In the present section, I wish to draw attention to the fact that in molecular spectroscopy, there exist other ways to use experimental information directly in order to draw conclusions on level–level correlations, and thereby on the applicability of the GOE. Subsequently, I discuss the question whether such "global indicators" also exist in nuclear physics. This would make the experimental work much less tedious.

In quantum beat experiments[21], pulsed laser light excites an atom within a band of excitation energies. The time–dependence of the deexcitation fluorescence light shows oscillations ("quantum beats"). They are due to the interference of amplitudes emitted from different levels within the energy band, and thus contain direct information on the eigenvalue spectrum. A second method[22] (anticrossing spectroscopy) uses the fact that under the influence of a magnetic field, the $m \neq 0$ components of triplet states with magnetic quantum number m can be made to move past a singlet state; at each such Landau–Zener anticrossing, the resonance fluorescence yield from the laser–exited singlet state has a minimum. Figure 3 shows the yield versus magnetic field strength for Methylglyoxal; the anticrossings are so dense that they overlap mutually. Nevertheless, the Fourier transform of the intensity autocorrelation function displays a "correlation hole"; this hole is evidence for GOE statistics within the triplet states[22,23]. We note that this information is available without examination of the individual anticrossings. This is the meaning of the expression "global indicators of chaos".

Unfortunately, I am not aware of the existence of such global indicators in nuclear physics. But I ask whether they exist, and whether enough time and thought has been spent to identify them.

One indicator has been suggested. Unfortunately, it is not as direct a measure of the two–point correlation function as are the observables in molecular physics described above. It has been argued[24] that the spreading width of the collective $E2$ strength may serve as such an indicator. The argument is this: Near the Yrast line, there exist individual rotational bands; collective $E2$ decay of one member of the band feeds another state in the same band. With increasing excitation energy (but fixed spin), the average spacing between different bands decreases and the effectiveness of the spin–dependent part of the intrinsic Hamiltonian in mixing states belonging to different bands, therefore increases. In consequence, collective $E2$ decay from a given state no longer populates only a single final state: The $E2$ strength

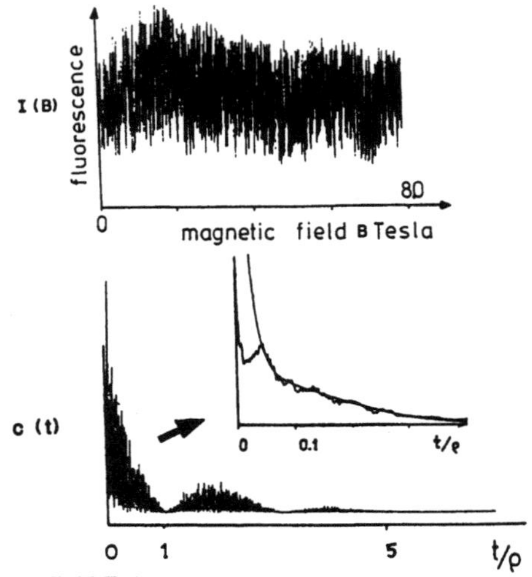

FIGURE 3

Resonance fluorescence yield versus magnetic field strength (upper part) and Fourier transform of the intensity autocorrelation function (lower part). The insert shows the "correlation hole" which is indicative of the long-range GOE correlations between levels with the same quantum numbers. (taken from ref. 22).

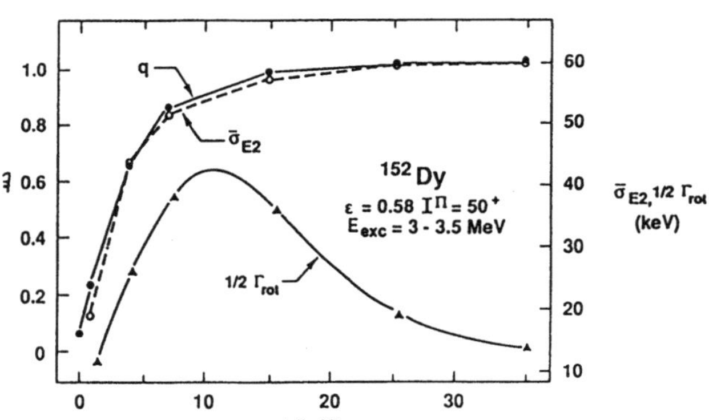

FIGURE 4

The chaoticity parameter q explained in the text (left scale), the average standard deviation of $E2$ strength, and the FWHM of $E2$ strength Γ_{rot} (right scale) versus the strength of the two-body interaction (taken from ref. 26).

acquires a spreading width $\Gamma^{\downarrow}$. In ref. 24 it was shown in the framework of a simple model that a value of $\Gamma^{\downarrow}$ amounting to several ten average level spacings signals the onset of chaos.

The suggestion that the spreading width of the collective $E2$ strength indicates the order-to-chaos transition[25,24] has been made quantitative in realistic model calculations performed by Åberg[26]. Diagonalization of a Hamiltonian containing deformed Nilsson states and a residual interaction yields eigenvalues from which the Δ_3 statistic can be determined. Defining the chaoticity parameter q by the decomposition $\Delta_3 = \Delta_3^{\text{Poisson}} \cdot (1-q) + \Delta_3^{\text{GOE}} \cdot q$, Åberg finds the very close parallelism between q and the fragmentation of $E2$ strength (here measured in terms of the standard deviation) shown in figure 4.

It remains to be seen whether experiments on high-spin states can measure the $E2$ strength fragmentation as a function of excitation energy.

4. STATISTICAL THEORY OF NUCLEAR REACTIONS, NUCLEAR DISSIPATION AND HAMILTONIAN TRANSPORT

Much progress has been made in recent years in the random-matrix modelling of statistical nuclear reactions. Starting from a random-matrix model for the Hamiltonian, it has been possible to derive without further ad-hoc assumptions or undue approximations energy-averaged cross-sections for compound and precompound reactions and, where necessary, higher moments and autocorrelation functions for cross-sections[2]. This work is relevant also for the analysis of data on tests of fundamental symmetries in compound-nucleus reactions[27].

In the present context, interest in such random-matrix modelling focusses on its justification in terms of chaotic motion: Are the statistical assumptions a mirror of chaotic nuclear dynamics? In the case of compound-nucleus reactions, the projection of the Hamiltonian onto the space of quasibound levels is modelled as a member of the GOE; this is consistent both with our knowledge of spectral fluctuation properties of isolated compound-nucleus resonances, and with the above-mentioned conjecture of Bohigas, Giannoni and Schmit. But what is the connection between this modelling and the newly emerging field[28] of classical and quantum chaotic scattering? And for which parameter values of the compound nucleus yields such modelling a scattering matrix which is simply a member of Dyson's circular ensemble? In the case of precompound reactions, it is commonly assumed that a hierarchy of states of increasing complexity exists. This hierarchy is modelled in terms of a chain of independent GOE's with nearest-neighbour hopping[2]. This type of modelling, at variance with the conjecture of Bohigas, Giannoni and Schmit, poses the question: Is there a dynamical justification for such partitioning of states into a hierarchy of classes? Does such partitioning correspond to a partitioning of classical phase space into regions such that in every region, almost all trajectories starting in this region fill it during a time which is small compared to the time it takes a trajectory on average to penetrate into neighbouring regions? This question connects to the problem of Hamiltonian transport, cf. below. In summary, we observe that in spite of successful modelling several questions concerning the connection to chaotic motion and to chaotic scattering, remain open.

Many aspects of deeply inelastic heavy-ion reactions have in the past been phenomenologically interpreted in terms of transport equations of the Master equation and of the Fokker-Planck-equation type. Is it possible to justify such approaches (in systems which are both conservative and isolated) in terms of chaotic motion? Aside from explicit studies such as that of ref. 19, the question has been discussed in general terms by W. Swiatecki[3]. This thoughtful and interesting paper lists many open questions both in relation to the nuclear level density, and to nuclear dissipation. These cannot be repeated her. Suffice it to say that Swiatecki suggests a connection between nuclear integrability and nuclear elasticity on the one hand, and between non-integrability and viscuous behaviour on the other.

I believe that the entire problem of nuclear dissipation must eventually be studied as a special case of Hamiltonian transport. This new discipline of dynamical theory, to the best of my knowledge at present developed only for bounded *classical* systems with two degrees of freedom[29,30], aims at finding a partitioning of phase space into countably many parts such that within each part, occupation probability dissipates on a time scale which is very short in comparision to the time it takes to transport occupation probability between different parts. If such a partitioning can be found, it is possible to replace Hamilton's equation of motion by a discrete Master equation which describes the change in time of the occupation probabilities of the parts of phase space defined by the partitioning. For the two degrees of freedom systems studied in refs. 29, 30 it has been possible to find the unique optimal partitioning. It is defined in terms of unstable periodic orbitals. It is a remarkable fact that these orbitals, which are so important in the semiclassical approximation to various chaotic quantum problems and in the study of scars, resurface also in the present context.

It seems that the theory of Hamiltonian transport is only at its beginning. The extension to more than two degrees of freedom, to scattering problems, and to quantum mechanics should give much additional insight into the nuclear physics problems mentioned above.

In the context of nuclear physics, this set of problems has been addressed in the framework of the time-dependent Hartree-Fock (TDHF) approach[31]. The authors express the TDHF equations as canonical classical equations of motion for TDHF trajectories. Collective coordinates are then defined in a self-consistent way, and the TDHF problem is transformed into a set of equations which describe the non-integrable interaction between collective and intrinsic degrees of freedom. This interaction manifests itself in the occurence of resonances. In Hamiltonian transport theory, resonances and their enumeration in terms of unstable periodic orbits form the basis of the partitioning of phase space referred to above, and I wonder whether the theory of ref. 31 can be developed further to yield a set of Master equations which describe the irreversible dissipative motion of the collective degrees of freedom.

5. CONCLUSIONS

I hope this (necessarily incomplete) review of recent developments has made apparent the enormous impact which the field of nuclear physics has received from the theory of classically chaotic dynamical systems and their quantum counterparts. The difficulties in making further progress, both experimentally and theoretically, are considerable. Two examples may illustrate this point: Experimentally, the collection of more data for statistical spectroscopy may not appear terribly rewarding in itself. The analysis of dissipative heavy-ion reactions is plagued by ambiguities. On the other hand, the nucleus as a self-bound quantum system is nearly unparalleled in posing a multitude of challenging questions. They relate to chaos in bounded and unbounded systems, to Hamiltonian transport, to the role of the exclusion principle, to giant resonances and large-amplitude collective motion, and to symmetry breaking in chaotic systems. I am confident that nuclear physics can profit much from future work on chaotic dynamical systems. I hope that the converse will also be true and that the questions discussed in this review may stimulate specific research on dynamical systems.

ACKNOWLEDGEMENT

I am grateful to many colleagues and friends, especially U. Smilansky, for discussions of the topics presented here, and to C. Lewenkopf and G. E. Mitchell for permission to use their results prior to publication.

REFERENCES

1) O. Bohigas and H. A. Weidenmüller, Ann. Rev. Nucl. Part. Sci. 38 (1988) 421

2) J. J. M. Verbaarschot, H. A. Weidenmüller and M. R. Zirnbauer, Phys. Rep. 129 (1985) 367
H. Nishioka, J. J. M. Verbaarschot, H. A. Weidenmüller and S. Yoshida, Ann. Phys. (N.Y.) 172 (1986) 67
H. Nishioka, H. A. Weidenmüller and S. Yoshida, Ann. Phys. (N.Y.) 183 (1988) 166

3) W. J. Swiatecki, Nucl. Phys. A 488 (1988) 375c

4) O. Bohigas, A.-J. Giannoni and C. Schmit, Phys. Rev. Lett. 52 (1984) 1

5) C. Lewenkopf, submitted to Phys. Rev. Lett.

6) H. Friedrich and D. Wintgen, Phys. Rep. 183 (1989) 37 and references therein

7) M. A. M. de Aguiar et al., Ann. Phys. 180 (1988) 167

8) R. V. Haq, A. Pandey and O. Bohigas, Phys. Rev. Lett. 48 (1982) 1086

9) J. F. Shriner, jr., G. E. Mitchell and E. G. Bilpuch, Phys. Rev. Lett. 59 (1987) 435

10) G. E. Mitchell, E. G. Bilpuch, P. M. Endt and J. R. Shriner, Phys. Rev. Lett. 61 (1988) 1473 and preprint (1989)

11) T. Guhr and H. A. Weidenmüller, Ann. Phys. (N.Y.) (in press)

12) H. L. Harney, A. Richter and H. A. Weidenmüller, Rev. Mod. Phys. 58 (1986) 607

13) T. von Egidy, H. H. Schmidt and A. N. Behkami, Nucl. Phys. A 481 (1988) 189 and references therein

14) A. Abul-Magd and H. A. Weidenmüller, Phys. Lett. 162 B (1985) 223

15) G. Mitchell et al., private communication (1990) and J. F. Shriner, G. E. Mitchell and T. Egidy, to be published

16) T. A. Brody, J. Flores, J. B. French, P. A. Mello, A. Pandey and S. S. M. Wong, Rev. Mod. Phys. 53 (1981) 385

17) D. Meredith, S. E. Koonin and M. R. Zirnbauer, Phys. Rev. A 35 (1988) 3499

18) R. Arvieu, F. Brut and J. Carbonell, Phys. Rev. A 35 (1987) 2389

19) B. Milek, W. Nörenberg and P. Rozmej, Z. Phys. 334 (1989) 233

20) W. Paar et al., preprint (1989)

21) S. Haroche, in "High-Resolution Laser Spectroscopy", Topics in Applied Physics, K. Shimoda, editor, Springer-Verlag Berlin (1976) p. 253

22) R. Jost and M. Lombardi, Lecture Notes in Phys. 263, Springer-Verlag Heidelberg (1986) p. 72 and references therein

23) T. Guhr and H. A. Weidenmüller, submitted to J. Chem. Phys.; U. Hartmann and H. A. Weidenmüller, in preparation

24) T. Guhr and H. A. Weidenmüller, Ann. Phys. (N.Y.) 193 (1989) 472

25) T. H. Seligman, J. J. M. Verbaarschot and H. A. Weidenmüller, Phys. Lett. 167 B (1986) 365

26) S. Åberg, preprint 1990

27) E. D. Davis and U. Hartmann. Phys. Rev. C (submitted)

28) B. Eckhardt and C. Jung, J. Phys. A Math. Gen. 19 (1986) L829

29) D. F. Escande, Phys. Rep. 121 (1985) 166

30) R. S. MacKay, J. D. Meiss and I. C. Percival, Physica 27 D (1987) 1

31) T. Marumori, Y. Hashimoto, K. Iwasawa and F. Sakata, preprint (1990)

Ericson Fluctuations Versus Conductance Fluctuations: Similarities and Differences*

H.A. Weidenmüller

Max-Planck-Institut für Kernphysik,
W-6900 Heidelberg, Fed. Rep. of Germany

Abstract: The universal conductance fluctuations of mesoscopic systems in the metallic regime are compared with Ericson fluctuations of compound nucleus cross sections. Both phenomena can be viewed as cross-section fluctuations in the domain of strongly overlapping resonances, the resonances having stochastic features. Differences arise because the conductance is a sum over many channels, and because the length of the mesoscopic system can be varied continuously.

1. The phenomena

In 1960, Ericson [1]) predicted the existence of stochastic fluctuations in compound-nucleus (CN) reaction cross sections. The advent of Tandem Van-de-Graaf accelerators with their high energy resolution led to the discovery of these fluctuations, and to much experimental and theoretical activity in their understanding [2]). Fig. 1 shows a nuclear-reaction cross section versus energy. The peaks and minima in this figure are not caused by individual, more or less isolated CN resonances. Indeed, it is known that at the energies here considered, the CN resonances have a mean spacing d which is several ten times smaller than their average width Γ. Therefore, the peaks (and minima) result from a constructive (or destructive) superposition of many overlapping resonances. The amplitudes of the resonances which superpose at any one energy are essentially random variables. This is because the dynamics of CN reactions is chaotic: In the domain of non-overlapping CN resonances ($\Gamma < d$), the distributions of level spacings and of partial neutron widths are consistent with distribution patterns of quantum systems which are chaotic in the classical limit [3]). By extrapolation, it is assumed that stochasticity governs the CN also at energies where $\Gamma \geqslant d$. The curve shown in fig. 1 therefore has the curious feature of being both reproducible and random. A measurement of the same reaction in the same energy interval with the same energy resolution will reproduce fig. 1. Nonetheless, the dependence of the cross section on energy in fig. 1 shows the features of a random process. Cross section fluctuations in the domain of strongly overlapping resonances carry the name Ericson fluctuations.

During the last decade, it has become possible to manufactue wires with lengths of the order of μm and widths of the order 0.01–0.1 μm, and two-dimensional semiconductor devices of similar sizes. At temperatures below 1 K or so, the mean free path for inelastic electron–phonon scattering is larger than the linear dimension of such a "mesoscopic" structure. On their way through the sample, electrons are

* Nuclear Physics **A518** (1990) 1–12 © 1990 – Elsevier Science Publishers B.V. (North-Holland)

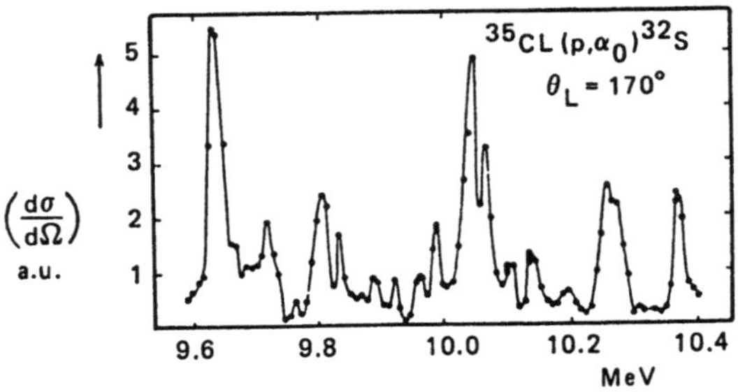

Fig. 1. A nuclear cross section displaying Ericson fluctuations versus proton bombarding energy. (Figure taken from ref. [2]).)

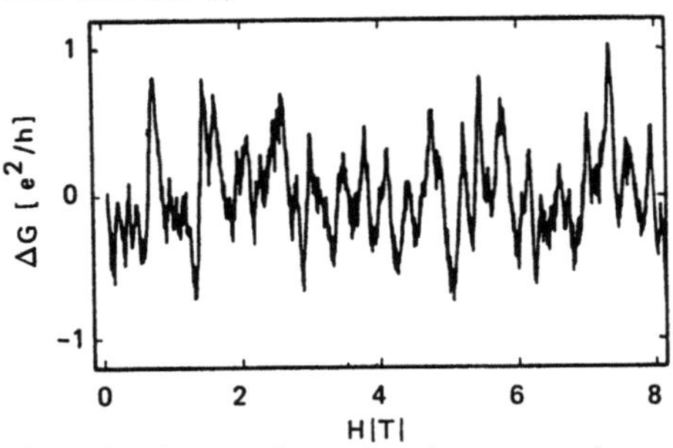

Fig. 2. Conductance fluctuations in an external magnetic field. (Figure taken from ref. [4]).)

scattered only elastically on the many random impurities. This situation is similar to the passage of light through a medium with a randomly varying index of refraction which gives rise to speckle patterns. Measurements of the conductance G of mesoscopic systems as a function of the Fermi energy E_F (which is changed by changing the gate voltage) or of an external magnetic field H (which serves to bend, classically speaking, the electron's trajectories and thereby causes a different set of scattering events on random scatterers) show a pattern of fluctuations which are rather similar to Ericson fluctuations [4]). An example is shown in fig. 2. On a given sample, the data are reproducible, although the pattern is seemingly random. However, different samples show different fluctuation patterns - hence the name "magnetofingerprints".

Fig. 2 shows only the *fluctuation* ΔG of the conductance G about its mean value $\bar{G}$. It is customary to express G in terms of the dimensionless conductance $g = hG/e^2$. The fluctuations Δg are found to be of order unity independent of sample length, sample geometry, and material composition of the sample, while typical values of $\bar{g}$ in mesoscopic systems and for good conductors range from a few hundred to a few thousand. This remarkable fact has led to the name "universal conductance fluctuations" for the phenomenon shown in fig. 2.

The present volume of Nuclear Physics, issued in honour of Torleif Ericson's 60th birthday, provides me with the opportunity to compare the two phenomena. As we proceed, we will discern striking similarities, but also essential differences. Is is my aim to keep the discussion free of all technical detail. It will nonethless carry us deeply into the modern field of "Chaotic quantum scattering".

In sect. 2, the observables related to the phenomena shown in figs. 1 and 2 are introduced, and their stochastic features are emphasized. Since the complete probability distribution for either observable is unknown, in sects. 3 and 4 we proceed to compare the mean values and fluctuations, respectively. Sect. 5 contains a short summary.

The present comparison, an outgrowth of my own work on both subjects, was triggered by remarks on the analogy between both types of phenomena made in ref. [5]).

2. The observables

For a comparison of both phenomena, it is important to clarify which observables they relate to. In the case of nuclear cross-section fluctuations, the ideal situation (approximately realized in fig. 1) is the one where a single channel dominates the cross section. This maximizes the fluctuations. In this case, the cross section is, except for kinematical factors, given by

$$\sigma_{ab} \sim |S_{ab}|^2 , \tag{1}$$

where $S_{ab}(E)$ is the element of the scattering matrix connecting channels a and b (with $a \neq b$), and E is the energy. To understand cross section fluctuations, it is obviously necessary to develop a theory for the fluctuations of $S_{ab}(E)$ versus E.

In the case of conductance fluctuations, we recall the relation between the conductance G and the dimensionless quantity $g = Gh/e^2$. For a disordered wire connected at either end to ideally conducting leads, g is given by the many-channel approximation to Landauer's formula,

$$g = 2 \sum_{ab} |S_{ab}^{\mathrm{LR}}(E_{\mathrm{F}})|^2 . \tag{2}$$

Here, L and R stand for the left and right lead, respectively. In either lead, there exists at the Fermi energy a multitude of transverse modes of the electron. Each of these modes defines a channel. The channels are labeled $a, b, \ldots$; $S_{ab}^{\mathrm{LR}}(E_{\mathrm{F}})$ is an element of the scattering matrix connecting channel a in the left lead with channel b in the right lead. Eq. (2) shows that g is the total transmission probability of the electron through the disordered sample. The factor 2 is due to spin.

A comparison of eqs. (1) and (2) displays obvious similarities and differences. In both cases, we deal with *bona fide* scattering problems; the scattering matrices in both cases fluctuate stochastically (either because of chaotic dynamics or because of random impurity scattering); in both cases, we deal with a situation with many ($\gg 1$) open channels. A significant difference lies in the absence of the channel summation in eq. (1). This enhances the fluctuations. Indeed, fig. 1 shows that the average cross section $\bar{\sigma}_{ab}$ and the fluctuations $[\overline{(\sigma_{ab} - \overline{\sigma_{ab}})^2}]^{1/2}$ are roughly equal. It was mentioned in sect. 1 that this is not the case for conductance fluctuations where $\bar{g}$ may easily equal several hundred while $[\overline{(g - \bar{g})^2}]^{1/2}$ is of order unity: The summation over many channels in eq. (2) acts like an averaging procedure and reduces the size of the fluctuations. In mesoscopic systems, it has been impossible for a long time to perform single-channel experiments of the Ericson type. In nuclei, it is conversely not customary (and would be very difficult) to study fluctuations of cross sections summed over many (but not over all) channels.

3. Mean values

In the domain of strongly overlapping CN resonances, the average value of $|S_{ab}|^2$, taken over an energy interval containing many fluctuations, is given by the Hauser–Feshbach formula $(a \neq b)$

$$\overline{|S_{ab}|^2} = T_a T_b / \sum_c T_c . \tag{3}$$

Here, $T_a = 1 - |\overline{S_{aa}}|^2$ with $0 \leq T_a \leq 1$ are "transmission coefficients" or "sticking probabilities", defined under the assumption that $\overline{S_{ab}} = 0$ for $a \neq b$. The set of transmission coefficients $\{T_a\}$ is a measure of the unitarity deficit $[1 - \sum_b |\overline{S_{ab}}|^2]$ of the average S-matrix. This fact suggests the physical interpretation of these coefficients: The energy-averaged S-matrix $\overline{S_{ab}}$ relates to the fast part of the nuclear reaction; the unitarity deficit of $\overline{S_{ab}}$ defines that part of the incident flux which spends a long time in the compound nucleus before decay occurs, and which thereby causes the appearance of stochastic fluctuations. Eq. (3) has a simple interpretation: The coefficient T_a is the probability of CN formation from channel a (hence "sticking probability"), while $T_b / \sum_c T_c$ is the branching ratio for decay into channel b.

Eq. (3) shows that formation and decay of the CN are independent. This statement - formulated as an hypothesis by N. Bohr in his famous 1937 paper [6]) introducing the concept of the CN - is only valid *on average*. Even so, two conditions must be met for this statement to hold. First, the decay time τ_{dec} of the CN must be large compared to the internal equilibration time τ_{equ} of the system. Second, the average width $\Gamma = \hbar / \tau_{\text{dec}}$ of the CN resonances must be large in comparison with their mean spacing d. We discuss these conditions quantitatively.

The decay width Γ is given by

$$\Gamma = \frac{d}{2\pi} \sum_a T_a . \tag{4}$$

The interpretation of this formula can be found in ref. [7]): A time-dependent wave packet in the space of CN resonances is quasiperiodic; it (nearly) returns to its original form after a time h/d; it therefore probes escaping into any one channel with frequency $\omega = d/h$. The escape width or decay width Γ is obtained by multiplying $\hbar\omega$ with the probability $\sum_a T_a$ of escaping into any one channel. The equilibration time τ_{equ} is related to a spreading width $\Gamma^{\downarrow} = \hbar / \tau_{\text{equ}}$; τ_{equ} is what solid-state physicists call the "diffusion time through the disordered sample"; in the context of nuclear physics, $\Gamma^{\downarrow}$ is typically estimated* to be 1 MeV. The condition $\tau_{\text{dec}} \gg \tau_{\text{equ}}$ or, equivalently, $\Gamma^{\downarrow} \gg \Gamma$, implies that CN configurations are thoroughly mixed before decay by particle emission sets in. It is intuitively obvious that thorough mixing (or "equilibration") is a prerequisite for independence of formation and decay of the CN. Quantitatively, fig. 1 suggests that the autocorrelation width of the fluctuations (which coincides with Γ, see sect. 4) is of the order 50 keV or so, and therefore very small compared to $\Gamma^{\downarrow}$.

The condition $\Gamma \gg d$ or $\sum_a T_a \gg 1$ is met when the resonances overlap strongly. In the opposite case, $\Gamma \ll d$, the CN cross section displays (nearly) isolated resonances

* This value is indicated by the spreading width of E1 giant resonances in nuclei and by the imaginary part of the optical-model potential a few MeV above threshold.

which behave stochastically, so that the cross section is a randomly fluctuating entity in this regime also. However, quantum mechanics implies that at each resonance, the cross section factorizes. As a consequence, the *average* cross section does not factorize exactly in the regime $\Gamma \ll d$. Put differently, the condition $\Gamma \gg d$ ensures that the decay time τ_{dec} is small compared to the Poincaré recurrence time h/d. Only then does the concept of equilibration have a meaning. Quantitatively, several tens of channels are open in the CN reaction of fig. 1, and the condition $\Gamma \gg d$ is well met.

The condition $\Gamma \gg d$ is embodied in eq. (3) which in fact is obtained as the leading term of an asymptotic expansion in powers of d/Γ, or of $(\sum_c T_c)^{-1}$. Analytical calculations of fluctuation properties of σ_{ab} and g use such asymptotic expansions; we return to this important point below.

As the bombarding energy increases, more and more channels open up, $\sum_a T_a$ increases, this increase outweighs the decrease of d, and Γ increases too, while $\Gamma^{\downarrow}$ remains roughly constant. At about 15–20 MeV bombarding energy, we have $\Gamma \simeq \Gamma^{\downarrow}$: Decay time and equilibration time are about equal; the CN decays before complete equilibration is attained. This is the domain of "preequilibrium" or "precompound" reactions which bear a close analogy to the conductance problem.

The average conductance $\bar{g}$ can be calculated in a model where a disordered wire is connected to two ideally conducting leads [8]). In the metallic regime $k_F l \gg 1$ (with k_F the Fermi wave number and l the elastic mean free path) eq. (2) yields for $\bar{g}$ the form

$$\bar{g} = 2\left(\sum_a T_a^R\right)\Pi^{LR}\left(\sum_b T_b^L\right) + \cdots . \tag{5}$$

As in eq. (2), the upper indices R, L refer to the right (left) lead, respectively, and the coefficients $T_a^c = 1 - |\overline{S_{aa}^{c,c}}|^2$ are "sticking probabilities" for the electron in the leads $c = \text{R, L}$ to "get stuck" in the disordered sample; cf. the discussion following eq. (3). Eq. (5) displays the term of leading order in an asymptotic expansion in inverse powers of $(\sum_{a,c} T_a^c)$; the dots indicate terms of higher order.

Except for the factor 2 and the summation over channels in eq. (5), the formal structure of the term on the r.h.s. is the same as in eq. (3). This is because the physical processes by which an electron enters or leaves the disordered sample are identical to formation and decay of the CN: In both cases, the interaction region (disordered sample or CN) is modelled by a random hamiltonian. Differences arise as we examine the coefficient Π^{LR} which describes the diffusive propagation of occupation probability from the left end of the disordered sample to its right end, or vice versa. It is given by

$$\Pi^{\text{LR}} = \left[\left(\sum_{a,c} T_a^c\right)\left(1 + \frac{\pi^2}{4}\left(1 - \frac{l}{L_x}\right)\frac{\Gamma}{E_c}\right)\right]^{-1} . \tag{6}$$

The quantities Γ and E_c appearing in the eq. (6) are identical with or at least closely related to the decay width Γ and the spreading width $\Gamma^{\downarrow}$, respectively, introduced in the context of Ericson fluctuations. Indeed, we have

$$\Gamma = \frac{\hbar}{\tau_{\text{dec}}} = \frac{d}{2\pi} \sum_{a,c} T_a^c , \tag{7}$$

where d is the mean level spacing in the disordered sample, and τ_{dec} the emission time of the electron from the disordered sample into the leads. Eq. (7) is identical to eq. (4). The "Thouless energy" E_c is defined in terms of the diffusion constant D and the length L_x of the sample,

$$E_c = \frac{\pi^2 \hbar D}{L_x^2} \simeq \frac{\hbar}{\tau_{diff}}, \tag{8}$$

where τ_{diff} is the diffusion time through the sample - analogous to the equilibration time τ_{equ} for Ericson fluctuations, so that E_c and $\Gamma^{\downarrow}$ also play analogous roles. We note that for $\Gamma \ll E_c$, eqs. (5) and (6) become identical in form to eq. (3) (again except for channel summations and the factor 2). This shows the very close similarity between Ericson and conductance fluctuations: In the latter case, we also deal with strongly overlapping resonances (in the disordered sample), as shown by eq. (7) and the fact that $\sum_{a,c} T_a^c \gg 1$. (The observation of conductance fluctuations in cases where only few channels are available has become possible only recently and is not discussed here). The value of $\bar{g}$ attained for $\Gamma \ll E_c$ has the same physical interpretation as the Hauser–Feshbach formula: The occupation probability uniformly fills the disordered sample ("equilibrates") before it decays by electron emission into the leads. Putting $T_a^c = 1$ for all (a, c), we note that for $\Gamma \ll E_c$, $\bar{g}$ attains the value Λ (with Λ the number of channels in either lead) rather than the maximum value 2Λ permitted by unitarity, cf. eq. (2). The reason is that the model of ref. [8]) does not allow for the existence of ballistic electrons. In the framework of a diffusion model, $\bar{g}$ cannot attain the "quantized contact resistance" value 2Λ.

Eq. (5) displays the "independence of formation and decay" of the "compound system", here defined by the statement that the electron is in the disordered sample. Eqs. (5) and (6) show that this independence holds irrespective of the value of Γ / E_c, in seeming contradiction to the satements made below eq. (4). The contradiction is resolved by observing that for geometrical reasons, electron emission into the left (right) lead is possible only from the left (right) end of the disordered wire whereas the CN may emit particles from any of its configurations. This last fact invalidates the factorized form (3) when the condition $\Gamma \gtrsim \Gamma^{\downarrow}$ is met.

The decisive difference between Ericson and conductance fluctuations lies in the fact that disordered samples of different lengths L_x can be produced. Since Γ and E_c depend differently on L_x, it is possible to attain, for the same material, different regimes: The "CN regime" where $\Gamma \leqslant E_c$, and the "precompound regime" where $E_c < \Gamma$. Indeed, eq. (8) shows that $E_c \sim L_x^{-2}$ while eq. (7) and the relation $d \sim L_x^{-1}$ show that $\Gamma \sim L_x^{-1}$. Most mesoscopic samples studied so far have lengths L_x such that $E_c > \Gamma$. Eqs. (5) and (6) show that in this regime, $\bar{g} \sim L_x^{-1}$, i.e. $\bar{g}$ is ohmic.

Fig. 3 shows a schematic plot of Γ. L_x, of $E_c \cdot L_x$ and of $d \cdot L_x$ versus length L_x. The points of intersection, $\Gamma = E_c$ and $d = E_c$, define length scales L_0 and L_{loc}, respectively. The first of these, L_0, defines the transition from the "CN regime" attained for $L \leqslant L_0$ to the "precompound" or "ohmic" regime where $L_0 < L$. The localization length L_{loc} is defined by the condition that Poincaré recurrence time and diffusion time coincide or, equivalently, that spreading width E_c and mean level spacing d are equal. This coincidence is made possible because $E_c \sim L_x^{-2}$. To a

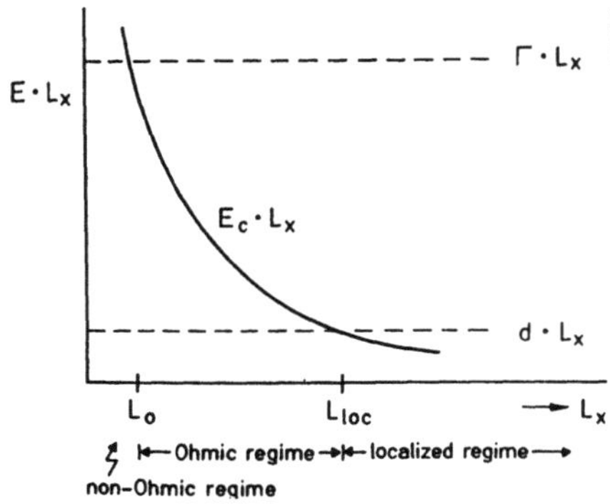

Fig. 3. Schematic plot of energy times length versus length L_x of the disordered sample.

nuclear physicist, it may seem curious that the spreading width *decreases* with the square of the sample length. But the fact is obvious when expressed in terms of the diffusion time through the sample: $\tau_{\rm diff} \sim L_x^2$ is an immediate consequence of the diffusion equation. For $L > L_{\rm loc}$ or $E_c < d$, wave functions are no longer mixed throughout the sample, and conductance is possible only through the expontential tails of individual wave functions: In the localized regime, the ohmic dependence of $\bar{g}$ on L_x is replaced by an exponential one, $\bar{g} \sim \exp(-L_x/L_{\rm loc})$.

In the nuclear physics context, these observations suggest the question whether localization effects could not be observed in precompound reactions. Indeed, the passage of occupation probability through a hierarchy of states of increasing complexity is akin to the passage of occupation probability through a disordered sample; increasing the length of the chain or the number of exciton states reached in the reaction is formally similar to increasing the length L_x of the disordered sample. Unfortunately, an increase in the number of exciton states participating in the reaction requires a rise in bombarding energy which in turn causes an exponential decrease in average level spacing. This shows that the equality $\Gamma^{\downarrow} \simeq d$ cannot be attained in nuclei.

4. Fluctuations

It might be expected that the elements S_{ab} of the S-matrix for CN reactions (which describe chaotic scattering) have a gaussian distribution, and this is indeed what theory predicts when many channels are open, and most $T_a \simeq 1$. More precisely: $S_{ab}(E)$ is a complex gaussian random process, uncorrelated with matrix elements carrying channel indices (a', b') of which at least one differs from both a and b. In the case that $\overline{S_{ab}} = 0$ for $a \neq b$, this implies that $|S_{ab}|^2$ has an exponential distribution. This theoretical expectation is borne out by the data: A plot of the frequency of occurrence of the cross section values in fig. 1, taken at fixed energy intervals, yields an exponential [2]). It follows immediately that

$$\overline{(|S_{ab}|^2)^2} - (\overline{|S_{ab}|^2})^2 = (\overline{|S_{ab}|^2})^2 . \tag{9}$$

This shows that the fluctuation $(\overline{\sigma_{ab}^2} - \overline{\sigma_{ab}}^2)^{1/2}$ of the cross section equals its mean value $\overline{\sigma_{ab}}$. Modifications of this result are due to two causes. Often more than one channel contributes to the cross section at fixed angle, and direct reactions cause $\overline{S_{ab}}$ to differ from zero for $a \neq b$. The influence of both modifications on the cross section fluctuations is well understood.

These statements apply to leading order in the asymptotic expansion which yields eq. (3), i.e. whenever $\sum_a T_a \gg 1$. In cases where this condition is violated, and where terms of higher order in the asymptotic expansion must be included, there have never been enough data on fluctuation properties of CN reactions to call for an extension of the theoretical analysis, except for the case $\Gamma \ll d$ where, however, interest was focussed on the spacing distribution of CN resonances. This situation has changed only recently in the context of tests of fundamental symmetries in CN reactions [9]).

Turning to conductance fluctuations, we naturally expect the elements S_{ab}^{LR} of the S-matrix in eq. (2) also to be complex gaussian random processes, and to leading order in the asymptotic expansion, this is indeed correct. We use this and the fact that S_{ab}^{LR} and S_{cd}^{LR} are uncorrelated (to the same order) for $(a, b) \neq (c, d)$ to work out the conductance fluctuations:

$$4 \sum_{abcd} \overline{|S_{ab}^{\mathrm{LR}}|^2 |S_{cd}^{\mathrm{LR}}|^2} - 4\left(\sum_{ab} \overline{|S_{ab}^{\mathrm{LR}}|^2} \right)^2 = 4\left(\sum_{ab} \overline{|S_{ab}^{\mathrm{LR}}|^2} \right)^2 . \tag{10}$$

With $\overline{|S_{ab}^{\mathrm{LR}}|^2} = T_a^{\mathrm{L}} \Pi^{\mathrm{LR}} T_b^{\mathrm{R}}$, this gives

$$4\left[\sum_a (T_a^{\mathrm{L}})^2 \right] \Pi^{\mathrm{LR}})^2 \left[\sum_b (T_b^{\mathrm{L}})^2 \right] . \tag{11}$$

With $T_d^c \simeq 1$ and Π^{LR} given by eq. (6), we find that for $\Gamma \leqslant E_c$, the fluctuations are indeed of order unity. We also note, however, that the fluctuations given by the expression (11) decrease rapidly with length since $\Pi^{\mathrm{LR}} \sim L_x^{-1}$ in the ohmic region, thus destroying both theoretical universality and, more importantly, the agreement with experiment.

What went wrong in the calculation of the expression (10)? Büttiker *et al.*, who used the reasoning just described, gave the answer [10]). The S-matrix elements are not gaussian, and S_{ab} is correlated with S_{cd} for $(a, b) \neq (c, d)$. Indeed, such correlations are found when the asymptotic expansion in powers of $(\sum_{ac} T_a^c)^{-1}$ is carried to higher orders than the first [5,11,8]). Also, terms of higher order are not negligible even though $\sum_{ac} T_a^c \gg 1$ because of the fourfold summation in eq. (10): Each summation essentially contributes a term $\sum_a T_a^c \gg 1$. And it is the contribution of such higher-order terms which accounts [11]) for the universality of the conductance fluctuations in the ohmic regime $E_c < \Gamma$.

We conclude that Ericson and conductance fluctuations, although both caused by a stochastic hamiltonian, are manifestations of chaotic quantum scattering in very different regimes. Ericson fluctuations occur in the regime $\Gamma \ll \Gamma^{\downarrow}$ and in cross section data involving few channels. They are direct manifestations of the gaussian distribution of the S-matrix elements. Conductance fluctuations, on the other hand, are usually measured in the regime $E_c > \Gamma$ (the ohmic regime). In this regime, the

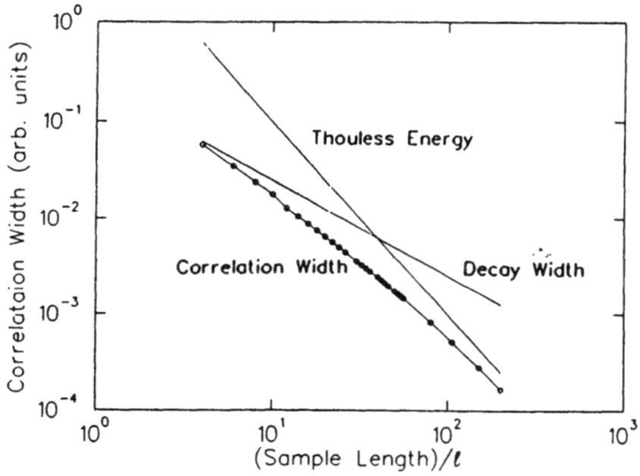

Fig. 4. Correlation width of a disordered sample with $T_a^c = 1$ for all a, c versus length in units of the elastic mean free path. (Taken from ref. [13]).)

gaussian term (10) contributes little to the fluctuations. The universality of the fluctuations is due to higher-order correlations between S-matrix elements. They dominate the fluctuations as L_x increases and approaches the localization length yielding a universal value for the fluctuations.

The dominance of higher-order correlations leads to the question: What is the distribution function for g as L_x approaches the localization length? It turns out [12]) that g is exponentially normal, with wide tails of the distribution that dominate the higher moments.

Besides the distribution of S-matrix elements, another measure of fluctuations is the autocorrelation function, $\overline{\sigma_{ab}(E)\sigma_{ab}(E+\Delta E)}-(\overline{\sigma_{ab}(E)})^2$ in the case of CN reactions, and $\overline{g(E_F)g(E_F+\Delta E)}-(\overline{g(E_F)})^2$ (with E_F the Fermi energy) in the case of conductance fluctuations. In both cases, the autocorrelation functions have (essentially) lorenzian shape, and are characterized by a width. In the case of Ericson fluctuations, this width coincides with the decay width Γ introduced in eq. (4). From the data in fig. 1, it is thus possible to infer the average CN lifetime $\hbar/\Gamma$. In the case of conductance fluctuations, the correlation width depends on L_x. For $L_x \leqslant L_0$, it is close to the decay width Γ of eq. (7), while for $L_x > L_0$, it is close to the spreading width E_c of eq. (8). This is shown quantitatively in fig. 4. The correlation width as determined by a fit to a theoretical calculation [13]) is plotted versus L_x; plots of Γ and E_c are also shown for comparison. A comparison of such results with the data has not yet been possible, for several reasons. First, it is difficult to manufacture macroscopically samples of different lengths. Second, fluctuations are more often investigated by applying an external magnetic field than by changing the gate voltage. Third, data are taken on samples with four-lead geometry, rather than with the two-lead geometry used in the model of ref. [13]).

5. Conclusions

The chaotic dynamics of the CN and the dynamics of disordered conductors are fundamentally similar. In particular, they give rise to similar fluctuation properties. This is the reason for the fundamental similarity between Ericson fluctuations and conductance fluctuations. Both phenomena can be viewed as random or chaotic scattering processes, governed by a scattering matrix in the domain of strongly overlapping resonances. The elements of this matrix are, to lowest order in an asymptotic expansion involving the inverse of the number of open channels, uncorrelated complex random variables with gaussian probability distribution.

Differences arise because Ericson fluctuations are most conspicuous when only few channels contribute to the reaction, whereas the conductance is necessarily a sum over many channels. Therefore, the r.m.s deviation of the nuclear cross section is of the order of the average cross section, while conductance fluctuations are only a small ripple on a large background.

A further difference arises because the CN is an equilibrated system while in typical mesoscopic systems, the diffusion time through the disordered sample is large compared to the emission time of the electron into the leads. This causes conductance fluctuations to be similar to properties of precompound rather than CN reactions, with obvious differences due to the geometry. Moreover, in a precompound or ohmic situation, the fluctuations would be strongly suppressed if the elements of the S-matrix were indeed strictly gaussian random variables. It is only through deviations from gaussian statistics, manifest in correlation terms of higher order in the asymptotic expansion, that the fluctuations survive and, for large length scales of the mesoscopic sample, attain their universal value.

A final difference is manifest in the autocorrelation functions of nuclear cross sections, and of the conductance. In the CN case of an equilibrated system, the correlation width is given by the inverse lifetime of the system for particle emission, while in typical mesoscopic systems, it is close to the inverse diffusion time through the system.

We have not discussed the passage of light- and sound-waves through disordered media. These phenomena provide another example of chaotic wave scattering [14]) and in some sense are intermediate between Ericson and conductance fluctuations.

References

1) T. Ericson, Phys. Rev. Lett. **5** (1960) 430
2) T. Ericson and T. Mayer-Kuckuk, Ann. Rev. Nucl. Sci. **16** (1966) 183
3) O. Bohigas and H.A. Weidenmüller, Ann. Rev. Nucl. Part. Sci. **38** (1988) 421
4) S. Washburn and R.A. Webb, Adv. Phys. **35** (1986) 375
5) P.A. Lee, A.D. Stone and H. Fukuyama, Phys. Rev. **B35** (1987) 1039
6) N. Bohr, Nature **137** (1936) 344
7) J.M. Blatt and V.F. Weiskopf, Theoretical nuclear physics (Wiley, New York, 1952) chapter VIII. 7C
8) S. Iida, H.A. Weidenmüller, and J.A. Zuk, Phys. Rev. Lett. **64** (1990) 583; Ann. of Phys., in press
9) E.D. Davis and D. Boosé, Z. Phys. **A332** (1989) 427
10) M. Büttiker, Y. Irmy, R. Landauer and S. Pinhas, Phys. Rev. **B31** (1985) 6207
11) B.L. Altshuler and B.I. Shkovskii, Sov. Phys. JETP **64** (1986) 127
12) B.L. Altshuler, V.E. Kravtsov and I.V. Lerner, Sov. Phys. JETP **64** (1986) 1352
13) A. Altland, private communication (1989) and to be published
14) See the special issue of J. Optical Soc. Ann. **2** (1985) 2096 on this field, and for more recent work S. Feng, C. Kane, P.A. Lee and A.D. Stone, Phys. Rev. Lett. **61** (1988) 834

Hot Nuclei – Landau Theory, Thermal Fluctuations and Dissipation

Y. Alhassid

Center for Theoretical Physics, Sloane Physics Laboratory, and A.W. Wright Nuclear Structure Laboratory, Yale University, New Haven, CT 06511, USA

Abstract. The basic ideas and theoretical methods used in the description of hot nuclei are reviewed. In particular, a macroscopic approach to shape transitions is discussed in the framework of the Landau theory in which the quadrupole shape degrees of freedom play the role of the order parameters. This theory describes the universal features of the nuclear shape evolution with temperature and spin. A unified description of fluctuations in all five quadrupole degrees of freedom is introduced and plays an important role in the calculation of physical observables. A macroscopic approach to the giant dipole resonance (GDR) in hot nuclei is developed. With all parameters fixed by the zero temperature nuclear properties, the theory predicts both the GDR cross-section and angular anisotropy of the γ-rays in very good agreement with recent experiments. The intrinsic shape fluctuations are the main cause for the resonance broadening at higher temperatures, while the orientation fluctuations are responsible for the observed attenuation in the angular anisotropy. Dissipation at finite temperature is discussed in the framework of a Langevin-like equation describing the time-dependent shape fluctuations. Non-adiabatic effects may cause motional narrowing of the resonance.

1. Introduction

The possibility of heating up a nucleus to a finite temperature opens a new direction in nuclear structure studies. The principal experimental techniques of forming such hot nuclei are via heavy ion fusion reactions with heavy projectiles. The relative kinetic energy in these collisions is deposited in internal excitations of the nucleus. If there is enough time, this energy will be redistributed among the large number of degrees of freedom of the heavy compound nucleus and

Eds: Y. Abe, H. Horiuchi, K. Matsuyanagi

would lead to an equilibrated hot nucleus [1-3]. A typical heavy nucleus can store up to several hundred MeV of excitation energy in its dense "reservoir" of excited states.

Numerous phenomena are encountered in the study of hot nuclei [4-12]. Various phase-transition-like phenomena are expected to occur as we heat the nucleus. At critical temperatures of T_c~0.5-1 MeV there is the disappearance of pairing [4], at T_c~1-3 MeV we have the shape transitions of well deformed nuclei associated with the melting of the shell structure [5-7], and at T_c~7-12 MeV there is the liquid-gas phase transition [8] above which the finite nucleus does not exist as such. The latter transition is hard to observe since Coulomb instabilities [8] start to develop at lower temperatures (T~5 MeV). Of great interest both experimentally [9-14] and theoretically [15-17] is the study of collective nuclear motion at finite temperature. In the following review we shall study mainly the shape transitions, the evolution of the nuclear shape with temperature and spin, and the collective nuclear motion at the corresponding range of temperatures and spins. However, many of the developed and discussed theoretical techniques are useful in the description of other phenomena in hot nuclei.

The main theoretical methods used to describe hot nuclei are taken from statistical mechanics. However, since the nuclear system is finite (typically in the order of 100 degrees of freedom) these techniques should be applied with care. In particular we shall see that thermal fluctuations play a major role even away from the critical point.

Recent advances in detector systems make it possible to study experimentally the properties of nuclei under conditions of high excitation and high spin. These properties are inferred from the measured spectra of γ-rays as well as of particles of low to medium mass that are emitted from the nuclear complex during its cooling process. A major probe of the shape of such hot nuclei is the giant dipole resonance [18-24] whose frequencies are known to depend on the nuclear shape.

The present review is organized as follows: In Sect. 2, we introduce the general concepts of hot nuclei: the equilibrium assumption and the definition of nuclear temperature and nuclear rotation. In Sect. 3, we discuss the various mean-field approximations used to treat hot rotating nuclei. The main discussion will focus on a recently developed macroscopic approach: the Landau theory of shape transitions [25,26]. Since the nuclear system is finite, fluctuations are important [27-32] and will be introduced in Sect. 4, using the framework of the Landau theory. In Sect. 5, we shall introduce a macroscopic theory of the giant dipole resonance (GDR) in hot nuclei [29,32]. Comparison will be made to experimentally observed quantities - the GDR absorption cross-section and the angular anisotropy [33,34] of the GDR γ-rays observed in the decay of the hot nucleus. Dissipation at finite temperature is also an important issue. It is discussed in Sect. 6 in the context of time-dependent fluctuations [35,36]. If these fluctuations are non-adiabatic, they will lead to motional narrowing [37,38] in the GDR from which it is possible to determine certain friction coefficients. Sect. 7 discusses briefly the effect of fluctuations on the B(E2) transitions in the continuum which were measured recently [39].

2. Hot Nuclei

2.1 The Equilibrium Assumption

At low energies the nuclear spectrum is discrete and a variety of theoretical models have been introduced to explain it. However, already for $E^{*}\gtrsim 6$ MeV the level density of a heavy nucleus is so high that its spectrum is quasi-continuous. Individual states are then hard to resolve experimentally or theoretically. Some renouncement of knowledge is required in order for the problem to become tractable again. The basic underlying assumption is the equilibrium assumption: all states at a given excitation energy, spin and any other conserved quantum numbers are equally probable. It is only under this assumption that a reduced global description of the nucleus in terms of a few macroscopic parameters can be achieved. The ensemble describing such a nucleus is the microcanonical ensemble.

Global equilibration means that various parts of the nucleus have enough time to reach complete relaxation and to explore all available phase space at the given excitation energy and spin. We shall refer to such a nucleus whose energy is shared among its many degrees of freedom as a hot nucleus. In a heavy ion reaction with a heavy projectile we expect that an important fraction of the fusion cross-section is the formation of an equilibrated nucleus. This is generally true for beam energy per nucleon of less than 5-7 MeV. It is of course an experimental task to identify that part of the cross-section which goes into the formation of an equilibrated hot nucleus [1-3].

2.2 Nuclear Temperature

The microcanonical ensemble at a given energy E which is proportional to $\delta(E-H)$ is not very convenient to work with. We therefore replace it by the canonical distribution $D \propto \exp(-H/T)$ at a certain temperature T. This temperature is determined by the condition that the average energy of the canonical ensemble is the given energy E. This leads to the following definition of nuclear temperature

$$\frac{1}{T} = \frac{d}{dE^*} \ln\rho(E^*) \ , \tag{2.1}$$

where $\rho(E^*)$ is the nuclear level density at excitation energy E^*.

The formal relation [40] between the microcanonical and canonical ensembles is established when the canonical partition function

$$Z(T) = \mathrm{Tr}\ e^{-H/T} \ , \tag{2.2}$$

is evaluated in terms of the level density (which is the microcanonical partition function $\rho(E) = \mathrm{Tr}\delta(E-H)$),

$$Z(T) = \int_0^\infty dE e^{-E/T} \rho(E) \ . \tag{2.3}$$

Relation (2.3) can be inverted via the inverse Laplace transform

$$\rho(E) = \frac{1}{2\pi i} \int_{-i\infty}^{i\infty} d\left(\frac{1}{T}\right) e^{E/T} Z(T) \ . \tag{2.4}$$

If (2.4) is evaluated by the saddle point approximation we obtain the relation (2.1). The exact transformation

(2.4) represents the fluctuations in the inverse temperature for a fixed value of E.

2.3 Nuclear Rotation

In the formation of a compound nucleus in a heavy ion reaction the nucleus usually acquired large angular momentum. The microcanonical equilibrium ensemble is then proportional to $\delta(E-H)P_J$, where P_J is the projection on states with a given spin J. It is again convenient to transform to the temperature representation where we define a partition function at a given spin J to be

$$Z(T,J) = \mathrm{Tr}\,(e^{-H/T} P_J) = \int_0^\infty dE e^{-E/T} \rho(E,J) \; . \qquad (2.5)$$

Here $\rho(E,J)$ is the level density at energy E and spin J. In terms of the density $\rho(E,M)$ with M being the spin projection [40]:

$$\rho(E,J) \approx - \frac{\partial}{\partial M} \rho(E,M) \Big|_{M=J+1/2} \; . \qquad (2.6)$$

Introducing the free energy in a frame rotating with angular velocity $\vec{\omega}$ [41]

$$F(T,\omega) = - T \ln \mathrm{Tr}\, e^{-(H-\vec{\omega}\cdot\vec{J})/T} \; , \qquad (2.7)$$

we can rewrite (2.5) in the form

$$Z(T,J) = \int_0^{4\pi T} \frac{i\omega d\omega}{4\pi T^2} e^{-\left[i(J+1/2)\omega + F(T,i\omega)\right]/T} \; . \qquad (2.8)$$

Thus by considering ω-fluctuations it is possible to calculate the partition function at a given spin. Usually (2.8) is evaluated by the saddle point approximation, where ω is determined from J by the relation

$$J + 1/2 = - \partial F(T,\omega)/\partial\omega \; . \qquad (2.9)$$

This accomplishes the transformation from the description in terms of the extensive variables E and J to the intensive conjugate variables T and ω. The free energy in the rotating frame (2.7) is the fundamental relation which contains all the thermodynamic information on the system. In the next section we shall discuss how to calculate such a quantity.

3. Mean-Field and Landau Theories

Since H in (2.7) is a many-body Hamiltonian, the evaluation of the exact many-body partition function is difficult and approximations must be introduced. Most of them are based on various versions of the mean-field approximation.

3.1 Mean-Field Approximations

In general we seek to calculate the free energy $F=F(T,\omega;\rho)$ as a function of some trial parameters ρ. The equilibrium configuration is then found by minimizing F with repect to ρ.

The basic variational principle from which such approaches are derived is the following: The exact canonical density matrix $D=\exp(-H(T))/Z(T)$ is obtained by minimizing the free energy functional

$$F[T,D] = \mathrm{Tr}(DH) + \mathrm{Tr}\,(D\ln D)\ , \qquad (3.1)$$

with respect to all possible density matrices satisfying

$$\mathrm{Tr}\,D = 1\ . \qquad (3.2)$$

If we restrict the space of variations to density matrices of the form

$$D_0 \propto \exp\,(-H_0/T)\ , \qquad (3.3)$$

where H_0 is a one-body Hamiltonian, we find

$$F[T,\rho] = \mathrm{tr}\,(t\rho + \frac{1}{2}\,\rho v\rho) + T\,\mathrm{tr}\,[\rho\ell n\rho + (1-\rho)\ell n(1-\rho)]\ . \qquad (3.4)$$

Here ρ is the one-body density matrix and t and v are the one-body kinetic energy and two-body potential energy, respectively.

Minimizing F in (3.4) with respect to ρ, we obtained the finite-temperature Hartree-Fock equation [6]

$$\rho = \{ 1 + \exp [(h_\rho - \mu)/T] \}^{-1} , \qquad (3.5)$$

where

$$h_\rho = t + v\rho , \qquad (3.6)$$

is the mean-field single-particle Hamiltonian.

If we wish to include pairing effects we can do it within the finite-temperature Hartree-Fock-Bogoliubov approximation [4].

If the phenomena we want to describe are the shape evolution with temperature and spin, a more restrictive choice of H_0 corresponds to that of the Nilsson Hamiltonian which is deformation-dependent but not self-consistent. The single-particle states with energies ϵ_i are assumed to be populated with the Fermi-Dirac occupation probabilities f_i

$$f_i = \frac{1}{1 + e^{(\epsilon_i - \mu)/T}} . \qquad (3.7)$$

To account for the correct average energy the Strutinsky method is used [7].

Another derivation of a mean-field approximation is based on the functional integral approach, where the many-body partition function is expressed as an integral over all "possible" one-body partition functions [42]:

$$Z(T) = \mathrm{Tr}\left(e^{-H/T}\right) = \int D[\sigma]\ e^{-F(T,\omega;\sigma)/T} . \qquad (3.8)$$

Here $\sigma(\tau)$ $(0 \le \tau \le T^{-1})$ is a single-particle density function which depends on an imaginary time-like parameter τ. For τ-independent σ's (the so-called "static approximation" [43,44]) the free energy $F(T,\omega;\sigma)$ is given by

$$F(T,\sigma) = - \frac{1}{2}\ \mathrm{tr}(\sigma v \sigma) - T \ln \mathrm{Tr}\left(e^{-H_\sigma/T}\right) . \qquad (3.9a)$$

Here H_σ is a sum of the single-particle Hamiltonians $h_\sigma^{(i)}$ where

$$h_\sigma = t + v\sigma . \qquad (3.9b)$$

Eqs. (3.9) describe the free energy of nucleons moving independently in a mean-field $v\sigma$. If the integral (3.8) is evaluated in the stationary phase approximation one recovers the Hartree-Fock equation (3.5). The advantage of the functional integral approach is that it allows us to go beyond the mean-field approximation by including fluctuations in σ. We shall return to this issue in Sect. 4.

To describe rotation of a hot nucleus we replace in (3.1) H by $H-\vec{\omega}\cdot\vec{J}$, which is the Hamiltonian in the frame rotating with angular velocity $\vec{\omega}$ [41]. Using one-body density matrices of the form (3.3) we obtain the cranked mean-field approximation where a Coriolis coupling $-\vec{\omega}\cdot\vec{j}$ is included in the one-body Hamiltonian (3.9b). The free energy is now $F = F(T,\vec{\omega};\rho)$ and can be expanded to second order in ω

$$F(T,w;\rho) = F(T,\omega=0;\rho) - \frac{1}{2} \sum_{i,j} I_{ij} (T;\rho)\, \omega_i \omega_j + \ldots \quad , \tag{3.10}$$

where I_{ij} is the symmetric moment of inertia tensor. It is found that

$$I_{ij} = \int_0^{1/T} < J_i(\tau)\; J_j(0) > d\tau \;, \tag{3.11}$$

where

$$< J_i(\tau)\; J_j(0) > = Z^{-1}\, \mathrm{Tr}\left[e^{-H_0/T} \left(e^{H_0 \tau}\; J_i\; e^{-H_0 \tau} \right) J_j \right] . \tag{3.12}$$

H_0 is the mean-field Hamiltonian characterized by the trial parameters ρ. If $|i>$ are the corresponding single-particle states with single-particle energies ϵ_i, then [26]

$$I_{zz} = \sum_i |<i|J_z|i>|^2 (-\partial f_i/\partial \epsilon_i) + \sum_{i\neq k} |<i|j_z|k>|^2 \left(\frac{f_k - f_i}{\epsilon_i - \epsilon_k} \right) . \tag{3.13}$$

The first contribution on the r.h.s. to I_{zz} is from the alignment of the single particle spins along the rotation axis while the second is from the collective rotation.

Open shell nuclei are usually found to be deformed in their ground state. This is a quantum mechanical effect related to the shell structure of the single-particle levels of the mean-field potential. Various mean-field calculations [6,7] indicate that when the temperature increases the nuclear shape is changing and eventually a transition to a spherical shape occurs at temperatures $T \approx 1$-2 MeV for deformed rare-earth nuclei. In the following we shall concentrate on the nuclear shape evolution with temperature and spin.

3.2 Landau Theory

The most relevant and universal features of any microscopic mean-field theory of shape transitions in hot nuclei can be described in the framework of the Landau theory [25,26]. The approach also offers a useful and economical parametrization of the results of microscopic calculations and singles out a small number of the most relevant combinations of the parameters on which the equilibrium shape depends.

The theory was introduced by Landau in 1937 [45] to describe symmetry breaking phase-transitions in statistical systems. The free energy which is a scalar is expanded in a set of trial parameters known as the order parameters. The equilibrium value of the order parameters is found by minimizing this free energy. In the symmetry-breaking phase these values of the order parameters are non-zero.

The most crucial trial parameters in the study of nuclear shapes are the quadrupole deformation parameters $\alpha_{2\mu}$ (μ=-2,-1,...,2). Thus instead of considering $F(T,\omega;\rho)$ we shall consider only an effective free energy $F(T,\omega;\alpha_{2\mu})$ by minimizing the former with respect to all other shape parameters. The symmetry-breaking phase is a deformed nucleus with $\alpha_{2\mu} \neq 0$. Thus the symmetry which is broken is the rotational invariance and the order parameters are $\alpha_{2\mu}$, the non-zero values of which characterize the degree of symmetry breaking, In the presence of rotation, the full rotational symmetry is explicitly broken by the preferred direction of the rotational axis. The remaining symmetry is that of rotations around the direction of $\vec{\omega}$, which is spontaneously broken in the phase-transition.

Since F must be a scalar, only rotationally invariant combinations of $\vec{\omega}$ and $\alpha_{2\mu}$ are allowed. The ω-independent invariants to fourth order are $(\alpha\times\alpha)^{(0)}$, $((\alpha\times\alpha)^{(2)}\times\alpha)^{(0)}$ and $(\alpha\times\alpha)^{(0)}\cdot(\alpha\times\alpha)^{(0)}$. The lowest invariants involving $\vec{\omega}$ are quadratic in ω such as

$((\omega \times \omega)^{(2)} \times \alpha)^{(0)}$, $((\omega \times \omega)^{(2)} \times (\alpha \times \alpha)^{(2)})^{(0)}$ etc. To second order in $\vec{\omega}$ one can arrange these terms in the form of the expansion (3.10). We can transform from the laboratory $\alpha_{2\mu}$ to the Hill-Wheeler intrinsic parameters (β,γ) and the Euler angles $\Omega=(\psi,\theta,\phi)$. The latter characterize the orientation of the nucleus' principal frame x'-y'-z' with respect to the rotation axis $\vec{\omega}$ which we choose along the z axis of the laboratory frame. The moment of inertia tensor I_{ij} is diagonal in the intrinsic principal frame and the components of $\vec{\omega}$ in that frame are

$$\begin{aligned} \omega_{x'} &= -\,\omega \cos\phi \sin\theta \\ \omega_{y'} &= \omega \sin\phi \sin\theta \\ \omega_{z'} &= \omega \cos\theta \quad . \end{aligned} \tag{3.14}$$

Thus (3.10) becomes

$$F(T,\omega;\alpha_{2\mu}) = F(T,\omega{=}0;\beta,\gamma) - \frac{1}{2}\left[I_{x'x'}\sin^2\theta\cos^2\phi + I_{y'y'}\sin^2\theta\sin^2\phi + I_{z'z'}\cos^2\theta\right]\omega^2 \ . \tag{3.15}$$

The above Landau expansion leads to the following

$$F(T,\omega{=}0;\beta,\gamma) = F_0(T) + A(T)\beta^2 - B(T)\,\beta^3\cos 3\gamma + + C(T)\beta^4 \ \ldots \quad , \tag{3.16}$$

and

$$\begin{aligned} I_{z'z'}(T;\beta,\gamma) &= I_0(T) - 2R(T)\,\beta\cos\gamma + 2I_1(T)\beta^2 + \\ &\quad + 2D(T)\,\beta^2\sin^2\gamma + \ldots \quad , \\ I_{x'x'}(T;\beta,\gamma) &= I_{z'z'}(T;\beta,\gamma - 2\pi/3) \ , \\ I_{y'y'}(T;\beta,\gamma) &= I_{z'z'}(T;\beta,\gamma + 2\pi/3) \quad . \end{aligned} \tag{3.17}$$

The temperature-dependent coefficients $F_0, A, B, C, I_0, R, I_1$

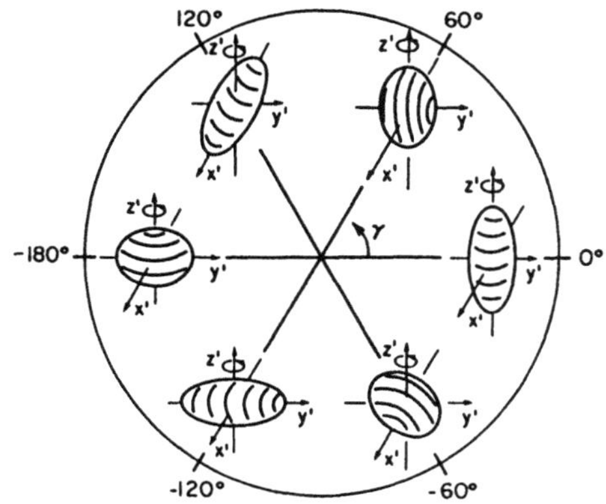

Fig. 1. The (β,γ) plane of shapes of deformed nuclei rotating around a princiapal axis z'.

and D are phenomenological parameters which are not determined by the Landau theory. However, the topography of the free energy surface (3.15) - (3.17) depends only on certain combinations of these parameters as we shall see below.

The equilibrium configuration is found by minimizing the free energy with respect to $\alpha_{2\mu}$. Minimizing first with respect to the orientation Ω, one obtains as a necessary condition for the minimum the standard result that the nucleus must be oriented such that one of its principal axes, say z', is directed along $\vec{\omega}$. We then allow (β,γ) to cover the full plane $(0 \leq \beta < \infty;$ $-180^\circ \leq \gamma \leq 180^\circ)$. This (β,γ) plane of shapes of deformed nuclei rotating around a principal axis z' is shown in Fig. 1. The sufficient conditions for a minimum require further that the rotation is around the axis with the largest moment of inertia, namely

$$I_{z'z'} > I_{x'x'}, I_{y'y'} \quad . \tag{3.18}$$

After minimizing with respect to the angles, the free energy (3.15) becomes

$$F(T,\omega;\alpha_{2\mu}) = F_0(T) + A(T)\beta^2 - B(T)\beta^3\cos3\gamma +$$

$$+ C(T)\beta^4 - \frac{1}{2} I_{z'z'}(T;\beta,\gamma)\omega^2 , \tag{3.19}$$

where $I_{z'z'}$ is given by (3.17).

For stability one must require C(T) > 0, and nuclei with deformed ground state have A(T) < 0 at low temperatures. The prolate-oblate asymmetry requires B(T) ≠ 0. For definiteness, we will discuss the more frequently occurring prolate case B(T) > 0. When T increases, A(T) changes sign at some temperature which is usually between 1 and 2 MeV for the deformed rare-earth nuclei. In the case D=0 (rigid body moment of inertia) the relevant dimensionless combinations are the reduced temperature

$$\tau = \frac{AC}{B^2} , \tag{3.20}$$

and ω/ω_c where

$$\omega_c = \frac{9}{16} (B/C) (B/R)^{1/2} , \tag{3.21}$$

is a critical angular velocity.

$$\tau = \frac{9}{8} \left[1 - (\omega/\omega_c - 1)^2\right] . \tag{3.22}$$

When τ increases above (3.22), the oblate shape becomes the only stable minimum. This is a second-order phase-transition where the order parameter (at the minimum) changes continuously. When $\omega/\omega_c<1$ (see l.h.s. of Fig. 2) and τ reaches (3.22), the oblate saddle point turns into a local minimum but the triaxial minimum is still separated from it. Instead, when τ increases past (3.22) a new triaxial saddle point emerges out of the oblate and moves towards the old. The two coalesce and disappear when τ reaches

$$\tau = \frac{9}{32} \left(1 + \frac{3}{4} \omega^2/\omega_c^2\right) . \tag{3.23}$$

Above the τ in (3.23) only the oblate minimum remains. The phase transition from the triaxial minimum to the oblate minimum occurs for τ between (3.22) and (3.23), when the free energies of these two minima are equal. This is then a first-order transition. The motion of the extrema is demonstrated in Fig. 2 for $\omega < \omega_c$ and for $\omega > \omega_c$. Fig. 3 is a phase diagram in the τ - ω/ω_c plane for D = 0 nuclei. The solid line for $\omega > \omega_c$ is the transition line of the second-order transitions. For $\omega < \omega_c$, there is a "coexistence" region whose boundaries are defined by (3.22) and (3.23). It is

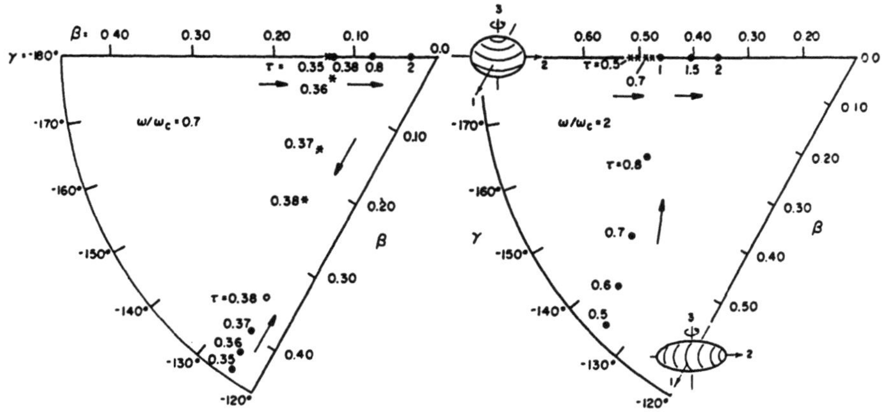

Fig. 2. Motion of the extrema in the β-γ plane with changing τ and fixed ω/ω_c. On the right a second-order transition is shown for ω/ω_c=2 with dots denoting a global minimum and asterisks a saddle point. On the left is a first-order transition for ω/ω_c=0.7 where open circles denote a local minimum.

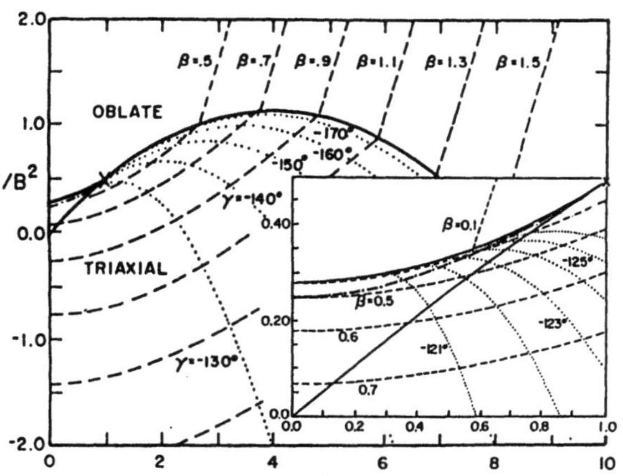

Fig. 3. Phase-diagram in the variables τ and ω/ω_c. The tricritical point is denoted by an ×. The solid line separates the triaxial phase from the non-collective phase. Insert: the first order transition region.

shown in more detail in the insert of Fig. 3. The point $\omega = \omega_c$, $\tau = \tau_c = 63/128$ separates the first order from the second order transitions and is known as a tricritical point [46]. Also shown in Fig. 3 are contour lines of β (in units of B/C) and γ. We see that near the tricritical point the shape changes rapidly from almost prolate to oblate.

In these reduced variables a universal phase diagram emerges. Minimizing (3.19) one finds that at negative values of τ, i.e. at low T and not too large ω, there are seven extrema in the entire (β,γ) plane. They are located symmetrically due to the $\gamma \leftrightarrow -\gamma$ symmetry of the free energy. For $D = 0$, the condition $I_{z'z'} > I_{x'x'}$, $I_{y'y'}$ (see Eq. (3.18)) selects those extrema which fall in the sector $|\gamma|>120^{\circ}$ only. At negative τ, there are three such extrema, one oblate ($\gamma = -180^{\circ}$) and two triaxial related by the $\gamma \leftrightarrow -\gamma$ symmetry. Only one of the latter should be considered, since they both give the same shape up to a rotation by 90° around the z axis. We choose to work in the sector $-180^{\circ} \leq \gamma \leq -120^{\circ}$. For $B > 0$ and $\tau < 0$, the triaxial configuration is the stable equilibrium shape and the oblate is a saddle point.

When τ increases towards positive values for fixed ω, one finds a behavior which depends on whether ω is smaller or larger than ω_c (see Eq. (3.21)). For fixed $\omega/\omega_c>1$ and τ increasing (see r.h.s. of Fig. 2), the triaxial minimum moves towards the oblate saddle point and coincides with it when τ reaches the value

3.3. **Microscopic Calculations**

The unfolding of the universal phase-diagram to the experimentally accessible variables, excitation energy and spin, requires the knowledge of the Landau parameters in the expansion (3.15)-(3.17). We have performed [47] microscopic calculations of the free energy surfaces of various nuclei and mapped them on the Landau expansion in order to determine the parameters A(T), B(T), C(T) etc.

The surfaces are calculated using cranked Nilsson-Strutinsky procedure for rotations along a principal axis z. The frequencies ω_k of the deformed harmonic potential well were parametrized according to Hill and Wheeler:

$$\omega_k = \omega_0 \exp\left[-\sqrt{\frac{5}{4\pi}}\,\beta\cos\left(\gamma - \frac{2\pi}{3}k\right)\right] . \qquad (3.24)$$

For ω=0 (no rotation) pairing is also included using a BCS monopole pairing force. In the liquid-drop energy we have allowed for large deformations by taking the exact Coulomb and surface energies expressed as elliptic

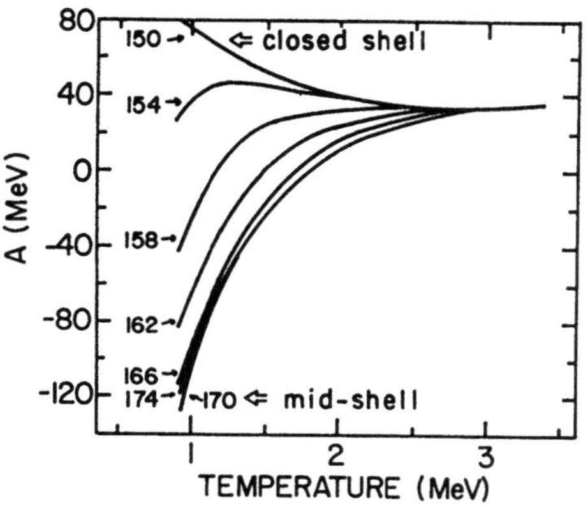

Fig. 4. The Landau parameter A(T) for various erbium isotopes as calculated microscopically. All curves converge to a common value ≈30 MeV for T ≳ 3 MeV.

integrals. For a given nucleus and a temperature T we can determine based on the expansion (3.15), the surfaces $F(T;\omega=0,\beta,\gamma)$ and $I_{zz}(T;\beta,\gamma)$. Using (3.16), we find F_0, A,B and C. Using (3.17) we determine the moment of inertia coefficients I_0,R,I_1 and D. The complete Landau expansion at any ω and $\alpha_{2\mu}$ is then given by (3.15). The Landau expansion (3.15)-(3.17) reproduces well the microscopic surfaces for T ≳ 1 MeV and moderate spins.

The Landau parameters have been determined [47] for all even-even rare-earth nuclei. They are found to obey simple systematics as a function of N and Z of the nucleus. Fig. 4 shows, for example, A(T) for different erbium isotopes in the temperature range T ≳ 0.8 MeV. A is the most crucial parameter in the Landau theory since the prolate to spherical shape transition at $\omega = 0$ occurs near the temperature where A changes sign. We see from Fig. 4 that the value of A is sensitive to shell effects. For mid-shell nuclei, which are strongly deformed in their ground state, A starts from negative values large in magnitude, and increases monotonically with T towards positive values. Such nuclei undergo shape transition from almost prolate to oblate at $\omega \neq 0$. For nuclei near shell closure, A starts positive and decreases monotonically while never becoming negative. Such nuclei start and stay spherical but become softer with increasing T. Above T ≈ 3 MeV all rare-earth nuclei have approximately the same A (≈30 MeV) due to the disappearance of shell effects. Figure 5 shows the

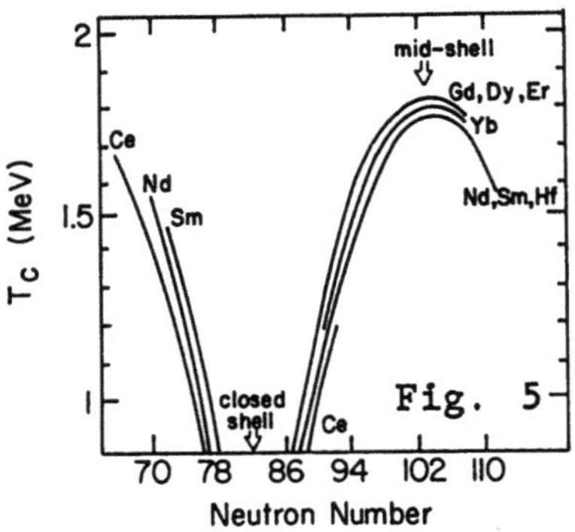

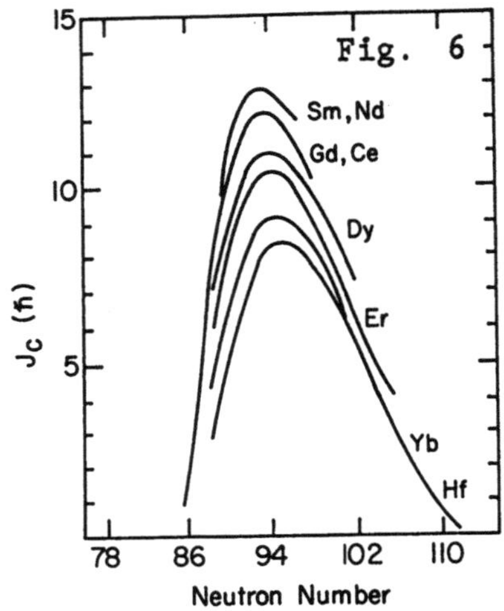

Fig. 5. The critical temperature T_c as a function of neutron number for even-even rare-earth nuclei. The maximal value of T_c occurs near the neutron mid-shell (N=104) for a given isotope family, and at the proton mid-shell (Z=66) among the various families.

Fig. 6. The angular momentum J_c at the tricritical point vs. neutron number for rare-earth nuclei.

critical temperature T_c versus neutron number for the even-even rare-earth nuclei. T_c is very close to the temperature at which $\tau = \tau_c$. On the phase-diagram of deformed nuclei the value of T_c determines the position of the line of triaxial to oblate shape transitions. The systematic dependence of T_c on the neutron and proton numbers is strikingly simple. For each family of isotopes between two closed neutron shells the values of T_c fall on an inverted parabola-like curve whose maximum is at mid-shell. The curve drops rapidly towards shell closure. This is of course consistent with the A systematics of Fig. 4. For various families of isotopes the parabola-like curves are arranged like onion shells where the innermost shells correspond to isotopes near proton shell closure at Z = 50 and 82, and the outermost shells are in the region of proton mid-shell, $_{66}$Dy. The largest critical temperature of $T_c \sim 1.85$ MeV is found in mid-neutron shell isotopes of Gd, Dy and Er.

Finally, we consider the angular momentum J_c which corresponds to $\omega = \omega_c$ i.e. $J_c = I_{zz}\omega_c$. The critical spin J_c is shown in Fig. 6 for various rare-earth isotope families. We note that since $\omega_c \sim B^{3/2}/C$ we expect

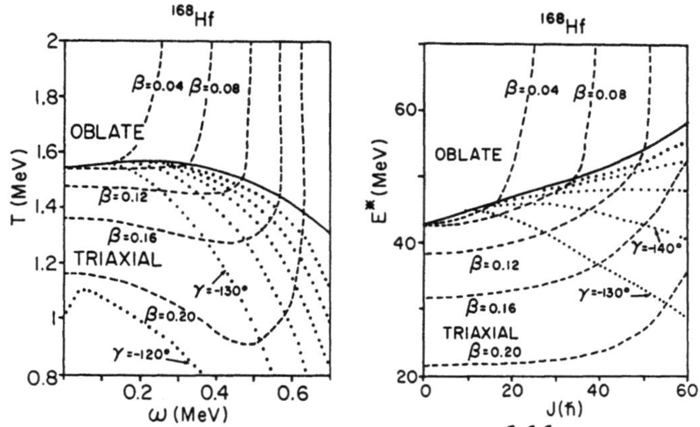

Fig. 7. Phase diagram for ^{166}Hf in intensive (T,ω) variables (left) and extensive (E^*,J) variables (right).

the J_c systematics to be dominated by the B-systematics. B on the other hand governs the prolate-oblate free energy difference which is $\Delta F = (24/256)\ (B^4/C^3)$. One can apply Hill and Wheeler's "thirds of the shell" rule [48] according to which prolate ground state deformation dominates in the first two-thirds of the shell while oblate deformation is typical for the last third. It is thus expected that $B \sim 0$ around two-thirds of a filled shell and is maximal around one-third of a filled shell. The behavior of J_c in Fig. 6 tends to support the foregoing analysis with J_c rising rapidly from neutron shell closure at 82 to a maximum in the neighborhood of the end of the first third of the shell (N=96) and then proceeding to fall towards zero at the end of the second third of the shell (N=112). An approximate proton "thirds of the shell" rule is also observed by inspecting the various curves in Fig. 6. The proton shell closure is 50 and 82, so that the outermost curves correspond to nuclei with 60-62 protons (Nd and Sm), while the innermost curves correspond to 70-72 protons (Yb and Hf). In general the values of J_c are relatively small ($\lesssim$ 10-15$\hbar$), indicating that the transitions are very close to being second-order.

Using the Landau parameters it is possible, for the nucleus under consideration, to unfold the universal phase-diagram to the physical variables (E^*,J). An example is shown in Fig. 7 for ^{166}Hf.

4. Fluctuations

Since the nuclear system is finite, fluctuations away from the mean-field configuration are important. According to Eq. (3.8) for the nuclear many-body partition function, all possible one-body potentials $v\sigma$ should be admitted each with a probability given by the Boltzmann factor

$$P\left[\sigma\right] \propto \exp\left[-F(T,\omega;\sigma)/T\right] , \tag{4.1}$$

with the free energy given by (3.9). For the finite nucleus P is maximal for $\sigma = \sigma_{eq}$ (the HF solution ρ of (3.5)) but is non-negligible for $\sigma \neq \sigma_{eq}$.

It is possible to evaluate the small amplitude fluctuations by approximating (4.1) by a Gaussian in σ-σ_{eq}, with a width given by the second variation of the exponent. We then obtain the temperature-dependent RPA approximation [42]. This approximation is not sufficient when some of the directions are "soft", in which case the integral (3.8) in these directions should be evaluated exactly [43].

We have developed a macroscopic approach to fluctuations [29,32] in the Landau framework in which fluctuations in the order parameters are treated exactly. In our case we assume a unified theory of shape fluctuations in which all five $\alpha_{2\mu}$ are included. The probability of finding a nucleus in a "state" with deformation $\alpha_{2\mu}$ is given by [49]

$$P\left[\alpha_{2\mu}\right] = Z^{-1} \exp\left[-F(T,\omega;\alpha_{2\mu})/T\right] , \tag{4.2}$$

where

$$Z(T,\omega) = \int D\left[\alpha\right] \exp(-F/T) , \tag{4.3}$$

is the nuclear partition function. The free energy F in (4.2) is given by the general Landau expansion (3.15)-(3.17). This theory takes into account large amplitude fluctuations in $\alpha_{2\mu}$. The phase-space volume element $D[\alpha_{2\mu}]$ is chosen to be the one which treats all five $\alpha_{2\mu}$ uniformly, namely the unitary invariant metric

$$D\left[\alpha_{2\mu}\right] = \prod_{\mu} d\alpha_{2\mu} = \beta^4 |\sin 3\gamma| d\beta d\gamma d\Omega . \tag{4.4}$$

It should be noted that our theory takes into account

fluctuations not only in the intrinsic shape β,γ but also in the nuclear orientation Ω relative to the rotation axis $\vec{\omega}$. Previous analyses of shape fluctuations have neglected the latter.

The importance of (4.2) is in its use of calculating expectation values of any physical observable X (which is shape-dependent)

$$< X > = \int D\left[\alpha_{2\mu}\right] P\left[\alpha_{2\mu}\right] X(\alpha) . \qquad (4.5)$$

In subsequent chapters we shall compare the value of $< X >$ obtained by (4.5) where shape fluctuations are included, with the value $X(\alpha_{eq})$ predicted by the mean-field theory, for several physical observables X. We shall see that they are quite different.

5. Giant Dipole Resonances in Hot Rotating Nuclei

The giant dipole resonance (GDR) built on nuclear excited states was observed in recent years at several laboratories, and is one of the major probes of shapes and properties of hot rotating nuclei [18-24].

In cold nuclei (T=0) the frequency of the giant resonance vibration is inversely proportional to the length of the semi-axis along which the vibration occurs [50,51]. Therefore in deformed nuclei the resonance splits and from the amount of splitting it was possible to determine its deformation. A similar technique was used to determine the shape of hot rotating nuclei [18-24]. However, due to shape fluctuations around the equilibrium shape, the relationship between the equilibrium nuclear shape and the GDR spectral shape is more complex. It has been recognized by several authors that thermal shape fluctuations play an important role in determining the width of the resonance [28,30,31].

Microscopic theories of GDR underestimate the width of the GDR [17]. The finite-temperature RPA also predicts that the GDR is rather insensitive to temperature. However, recent measurements indicate strong dependence of the width on temperature for spherical nuclei [18]. By using a phenonemological escape width [17], one can increase the calculated width, but the spectral shape is still in disagreement with the observed one. We have developed a macroscopic approach [29,32] to the GDR in hot nuclei in the framework of the Landau theory. All parameters are determined from the properties of the T = 0 GDR. In this theory unified large-amplitude fluctuations in the order parameters ($\alpha_{2\mu}$) play an important role. It is the first theory which is able to reproduce both the observed cross-section and the angular anisotropy of the GDR γ-rays emitted from the hot nucleus [33,34].

5.1 Macroscopic Theory

In this section we shall derive an exact expression for the GDR absorption cross-section $\sigma(\epsilon)$ and the angular anisotropy coefficient $a_2(\epsilon)$ (see definition below) in terms of the dipole temporal autocorrelation function evaluated at equilibrium.

The differential cross-section for a nucleus of energy E and spin J to emit an electric dipole γ-ray of energy ϵ and direction Θ with respect to an axis to be specified below is

$$\frac{d\Gamma_{em}}{d\epsilon d\Omega} = \frac{1}{2\pi\hbar}\left(\frac{\epsilon}{\hbar c}\right)^3 \frac{1}{\rho((\epsilon,J)}$$

$$\sum_{\substack{i,f\\ \mu,M_i}} |\langle f\ J_f M_f|D_\mu|i\ J\ M_i\rangle|^2\ \delta(E-E_i)\ \delta(E'-E_f)\ F_\mu(\Theta)\ , \tag{5.1}$$

where $E' = E-\epsilon$. The sum represents an average over all initial states of energy E and spin J and a sum over all final states of energy E'. Here $\rho(E,J)$ is the initial level density and D_μ is the dipole operator. The angular functions $F_\mu(\Theta)$ are given by

$$F_\mu(\Theta) = |d^1_{\mu,1}(\Theta)|^2 + |d^1_{\mu,-1}(\Theta)|^2\ , \tag{5.2}$$

where d^1 are the Wigner matrices for spin 1. Eq. (5.1) can be rewritten in the following way:

$$\frac{d\Gamma_{em}}{d\epsilon d\Omega} = \frac{1}{(2\pi\hbar)^2}\left(\frac{\epsilon}{\hbar c}\right)^3 \sum_\mu \int_{-\infty}^{\infty} dt e^{i\epsilon t/\hbar} \langle D^\dagger_\mu(t)\ D_\mu(0)\rangle_{E,J}\ F_\mu(\Theta)\ , \tag{5.3}$$

where $D_\mu(t)$ is the dipole operator in the Heisenberg representation and the average is over the microcanonical ensemble

$$< X >_{E,J} = \mathrm{Tr}\left[X\,\delta(E-H)\,P_J\right] / \mathrm{Tr}\left[\delta(E-H)\,P_J\right] . \quad (5.4)$$

In Eq. (5.4) P_J is the projection on states with spin J.

As usual we replace in (5.3) the microcanonical average by a canonical average $<D^{\dagger}_{\mu}(t)D_{\mu}(0)>_{T,\omega}$ where

$$< X >_{T,\omega} = \mathrm{Tr}\left[Xe^{-(H-\vec{\omega}\cdot\vec{J})/T}\right] / \mathrm{Tr}\ e^{-(H-\vec{\omega}\cdot\vec{J})/T} . \quad (5.5)$$

In doing so, we have chosen a preferred direction $\vec{\omega}$. Choosing this direction to be the z-axis, we are measuring the γ-ray with respect to the rotation axis $\vec{\omega}$, i.e. approximately with respect to the spin direction. From Eqs. (5.2) and (5.3) we find that the angular distribution of the γ-rays can be written as

$$\hbar\,\frac{d\Gamma_{em}}{d\epsilon\, d\Omega} = \frac{1}{4\pi}\left(\frac{\epsilon}{\pi\hbar c}\right)^2 \sigma(\epsilon)\left[1 + a_2(\epsilon)\,P_2(\Theta)\right] , \quad (5.6)$$

where

$$\sigma(\epsilon) = \frac{2\pi\epsilon}{3\hbar^2 c}\int_{-\infty}^{\infty} dt e^{i\epsilon t/\hbar}\ <D^{\dagger}_{\mu}(t)\ D_{\mu}(0)> , \quad (5.7)$$

and

$$a_2(\epsilon) = \frac{1}{2} - \frac{3}{2}\,\frac{\int_{-\infty}^{\infty} dt\ e^{i\epsilon t/\hbar}\ <D^{\dagger}_0(t)\ D_0(t)>}{\int_{-\infty}^{\infty} dt\ e^{i\epsilon t/\hbar}\ \sum_{\mu} <D^{\dagger}_{\mu}(t)\ D_{\mu}(0)>}$$

$\sigma(\epsilon)$ is the GDR absorption cross-section and a_2 is known as the angular anisotropy parameter. If the angular distribution is measured with respect to the beam axis, then, assuming it is perpendicular to the spin direction, we have to multiply (5.8) by (-1/2). In the following and in all calculations we shall use this a_2 unless otherwise specified.

5.2. Fixed Deformation

In order to calculate the dipole correlation functions in (5.7) and (5.8) we first need to evaluate them for a constant deformation $\alpha_{2\mu}$. In Section 5.3 we will show how to take into account the fluctuations of $\alpha_{2\mu}$ at finite temperature. In the absence of rotations ($\omega = 0$) the tensor $\langle D_\mu^\dagger(t) D_\mu(0)\rangle$ is diagonal in the intrinsic frame of the nucleus. In that frame $\alpha_{2\mu}$ is also diagonal and is defined in terms of the Hill-Wheeler coordinates β,γ and the Euler angles Ω which determine the orientation of the nucleus relative to the rotation axis $\vec{\omega}$. Rotation matrices relate the correlation tensor in the intrinsic frame to that in the laboratory frame. We adopt the common assumption that the resonant energies of the three giant dipole oscillatory modes are inversely proportional to the semi-axis lengths of the ellipsoid representing the nuclear shape [50,51]

$$E_j^0 = E_0 \frac{R_0}{R_j} = E_0 \exp\left[-\sqrt{\frac{5}{4\pi}}\, \beta \cos\left(\gamma - \frac{2\pi}{3} j\right)\right]; \quad (j=1,2,3) \,, \tag{5.9}$$

and that the Fourier transform of the correlation tensor (multiplied by the additional factor of ϵ in (5.7)) is proportional to a Lorentzian,

$$f_j(\epsilon) = \frac{\Gamma_j \epsilon^2}{\left(\epsilon^2 - E_j^2\right)^2 + \Gamma_j^2\, \epsilon^2} \,. \tag{5.10}$$

In (5.10) the natural frequencies are $E_j(\omega = 0) = E_j^0$ as given by (5.9) and Γ_j are the decay widths along the j-th intrinsic axis; E_0 is a constant which varies from nucleus to nucleus. Equation (5.9) follows from the nuclear hydrodynamics model where giant resonances are oscillations in the proton and neutron fluid densities and thus have wavelengths nearly proportional to the length of the semi-axis along which the oscillations occur [50,51]. Equation (5.10) follows from assuming that the dipole in the intrinsic principal frame is damped as is the displacement of a harmonic oscillator in an external field. The strength of the Lorentzian (5.10) is fixed by the classical sum rule [50].

In the presence of rotation ($\omega \neq 0$) the Coriolis and centrifugal forces shift the normal frequencies from E_j^0 to $E_j(\vec{\omega})$ and make the correlation tensor nondiagonal in the intrinsic frame. However, there exists a canonical transformation to normal coordinates in which the correlation tensor is diagonal. To find that transformation we model the giant dipole oscillator by a rotating deformed three-dimensional harmonic oscillator. The Hamiltonian in the intrinsic rotating frame is

$$H(\vec{d},\vec{p}) = \frac{1}{2}\,\vec{p}^{\,2} + \frac{1}{2}\sum_i E_i^{o2}\, d_i^2 - \vec{\omega}\cdot(\vec{d}\times\vec{p}) \quad , \qquad (5.11)$$

where $\vec{d}$ is the giant dipole operator in the intrinsic frame and $\vec{p}$ is its conjugate momentum. In [32] we have shown that the correlation function in this frame can be written in terms of the raising and lowering operators for the three normal modes, $a_s^\dagger$ and a_s, and a matrix $\mathfrak{M}$ which relates the normal coordinates to the intrinsic ones:

$$\langle d_\sigma^\dagger(t)\; d_\tau(0)\rangle = \sum_{st} \mathfrak{M}_{\sigma s}^*\, \mathfrak{M}_{\tau t}\, \langle a_s(t)\; a_t^\dagger(0)\rangle \; . \qquad (5.12)$$

The matrix $\mathfrak{M}$ depends on ω, β and γ (through E_j^0) and Ω. As in the nonrotating case, the Fourier transform of the correlation tensor in the "normal" frame, $\langle a_s(t)\; a_t^\dagger(0)\rangle$, is assumed to be diagonal and to have the Lorentzian form (5.10) but now with shifted E_j.

The relationship between the dipole operator in the laboratory frame (D_μ) and in the intrinsic rotating frame (d_μ) is

$$D_\mu(t) = e^{-i\vec{\omega}t\cdot\vec{J}/\hbar} \sum_\lambda \mathcal{D}_{\mu\lambda}(\Omega)\; d_\lambda(t)\; e^{+i\vec{\omega}t\cdot\vec{J}/\hbar} \quad , \qquad (5.13)$$

where $\Omega = (\psi,\theta,\phi)$ are the Euler angles and $\mathcal{D}$ is the Wigner rotation matrix for spin 1.

We find that the cross-section in the laboratory frame is given by

$$\sigma(\epsilon;\beta,\gamma,\Omega) = \frac{4\pi e^2\hbar}{mc}\,\frac{Z\,N}{A}\sum_{j=1}^{3}\Big[S_j^{(0)}\, f_j(\epsilon) + S_j^{(-)}\, f_j(\epsilon+\hbar\omega) + S_j^{(+)}\, f_j(\epsilon-\hbar\omega)\Big] \quad , \qquad (5.14)$$

while the angular anisotropy is

$$a_2(\epsilon;\beta,\gamma,\Omega) = \frac{3}{4}\,\frac{\sum_j S_j^{(0)}\, f_j(\epsilon)}{\sum_j S_j^{(0)}\, f_j(\epsilon) + S_j^{(-)}\, f_j(\epsilon+\hbar\omega) + S_j^{(+)}\, f_j(\epsilon-\hbar\omega)} - \frac{1}{4}. \qquad (5.15)$$

Here $S_j^{(0,\pm)}$ are the resonance strengths in units of the classical sum rule:

$$S_j^{(\pm)} = \frac{1}{6}\sum_{\mu\nu}\left[(\delta_{\mu\nu}-n_\mu n_\nu)\mathrm{Re}(\mathfrak{M}^*_{\mu j}\mathfrak{M}_{\nu j}) \pm \sum_\lambda \epsilon_{\lambda\mu\nu} n_\lambda \mathrm{Im}(\mathfrak{M}^*_{\mu j}\mathfrak{M}_{\nu j})\right]$$

$$S_j^{(0)} = \frac{1}{3}\sum_{\mu\nu} n_\mu n_\nu \ \mathrm{Re}(\mathfrak{M}^*_{\mu j}\mathfrak{M}_{\nu j}) \ . \tag{5.16}$$

Here $\hat{n}$ is a unit vector along the rotation axis and n_μ are its components in the principal intrinsic frame. $\epsilon_{\lambda\mu\nu}$ is the completely antisymmetric tensor. From (5.14) we see that σ is generally the sum of nine Lorentzians in the laboratory frame (which are non-degenerate for triaxial nuclei rotating around an axis which is not principal). For noncollective nuclear rotations all but two sets of the Lorentzians coincide; all but five sets of lines coincide when the angular velocity is parallel to a principal axis. For rare-earths the rotational splitting is usually small ($\omega \lesssim 0.5$ MeV) compared to the widths of the Lorentzians. For lighter nuclei it may be more significant—due to the fact that ω can be as large as 2-3 MeV for moderate spins.

The cross-section (5.14) and a_2 in (5.15) will be completely specified as soon as the widths Γ_j in (5.10) are defined. We use for these widths a power-law [52] dependence on the resonance energy

$$\Gamma_j = \Gamma_0 \left(\frac{E_j}{E_0}\right)^\delta , \tag{5.17}$$

where Γ_0 depends on the nucleus under consideration. Equation (5.17) is consistent with the experimental dependence of the ground-state GDR widths on deformation in heavy nuclei. It can also be derived using surface dissipation models. In such a theory the width, in units of Γ_0, is reduced to a purely geometrical quantity expressed as an elliptic integral [53]. It is well approximated by a power law with $\delta = 1.6$. The only free parameters (E_0, Γ_0, δ) which are assumed to depend weakly on T are determined from the zero temperature properties of the nucleus as discussed in [32].

If the effect of the Coriolis force on the eigenfrequencies is neglected ($\omega \to 0$), it is possible to

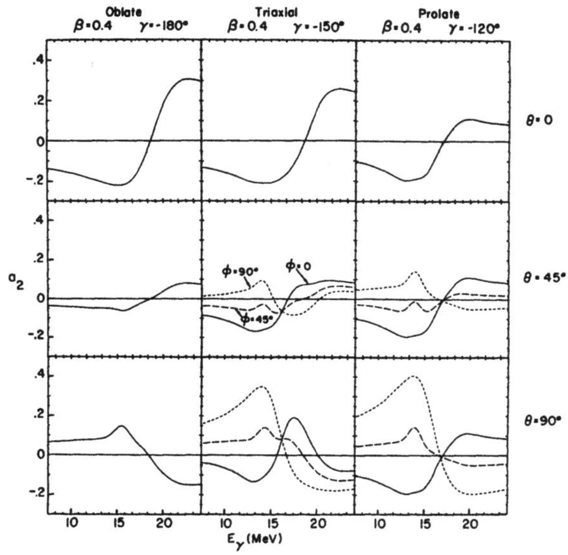

Fig. 8. The dependence of a_2 on orientation (θ,ϕ) for different intrinsic shapes.

derive an analytic expression for $\sigma(\epsilon)$ and a_2 as a function of the orientation (θ,ϕ). In this approximation, the correlation tensor $\langle d_i^\dagger(t)d_j(0)\rangle$ $(i,j=1,2,3)$ is diagonal and from Eqs. (5.7) and (5.8) we find

$$\sigma \propto f_x + f_y + f_z, \tag{5.18}$$

and

$$a = -\frac{1}{2}\,\frac{(f_x + f_y)/2 - f_z}{f}\,\frac{3\cos^2\theta - 1}{2} + \frac{3}{8}\,\frac{f_x - f_y}{f}\,\sin^2\theta\,\cos 2\phi\ , \tag{5.19}$$

where $f = f_x + f_y + f_z$ and f_i $(i=x,y,z)$ are the Fourier transforms of $\langle d_i^\dagger(t)d_i(0)\rangle$ as given by Eq. (5.10). Notice that a_2 in (5.19) is the one with respect to the beam axis.

Here a_2 is independent of the Euler angle ψ. For axially symmetric shapes (with symmetry axis z) $f_x = f_y$ and a_2 is then also independent of ϕ. It is seen that while σ is independent of orientation (in this approximation) a_2 is very sensitive to orientation. This is demonstrated in Fig. 8.

5.3 Thermal Fluctuations (Adiabatic Theory)

If the fluctuations in the shape are slow enough then the equilibrium averages in (5.7) and (5.8) are done according to (4.5) by averaging the contribution from a fixed $\alpha_{2\mu}$ over all possible intrinsic shapes and shape orientations. This is demonstrated in Fig. 9 for the GDR cross-section.

To compare our theory with experiment we have applied it [33] to three cases for which precision measurements were recently taken [34]: ^{90}Zr at T = 1.6 MeV, J = 10$\hbar$; ^{90}Zr at T = 1.7 MeV, J = 22$\hbar$; and ^{92}Mo at T = 2 MeV, J = 33$\hbar$. The free energy surfaces were constructed from a cranked Nilsson-Strutinsky Hamiltonian. The cranking calculations were done only for $\vec{\omega}$ parallel to a principal axis from which the moments of inertia were determined as a function of β and γ. The free energy for a general orientation Ω is then determined by the expression (3.15). We have calculated the phase-diagrams of these nuclei and found that at the equilibrium configuration ω = 0.47, 1.03 and 1.36 MeV respectively in each of the above three cases. The equilibrium shape is a non-collective oblate (γ = -180°; θ = 0) whose deformation increases with spin; β = 0.02, 0.08 and 0.16, respectively. When the metric (4.4) is included with the distribution (4.2), the resulting most probable shapes are triaxial ($\gamma \approx$ - 150°) and their deformation is significantly larger; β = 0.24, 0.31 and 0.46, for the above three cases. From the ground-state GDR data in ^{90}Zr and ^{92}Mo we have determined E_0 = 16.82 MeV and Γ_0 = 5 MeV. Both σ and a_2 were then calculated from (5.7) and (5.8) where fluctuations in the intrinsic deformation (β,γ) were included as well as those in the orientation (θ,ϕ).

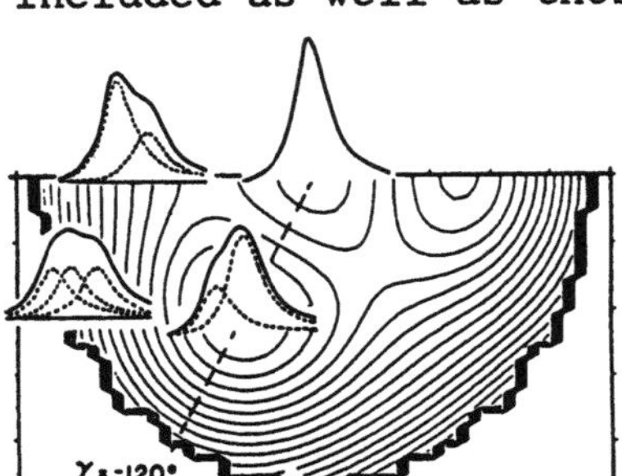

Fig. 9. GDR cross-sections are shown at various deformations in the (β,γ) plane. The observed cross-section is obtained by averaging according to (4.5).

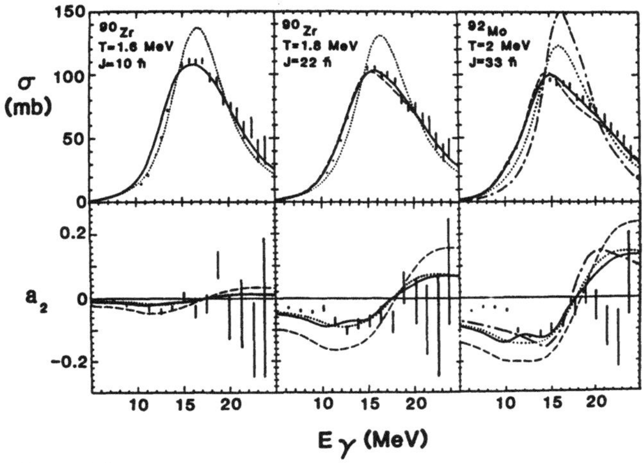

Fig. 10. The GDR absorption cross-section σ (top row) and the angular anisotropies a_2 (bottom row) versus γ-ray energy E_γ for ^{90}Zr and ^{92}Mo. The error bars are the experimental data. The following calculations are shown: with orientation fluctuations (solid lines) and without orientation fluctuations (dashed lines) using the metric (4.4); without the orientation fluctuations using the metric (5.20) (dotted lines) and the mean-field results (dashed-dotted lines in the right column).

The results are shown by the solid curves in Fig. 10. They agree very well with the experiment [34] shown by the error bars. In particular, the theory reproduces accurately the observed broadening of σ at finite temperature. In judging the quality of the agreement between the calculated and experimental a_2, the region $E_\gamma \lesssim 11$ MeV should be disregarded since there the γ-rays from daughter nuclei contaminate the signal. The latter have lower spin and energy and tend to drive the observed a_2 to zero. Also, at the high energy side we are at the tail of the resonance and the error bars are large. Thus the range to consider in Fig. 10 for the a_2 fit is $11 \lesssim 20$ MeV. The dashed lines in Fig. 10 are the results of similar calculations but with no orientation fluctuations. The metric (4.4) is used. Comparing them with the solid lines, we see that the effect of the orientation fluctuations on the cross-section is small. Only in the high spin case ($J=33\ \hbar$) is this effect measurable, and provides a correction at the high energy side of the resonance. However, the effect of the orientation fluctuations on a_2 is large; it causes a considerable attenuation in a_2 in agreement with the experiments.

The dotted lines in Fig. 10 show the results of the calculations where the fluctuations are evaluated with a metric used by other authors [28,30,31]

$$D[\alpha] = \beta \, d\beta d\gamma \ , \qquad (5.20)$$

which does not include orientation fluctuations. We see that these results are in strong disagreement with the measured GDR cross-sections. The GDR widths are considerably underestimated by the calculations and their shape is different from the experimental ones. The a_2's on the other hand are close to those calculated with the metric (4.4) and with orientation fluctuations. If a temperature-dependent width is added to fit the experimental width of the GDR cross-sections, then a disagreement with a_2 will result. Also, if orientation fluctuations are included with the metric (5.22), then a_2 will be strongly attenuated relative to the experiment. Thus only a theory with intrinsic shape and shape orientation fluctuations according to the unitary metric (4.4) can reproduce both the observed cross-sections and a_2's. The good agreement between experiment and theory is also an indirect confirmation of the existence at finite temperatures of non-collective oblate shapes whose deformation increases with spin.

We have seen that the choice of the metric plays an important role when compared with the experimental results. The main difference between the two metrics is that the unitary metric (5.4) favors large and triaxial deformations far more than Eq. (5.20). This is illustrated in Fig. 11 which shows for both metrics the probability density for an intrinsic deformation in cartesian coordinates $\beta\cos\gamma$ and $\beta\sin\gamma$. Notice the peaks along triaxial shapes in the case of the unitary metric. The difference between the two is more noticeable at higher temperatures.

To understand the effect of orientation fluctuations we use the approximation of Sect. 5.2. Since σ is insensitive to orientation, it is not affected much by orientation fluctuations. However when we average a_2 in Eq. (5.19) over orientation, we obtain (for axial nuclei)

$$a_2 = \frac{f_\perp - f_{||}}{2f_\perp + f_{||}} \, r \ , \qquad (5.21)$$

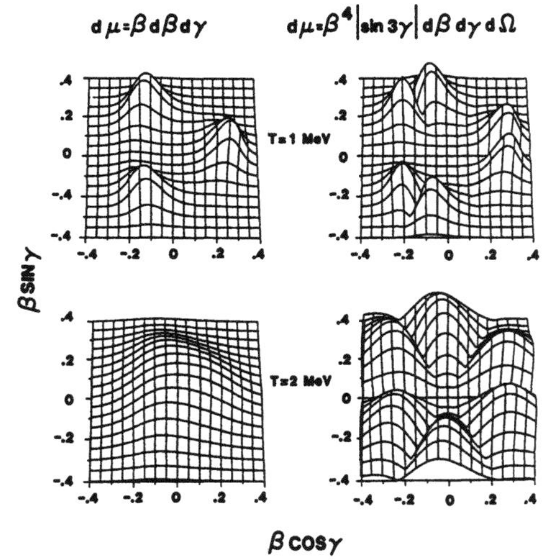

Fig. 11. Comparison of the unitary metric (5.4) and the metric $\beta d\beta d\gamma$ (Eq. (5.20)). Probability densities are shown for ^{166}Er at T=1 and 2 Mev (ω=0). Left: Using the metric (5.20); right: using the unitary metric.

where $\parallel$ and $\perp$ denote the direction parallel and perpendicular to the symmetry axis. r is an attenuation factor given by

$$r = \left\langle \frac{3\cos^2\theta - 1}{2} \right\rangle =$$

$$= \frac{3}{4x} e^x \Big/ \int_0^1 e^{xz^2} dz - \frac{3}{4x} - \frac{1}{2} , \qquad (5.22)$$

where

$$x = (I_{\parallel} - I_{\perp})\, \omega^2/2T . \qquad (5.23)$$

The attenuation factor r is shown vs. x in Fig. 12. Notice that the attenuation is smaller (i.e. r closer to 1) when $|x|$ is larger. From (5.23) this happens when the spin is higher (and there are less fluctuations in orientation) or when $\Delta I=|I_{\parallel}-I_{\perp}|$ is larger (at larger deformation).

The value of a_2 is also sensitive to the intrinsic shape, and its magnitude increases with deformation (for a spherical nucleus $a_2 = 0$). Intrinsic shape fluctuations tend to drive the most probable deformation to higher values (compared with the equilibrium deformation), for which the resulting anisotropy is

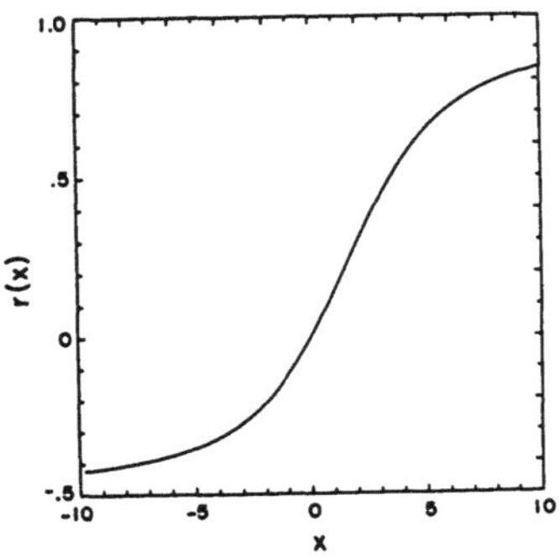

Fig. 12. Attenuation factor r (Eq. (5.22) vs. x (Eq. (5.23)). For prolate shapes $x < 0$ and for oblate shapes $x > 0$.

larger. For example ^{90}Zr at $T = 1.6$ MeV and $J = 22\hbar$ has an equilibrium deformation of $\beta \approx 0.08$ but a most probable $\beta \approx 0.31$. The enhancement of a_2 due to intrinsic shape fluctuations is counteracted by its suppression due to orientation fluctuations. Thus the observed a_2 is an indirect indication of the equilibrium shape. In the experiment of Fig. 10 it was therefore possible to demonstrate liquid-drop-like shape changes from spherical to oblate with increasing spin [33,34].

5.4 Sensitivity of the GDR to Shape

As shown in Sect. 5.3 our theory provides a good description of the available data. Other examples for the GDR cross-section in the strongly deformed rare-earth region are shown in Fig. 13. From the $T = 0$ GDR cross-section of ^{166}Er we have determined $E_0 = 14.4$ MeV and $\Gamma_0 = 3.64$ MeV. We have calculated the GDR cross-section for ^{166}Er and ^{160}Er for the temperatures and spins shown in the figure. The solid lines are from the experimental data. Again, when the unitary metric is used (dashed lines) we obtain good agreement with the data while for the metric (5.20) (dotted lines) the agreement is poor.

Having established the ability of our theory to provide a quantitative account of the experimental situation, we will now discuss the question of how sensitive the GDR is as a probe of the shape of hot nuclei. Indeed, with increasing T, the thermal fluctuations increase due to two factors: the explicit appearance of T in the denominator of the exponent in

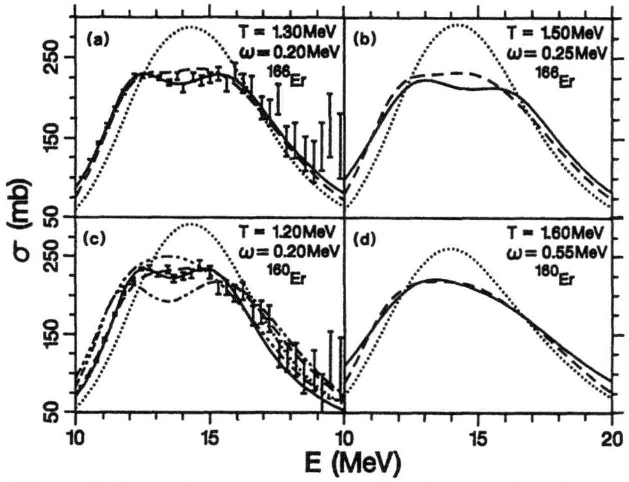

Fig. 13. Comparison of the calculated photoabsorption GDR cross-section (dashed-lines) with the results of CASCADE code fits to the experimental ones (solid lines). (a) ^{166}Er at T = 1.30 MeV, ω = 0.20 MeV (corresponding to an excitation energy E^* = 49.2 MeV in [14]). (b) ^{166}Er at T = 1.50 MeV, ω = 0.20 (E^* = 43.2 MeV in [14]). (d) ^{160}Er at T = 1.60 MeV, ω = 0.55 MeV (E^* = 90.3 MeV in [20]). The dotted lines represent the calculated GDR with the $\beta d\beta d\gamma$ metric. The error bars in (a) and (c) are only suggestive of the accuracy of the experimental measurements and do not represent the actual data.

(4.2), and the T dependence of the free energy surface. It is the last factor which is associated with the shape changes caused by heating and rotating a deformed nucleus. In order to see the sensitivity of the GDR results to these changes, we show in Fig. 13 (for ^{166}Er at T = 1.2 MeV and ω = 0.2 MeV) the GDR cross-sectionobtained by using the "wrong" surfaces F in Eq. (4.2): the T = 0 (the "dot-dot-dashed-dashed" line) and the T = 3 MeV (the "dot-dot-dot-dashed" line) surfaces. The experimental results clearly indicate that at T = 1.2 MeV ^{160}Er, though still prolate [14], has a softer energy surface than at T = 0. However, at higher temperatures (T $\gtrsim$ 1.6 MeV) the GDR becomes much less sensitive to variations in the energy surface [32].

In Fig. 14, we have plotted the cross-section's full width at half maximum (FWHM) as contour lines in the T-ω (or T-J) plane for ^{166}Er (a typical deformed nucleus)

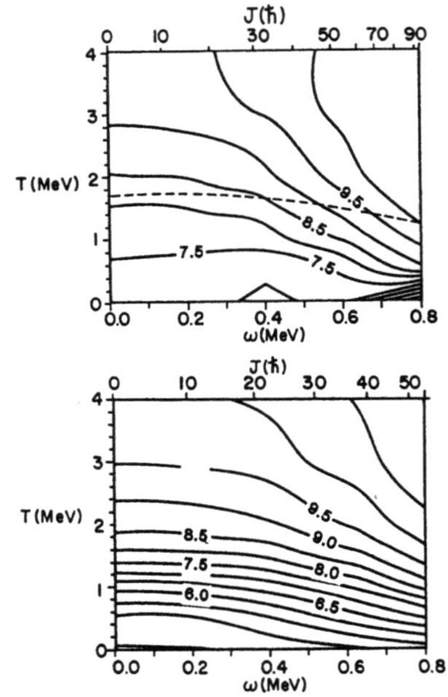

Fig. 14. FWHM contour lines (MeV) in the T-ω (or T-J) plane of the GDR for ^{166}Er (top) and ^{140}Ce (bottom). The shape transition line is shown by the dashed line in the upper figure.

and for ^{140}Ce (a typical spherical nucleus). The latter nucleus shows a more dramatic change with temperature: its width at T = 2 MeV is about twice that at T = 0 MeV in accord with experiment [18]). This occurs because its free-energy surface "softens" as T rises from 0 to 2 MeV and because it has as a spherical nucleus only a single Lorentzian component at T = 0. Erbium has a deformed ground state and its FWHM changes more slowly for T ≤ 2 MeV.

Another interesting phenomenon is that for both nuclei the widths increase much more slowly at temperatures above 3 MeV. This is associated with the disappearance of shell effects in this temperature region.

6. Dissipation and Time-dependent Fluctuations in Hot Nuclei

Though the static fluctuation theory works reasonably well for many of the observed GDR cross-sections, there are several cases, in particular at higher-temperatures

(T ≥ 2 MeV) where the above theory [35] overestimates the observed width of the GDR [20]. It was recently suggested that in analogy with nuclear magnetic resonance in condensed matter systems [54] and with rotational damping of B(E2) transitions in the continuum [37], the GDR could display a phenomenon known as "motional narrowing" [38]. The physical reason for this is that the nucleus is hopping so fast between its various shapes that the GDR does not have enough time to probe separately each nuclear shape. Thus there is less broadening of the GDR due to its coupling with the low-lying quadrupole collective states. This idea was incorporated in [38] within the framework of a microscopic model for the strength function which is based on a description of the compound nucleus by random matrix theory. Although the model is successful in explaining motional narrowing, it is rather complicated and very difficult to solve at intermediate situations where the time scale for hopping between shapes is comparable to that associated with the spread of the GDR frequencies from shape fluctuations. Furthermore, this theory is proper for a single resonance and does not take into account the coupling between the various resonances of a deformed nucleus. In particular, when the Gaussian orthogonal ensemble (GOE) of random matrices is used and under certain approximations, one always obtains in the adiabatic limit the Wigner semicircular distribution. This distribution is inconsistent with those obtained in previous adiabatic theories.

We have introduced a relatively simple macroscopic theory of time-dependent fluctuations [35,36], within the framework of the Landau theory, which generalizes our previous theory of fluctuations to non-adiabatic situations. The dynamics of the quadrupole shape fluctuations are described by a model analogous to Brownian motion, where the quadrupole degrees of freedom play the role of the heavy particle immersed in a fluid and all the other nuclear degrees of freedom play the role of the surrounding fluid molecules. We assume that at the temperatures of interest the quadrupole motion is overdamped. The thermodynamical driving force (to equilibrium) is proportional to the free energy gradient (taken with respect to $\alpha_{2\mu}$), and the random force describes the interaction of the quadrupole shape parameters $\alpha_{2\mu}$ with all other degrees of freedom. Since the Landau expansion of the free energy contains terms up to fourth order in the deformation, the equation of motion for $\alpha_{2\mu}$ has the form of a non-linear Langevin equation [55]. When compared with the theory of Sect. 5, the current model contains only one additional parameter; it is proportional to the mean relaxation rate of the quadrupole motion. This parameter spans the full range between the adiabatic limit and the sudden limit. In the latter limit the time scale for changes in the quadrupole shape is much shorter compared to the time scale associated with the spread in the GDR frequencies over a typical range of deformations. The other parameter in the theory, i.e. the magnitude of the correlation of the random force, is completely determined in terms of the relaxation parameter through the fluctuation-dissipation theorem. This theorem must be satisfied in order for the equilibrium distribution to be a solution of the equation of motion. Indeed, what we have is a picture of a dynamical equilibrium of the nucleus, in which the shape fluctuates in time but the distribution of shapes at any given moment is the equilibrium distribution.

6.1 Langevin Equation

Assuming the quadrupole vibrations to be overdamped at the temperatures of interest, we model the quadrupole fluctuations by an equation of the Brownian motion type

$$\dot{\alpha}_{2\mu} = -\frac{1}{\chi}\frac{\partial F}{\partial \alpha^*_{2\mu}} + f_{2\mu}(t) \quad . \tag{6.1}$$

$F(T,\omega,\alpha_{2\mu})$ is the free energy surface at temperature T and angular velocity ω, so that $-\partial F/\partial\alpha^*_{2\mu}$ plays the role of an external average driving force at finite temperature. The quantity $f_{2\mu}(t)$ is a random force which describes the interaction of the quadrupole shape degrees of freedom with all the others. It is assumed gaussian and stationary, satisfying the relations

$$\begin{aligned} &< f_{2\mu}(t) > = 0 \\ &< f_{2\mu}(t)\, f^*_{2\mu'}(t') > = \xi\ \delta(t-t')\ \delta_{\mu\mu'} \quad , \end{aligned} \tag{6.2}$$

where ξ characterizes the magnitude of the correlation. The correlations in (6.2) express the physical assumption that the forces $f_{2\mu}$ vary rapidly on the time scale of the variation of $< \alpha_{2\mu} >$. Notice that, similar to $\alpha_{2\mu}$, $f_{2\mu}$ are complex but satisfy the reality condition

$$f^*_{2\mu}(t) = (-)^{\mu} f_{2-\mu}(t) \ . \qquad (6.3)$$

Relations (6.2) imply that the real and imaginary parts of $f_{2\mu}$ are uncorrelated. Taking the average of (6.1) we find

$$< \dot{\alpha}_{2\mu} > = - \frac{1}{\chi} < \frac{\partial F}{\partial \alpha^*_{2\mu}} > \ . \qquad (6.4)$$

For a quadratic F this describes the relaxation of the average deformation to its equilibrium value defined by $< \partial F/\partial \alpha^*_{2\mu} > = 0$. The parameter χ of the model is then proportional to the mean relaxation time of $\alpha_{2\mu}$ to its equilibrium value. The first order equation (6.1) is correct if the quadrupole motion is strongly overdamped. Otherwise we have to use a second-order equation which contains an additional mass parameter. The reduction from the second to first-order equation is discussed elsewhere [56]. Also $1/\chi$ in Eq. (6.1) and the correlation function (6.2) can be more generally an anisotropic tensor $(1/\chi)_{\mu\mu'}$ which depends on $\alpha_{2\mu}$. In this paper we have taken the simplest possible form which is isotropic. If, for example, the relaxation times for the nuclear orientation are different from those of the intrinsic shape we have to consider an anisotropic $1/\chi$ in (6.1).

Equation (6.1) is a non-linear Langevin equation. The (non-linear) driving force is found by taking the derivatives of the Landau expansion of F analytically in the rotating frame. Notice that it is important to consider this expansion for a general $\alpha_{2\mu}$ and not just for situations in which a principal axis of the nucleus is parallel to $\vec{\omega}$. Indeed, time-dependent fluctuations in the orientation of the nucleus are possible. We find

$$\frac{\partial F}{\partial \alpha^*} = 2A\, \alpha_{2\mu} + 3B\, b_{\mu\nu}\, \alpha_{2,\mu-\nu}\, \alpha_{2\nu} + 4C(-)^{\nu}\, \alpha_{2\nu}\alpha_{2-\nu}\alpha_{2\mu} \ ,$$

$$- \left[2(I_1+D)\right] \alpha_{2\mu}\, \delta_{\mu,\pm 2} + (2I_1+D)\alpha_{2\mu}\delta_{\mu,\pm 1} - (R-2I_1\alpha_{2\mu})\delta_{\mu,0}\Big]\omega^2$$

$$(6.5)$$

where $b_{\mu\nu}$ is proportional to a Clebsch-Jordan coefficient, $b_{\mu\nu}=\sqrt{7/2}\ (2\ \mu-\nu\ 2\ \nu|2\ \mu)$. Here $\vec{\omega}$ is chosen parallel to the z-axis in the laboratory frame, and A,B,C,... are the Landau parameters of Sect. 3.

The Langevin equation (6.1) defines a stochastic Markov process which determines an ensemble of "trajectories" $\{\alpha_{2\mu}(t)\}$. At any time t we can then construct a distribution $P(\alpha_{2\mu},t)$ of the shapes such that $P(\alpha)\Pi_\mu d\alpha_{2\mu}$ is the probability of finding a shape α in the volume element $\Pi_\mu d\alpha_{2\mu}$ around $\alpha_{2\mu}$. The equation of motion of P is discussed in the next section.

6.2 Fokker-Planck Equation

Since the process $\alpha_{2\mu}(t)$ in (6.1) is Markovian, it obeys a master equation whose form can be determined from the so-called jump moments. Starting from a given shape $\alpha_{2\mu}$, the average jump (first moment) is calculated from (6.1) to be

$$< \Delta\alpha_{2\mu} > = -\frac{1}{\chi}\frac{\partial F}{\partial\alpha^*_{2\mu}}\Delta t \quad , \tag{6.6}$$

while the second moments are

$$< \Delta\alpha_{2\mu}\Delta\alpha_{2\mu'} > = \xi\ \delta_{\mu\mu'}\Delta t \quad . \tag{6.7}$$

The distribution $P(\alpha_{2\mu},t)$ therefore satisfies a Fokker-Planck equation [57]

$$\frac{\partial P}{\partial t} = -\frac{\partial}{\partial\alpha_{2\mu}}\left[-\frac{1}{\chi}\frac{\partial F}{\partial\alpha^*_{2\mu}}P\right] + \frac{1}{2}\xi\frac{\partial^2 P}{\partial\alpha_{2\mu}\partial\alpha^*_{2\mu}} \quad . \tag{6.8}$$

The first term on the right of (6.8) is known as the "drift" term while the second is the "diffusion" term. ξ then has the meaning of the macroscopic diffusion constant and (6.7) is just the Einstein relation which relates ξ to the microscopic jumps of $\alpha_{2\mu}$.

6.3 Fluctuation-Dissipation Theorem

Any solution to the Fokker-Planck equation converges to the stationary solution $P_{st}(\alpha_{2\mu})$ which satisfies

$$\frac{\partial}{\partial\alpha_{2\mu}}\left[\frac{1}{\chi}\frac{\partial F}{\partial\alpha^*_{2\mu}}P_{st} + \frac{1}{2}\xi\frac{\partial P_{st}}{\partial\alpha^*_{2\mu}}\right] = 0 \quad . \tag{6.9}$$

A solution to (6.9) is

$$P_{st} \propto e^{-\frac{2}{\chi\xi} F} . \qquad (6.10)$$

Since (6.10) should coincide with the equilibrium distribution

$$P_{eq} \propto e^{-F/T} , \qquad (6.11)$$

we conclude that

$$\xi = \frac{2T}{\chi} . \qquad (6.12)$$

Eq. (6.12) determines the correlation function of the random force in (6.1). Equation (6.12) is just a special case of the fluctuation-dissipation theorem which connects the fluctuations (of the random force) at equilibrium with a dissipation parameter characterizing the relaxation to equilibrium. Relation (6.12) determines ξ in our model (6.1), so that the only undetermined parameter of the model is χ.

In terms of P_{eq}, Eq. (6.1) can be rewritten as

$$\frac{\partial P}{\partial t} = \frac{\partial}{\partial \alpha_{2\mu}} \left[P_{eq} \frac{\partial}{\partial \alpha^*_{2\mu}} \left(\frac{1}{2} \xi \frac{P}{P_{eq}} \right) \right] . \qquad (6.13)$$

To show that any solution of (6.1) converges to P_{eq}, we define the relative information content [58] H of P with respect to P_{eq} by

$$H = \int D[\alpha] \; P \ln \left(\frac{P}{P_{eq}} \right) , \qquad (6.14)$$

where $D[\alpha] = \prod_\mu d\alpha_{2\mu}$. Using (6.13) we find that

$$\frac{dH}{dt} = \int D[\alpha] \frac{\partial P}{\partial t} \ln \left(\frac{P}{P_{eq}} \right) =$$

$$= - \frac{1}{2} \xi \int D[\alpha] \frac{P}{P_{eq}} \sum_\mu \left| \frac{\partial}{\partial \alpha_{2\mu}} \left(\frac{P}{P_{eq}} \right) \right|^2 \le 0 , \qquad (6.15)$$

is always non-positive. Thus H decreases monotonically in time unless $P = P_{eq}$, for which $dH/dt = 0$. For $P \neq P_{eq}$ we have $dH/dt < 0$, and H will

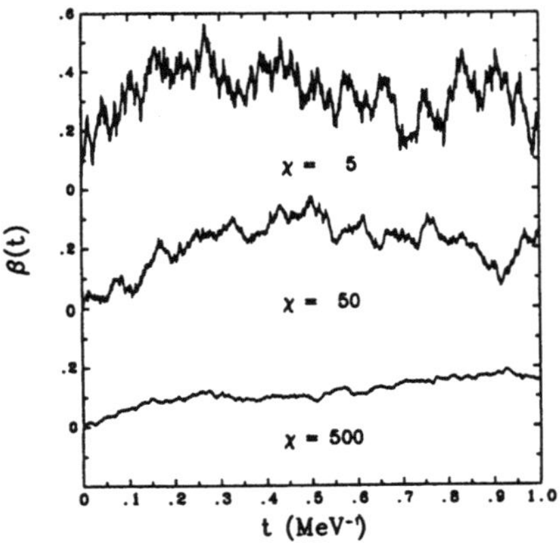

Fig. 15. Typical shape trajectories β vs. time t for ^{112}Sn at T = 1.8 MeV and ω = 0.535 MeV, obtained from the solution of (6.1). The three trajectories shown correspond to various values of χ: χ = 500 (adiabatic limit), χ = 50 and χ = 5 (sudden limit).

decrease gradually to zero so that any P will eventually decay to P_{eq} (where H = 0). In fact H is a measure of the entropy difference of the equilibrium distribution relative to the instantaneous distribution P.

If at time t = 0, P = P_{eq}, then initially H = 0 and therefore H = 0 for all times t so that P = P_{eq} for any t. This is the stationary solution of the Fokker-Planck equation (6.8). The corresponding ensemble of solutions $\{\alpha_{2\mu}(t)\}$ of the Langevin equation is obtained when the initial shapes $\{\alpha_{2\mu}(0)\}$ are distributed according to P_{eq}. This is a stationary Markov process, which describes an equilibrated nucleus at temperature T whose shape fluctuates but its shape distribution is constant. In what follows $\{\alpha_{2\mu}(t)\}$ will denote the above process.

In Sect. 6.4 we shall explain how the ensemble $\{\alpha_{2\mu}(t)\}$ can be constructed by Monte-Carlo techniques. In Fig. 15 we show three typical "trajectories" $\beta(t)$ obtained in a Monte-Carlo solution of (6.1) for three different values of the parameter χ. Since χ is proportional to the relaxation time of $\beta(t)$, the decay of $\beta(t)$ to its equilibrium value is faster as χ gets smaller. According to (6.12), the variance of the random force, which is proportional to ξ, is inversely proportional to χ. Thus as χ gets smaller the r.m.s. value of the random force gets larger and the motion is less smooth and more erratic when shown on the same time scale.

6.4 Non-Adiabatic Theory of the GDR

The giant dipole resonance absorption cross-section $\sigma(\epsilon)$ and angular anisotropy $a_2(\epsilon)$ can be calculated from the Fourier transform of the temporal autocorrelation functions of the dipole operator $\vec{D}$ at equilibrium (see Eqs. (5.7) and (5.8)). Such a correlation function can be determined from an equation of motion for $\vec{D}$. In the following we will assume that the giant vibrations are described by a three-dimensional damped harmonic oscillator rotating with angular velocity $\vec{\omega}$. The vibration frequencies and intrinsic damping widths are assumed to depend on the quadrupole deformation asin the adiabatic theory of Sect. 5. Since the quadrupole deformation is fluctuating in time according to a stochastic process, the dipole equation also describes a stochastic process which is coupled to the Langevin equation.

Here we shall denote by $\vec{D}$ the dipole operator in the frame which is rotating with constant angular velocity $\vec{\omega}$ and by $\vec{P}$ the canonical conjugate momentum. Since the nuclear shape and orientation are fluctuating, this frame is usually not the intrinsic frame of the nucleus. We have

$$\dot{\vec{D}} = \vec{P} - \vec{\omega} \times \vec{D} - \frac{1}{2} \Gamma \vec{D}$$

$$\dot{\vec{P}} = - E^2 \vec{D} - \vec{\omega} \times \vec{P} - \frac{1}{2} \Gamma \vec{P} , \qquad (6.16)$$

where E and Γ are the frequency and damping matrices, respectively. E is the matrix which in the principal intrinsic frame is diagonal with elements given by Eq. (5.9).

The transformation from the rotating frame to the principal intrinsic frame is the one which diagonalizes $\alpha_{2\mu}$. Thus in the rotating frame

$$E = E_o \exp \left(- \sqrt{\frac{5}{4\pi}} Q \right) , \qquad (6.17)$$

where the matrix Q in cartesian coordinates is

$$Q = \begin{pmatrix} -\alpha_{20}/2 + \sqrt{3/2}\,\mathrm{Re}\,\alpha_{22} & \sqrt{3/2}\,\mathrm{Im}\,\alpha_{22} & -\sqrt{3/2}\,\mathrm{Re}\,\alpha_{21} \\ \sqrt{3/2}\,\mathrm{Im}\,\alpha_{22} & -\alpha_{20}/2 - \sqrt{3/2}\,\mathrm{Re}\,\alpha_{22} & -\sqrt{3/2}\,\mathrm{Im}\,\alpha_{21} \\ -3/2\,\mathrm{Re}\,\alpha_{21} & -\sqrt{3/2}\,\mathrm{Im}\,\alpha_{21} & \alpha_{20} \end{pmatrix} \tag{6.18}$$

The damping matrix Γ is then given from the power law (5.17)

$$\Gamma = \Gamma_o \left(\frac{E}{E_o}\right)^{\delta} = \Gamma_o \exp\left[-\delta\sqrt{\frac{5}{4\pi}}\,Q\right] , \tag{6.19}$$

as in the adiabatic case.

The GDR equations of motion (6.16) are coupled to the quadrupole equation (6.1) through the α dependence of E and Γ. They thus also describe a stochastic process. For any given solution $\alpha_{2\mu}(t)$ of (6.1) we can solve (6.16) for $\vec{D}(t)$. The transformation from $\vec{D}(0)$, $\vec{P}(0)$ to $\vec{D}(t)$, $\vec{P}(t)$ (for a given $\alpha_{2\mu}(t)$) is linear so that it is sufficient to solve (6.16) for six possible independent initial conditions for $\vec{D}(0)$ and $\vec{P}(0)$. The correlation function $< \vec{D}(t)\vec{D}(0) >$ can then be calculated by averaging over the whole ensemble $\{\alpha_{2\mu}(t)\}$ if the initial dipole correlations (at $t = 0$) are known.

In deriving the initial dipole correlation we are guided by the limit where $\alpha_{2\mu}$ are constant (independent of time). In this limit the correlation function (for fixed $\alpha_{2\mu}$) should give the same contribution to the GDR cross-section that we have assumed in the adiabatic fluctuation theory. This initial correlation function is found to be

$$< \vec{D}(0)\ \vec{D}(0) > = \frac{1}{2}\,E^{-1} , \tag{6.20a}$$

which is simply the ground state expectation of $\vec{D}^2$. The other relevant initial condition

$$< \vec{P}(0\ \vec{D}(0) > = -\ \frac{i}{2} , \tag{6.20b}$$

also follows from considering the ground-state giant resonance state vector.

The degree of adiabaticity of the process is determined by the parameter χ. To see that, we define the adiabaticity parameter η as the ratio between the frequency spread ΔE of the GDR due to variations in the static deformation and the mean relaxation rate λ of the quadrupole motion

$$\eta = \frac{\Delta E}{\lambda} \quad . \tag{6.21}$$

We can estimate the frequency spread from

$$\Delta E = \sqrt{5/4\pi}\; E_o\, \Delta\beta \ . \tag{6.22}$$

In (6.22), $(\Delta\beta)^2$ is the variance of β at equilibrium. λ in (6.21) is given by $\lambda = 1/t_c$, where the relaxation time t_c can be shown to be [36]

$$t_c \approx \chi\, \frac{\langle\Delta\alpha_{2\mu}\ \Delta\alpha^*_{2\mu}\rangle}{5T} \quad . \tag{6.23}$$

We then find

$$\eta \approx \frac{E_o}{\sqrt{20\pi}}\, \frac{\chi}{T}\, (\Delta\beta)\ \langle\Delta\alpha_{2\mu}\ \Delta\alpha^*_{2\mu}\rangle \ . \tag{6.24}$$

Thus, η is proportional to the parameter χ of Eq. (6.1). The adiabatic limit corresponds to $\eta \gg 1$ $(\chi \to \infty)$. The opposite limit $\eta \ll 1$ is referred to as the sudden limit, where the quadrupole deformation fluctuates very rapidly. In the adiabatic limit $(\eta \gg 1)$ we can assume the quadrupole deformation $\alpha_{2\mu}$ in the GDR equations (6.16) to be frozen at its initial value $\alpha_{2\mu}(0)$, so that the Fourier transform of the dipole correlation function (for a fixed $\alpha_{2\mu}(0)$) is a superposition of Breit-Wigner curves. The actual GDR absorption cross-section becomes then the average over the initial distribution, i.e. the equilibrium distribution

$$\sigma_{abs}(\epsilon;T,\omega) = \frac{\int D[\alpha]\ e^{-F/T}\ \sigma(\epsilon;\alpha_{2\mu})}{\int D[\alpha]\ e^{-F/T}} \quad , \tag{6.25}$$

with $D[\alpha] = \Pi_\mu\, d\alpha_{2\mu} = \beta^4 |\sin 3\gamma|\ d\beta\ d\gamma\ d\Omega$. Eq. (6.25) is identical with our previous adiabatic model of Sect. 5.3. Furthermore, the unitary metric (4.4) emerges as the one which should be used in the adiabatic limit if (6.1) describes the correct dynamical evolution of $\alpha_{2\mu}$.

To solve the stochastic equations (6.1) and (6.16), and to determine the dipole correlation function in the general case, we proceed as follows:

(i) We choose an initial ensemble of quadrupole deformations $\{\alpha_{2\mu}(0)\}$ which is distributed according to the equilibrium ensemble $\exp(-F/T)$.

(ii) For each $\alpha_{2\mu}(0)$ we solve (6.1) by Monte-Carlo techniques [36]. We use a second order stochastic Runge-Kutta method such that

$$\alpha_{2\mu}(t+\Delta t) = \alpha_{2\mu}(t) + g_{2\mu}\,\Delta t + \sqrt{\Delta t\,\xi}\;Y_{2\mu}\ , \qquad (6.26)$$

where $g_{2\mu}$ is an average of $-\chi^{-1}\partial F/\partial\alpha^*_{2\mu}$, and the real and imaginary parts of $Y_{2\mu}$ are five independent standard normal random variables. We obtain an ensemble of "shape trajectories" $\{\alpha_{2\mu}(t)\}$ which is equilibrated at any time.

(iii) For each trajectory $\alpha_{2\mu}(t)$ we solve (6.16) for $\vec{D}(t)$ in terms of $\vec{D}(0)$ and $\vec{P}(0)$. The correlation $<\vec{D}(t)\vec{D}(0)>$ is then calculated by averaging over the ensemble $\{\alpha_{2\mu}(t)\}$ using the appropriate (quantum mechanical) initial correlation functions (6.20).

Non-adiabatic effects are seen most clearly when one assumes a zero intrinsic width (i.e. $\Gamma_0 = 0$), so that broadening of the resonance arises only from the coupling to the quadrupole degrees of freedom. We thus consider such an hypothetical ^{166}Er nucleus at T = 1.5 MeV and $\omega = 0$. The solid lines in Fig. 16 are the fit to the Monte Carlo calculations (error bars) of the GDR absorption cross-section and the dotted line is the adiabatic model. At $\chi = 750$ we are close to the adiabatic limit where three peaks are seen. At smaller χ the two peaks on the right coalesce and get narrower. Then the left peak starts to move to the right while disappearing and in the sudden limit we have a single narrow Lorentzian. Thus, though the general effect is that of motional narrowing as discussed in [38], the actual spectral shape of the resonance is also sensitive to η. Note that as the process becomes more sudden it is necessary to take a smaller time step Δt since the quadrupole fluctuations are more erratic.

Realistic calculations ($\Gamma_0 \neq 0$) are shown in Fig. 17 for ^{112}Sn where our adiabatic model ($\chi = 500$) overestimates the experimental widths [20] (dotted

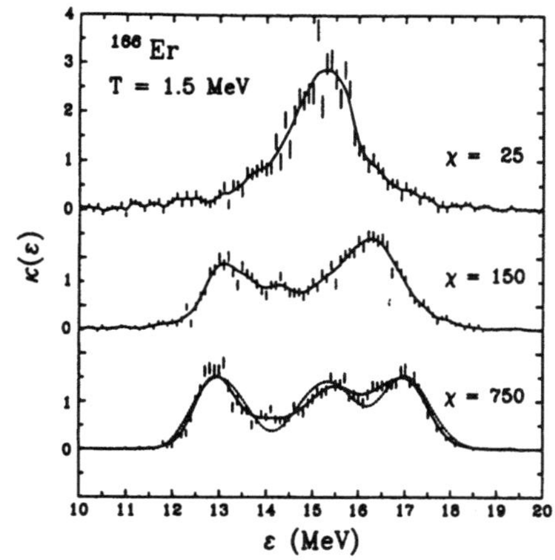

Fig. 16. The Fourier transforms (solid lines) of the dipole correlation function found from the solution of the stochastic equations (6.1) and (6.16) for various values of χ (χ=750, 150, and 25). For the purpose of demonstrating the effects of nonadiabaticity we have chosen a hypothetical case where Γ_0=0 (i.e. no intrinsic damping of the dipole) for ^{166}Er at T=1.5 MeV. The bars are the statistical errors associated with the Monte-Carlo calculations and the dashed lines present the adiabatic model of Sect. 5. Notice that various peaks coalesce and get narrower as the process becomes less adiabatic (χ get smaller).

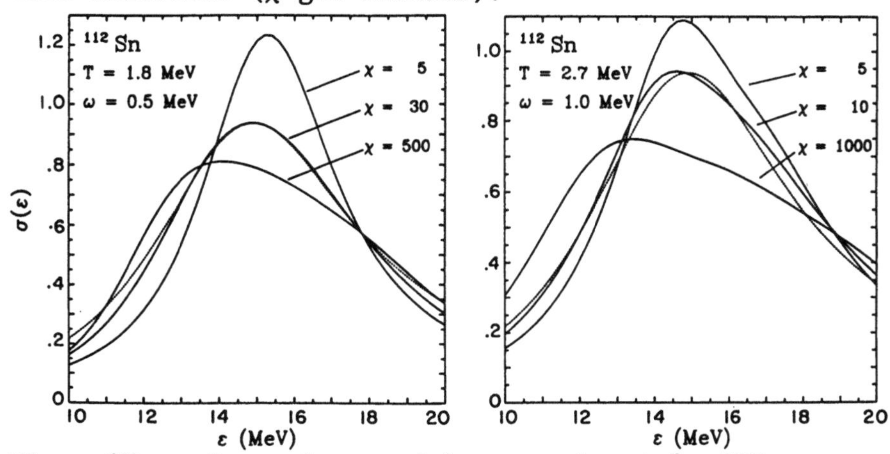

Fig. 17. Comparison with experimental GDR cross-sections for ^{112}Sn [20]. The dotted lines shown are the experimental cross-sections. The solid lines are the results of our stochastic model for several values of χ. The higher and the lower values of χ are the adiabatic and sudden limits, respectively. The intermediate values of χ show the best fit to the data.

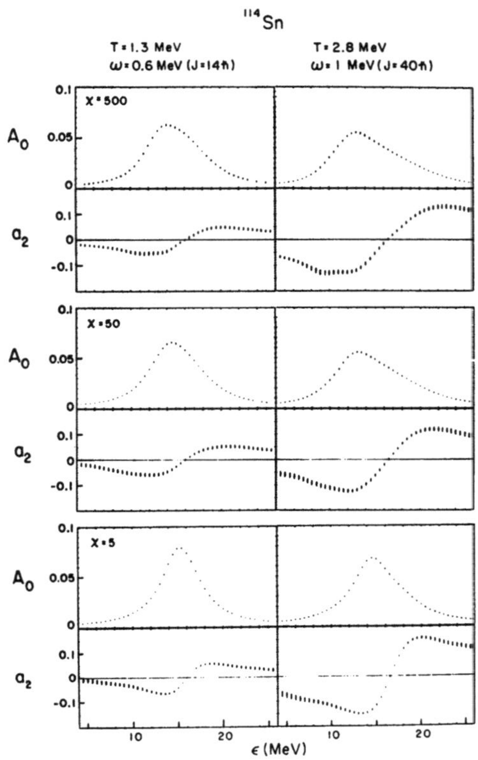

Fig. 18. Non-adiabatic effects on $\sigma \propto A_0$ and a_2 for ^{114}Sn at low spin (left side) and high spin (right side). Both σ and a_2 are shown for various values of χ: $\chi = 500$ (adiabatic), 50 and 5 (sudden). Notice the motional narrowing in σ and the stronger peaks in a_2 when the sudden limit is approached.

lines). We used $E_0 = 15.2$ MeV, $\Gamma_0 = 3.76$ MeV and $\delta = 1.6$, and calculated the GDR absorption cross-section for several values of χ. As χ decreases the resonance gets narrower and its structure changes. The valueswhich fit the experiment the closest are shown on the figure and correspond to intermediate η's.

In the sudden limit $\eta \ll 1$ $(\chi \to 0)$ it is possible to reduce the stochastic equation (6.16) to an equation of motion for $\langle\vec{D}\rangle$ which is basically that of a damped rotating oscillator with some effective frequency and damping width. The effective damping $\tilde{\Gamma}$ is estimated to be (in the $\Gamma_0 = 0$ case)

$$\tilde{\Gamma} \approx \frac{(\Delta E)^2}{\lambda} = \eta\ \Delta E \ll \Delta E \qquad (6.27)$$

which is narrower by a factor η than the width in the adiabatic limit. This is exactly the motional narrowing effect discussed in [38]. In realistic calculations Γ_0 has to be added to the r.h.s. of (6.29). The effects of non-adiabaticity on a_2 are shown in Fig. 18 together with those on σ. Two cases are shown for ^{114}Sn, at low (J = 14 $\hbar$) and high (J = 40 $\hbar$) spins. In the sudden limit the minimum and maximum of $a_2(\epsilon)$ become more sharply peaked and larger in magnitude. They are also closer to the central energy E_0. These effects are consistent with the motional narrowing in $\sigma(\epsilon)$.

We have seen earlier that for ^{112}Sn the GDR cross-section is consistent with an intermediate situation between the adiabatic and sudden limits. Thus we expect a_2 to look more like the intermediate cases in Fig. 18. Unfortunately the a_2's for ^{112}Sn were not measured yet. It will be very interesting to determine them experimentally. This will provide a more crucial test of non-adiabatic effects in the GDR.

7. Quasi-Continuum E2 Transition

At lower excitation energies ($E^* \leq 8$ MeV), the E1 decay rate is lower and E2 transitions above Yrast dominate. The quasicontinuum E2 spectrum has been recently observed [39] by subtracting the discrete lines and the statistical γ-rays, and then decomposing the remaining quasicontinuum into dipole and quadrupole parts. The observed E2 bump is interpreted as the result of E2 transitions within dense rotational bands in the continuum [39]. These E2 transitions are thus a possible probe of the properties of "warm" nuclei whose excitation energy is below the neutron separation energy. An interesting observation was made through the analysis of the experimental data [39] of warm transitional dysprosium isotopes (^{152,154,156}Dy): a strong collective B(E2) of about 300 W.u. was required to reproduce the measured Doppler shift and the spectral shape of the E2 peak.

For transitional nuclei, the shape transition temperature is low [47] ($T_{cr} \sim 0.5 - 1$ MeV). Figure 19 shows the phase diagram in the energy-spin plane, where only the transition lines ($\gamma = -180^\circ$) are displayed. The transitions at J = 0 happen at $E^* \approx 12$, 21 and 30 MeV for ^{152,154,156}Dy, respectively. In the experiment

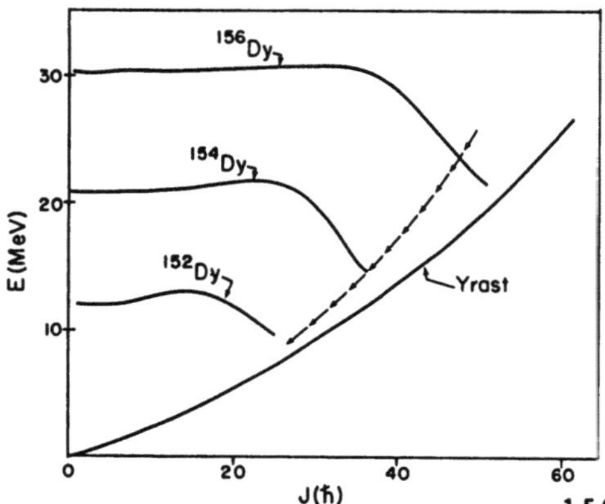

Fig. 19. Phase diagram of ^{152,154,156}Dy in the energy-spin plane. Shown are the Yrast line and the transition lines. The arrows indicate an approximate most probable γ-ray decay pathway.

of [39] the γ-cascades start at a given entry point and proceed along different paths determined by the competition between E2 and E1. The most probable path is shown in the above figure by arrows. To estimate B(E2) for a general triaxial rotating shape we assume an asymmetric rotor. If its angular momentum J is large it has in addition to J an approximate good ("wobbling") quantum number n. The E2 transitions proceed along the bands n=const. with strength [59]

$$B(E2;\ n,J \rightarrow n,J-2) \approx \frac{5}{16}\pi\, e^2 Q_{22}^2 , \qquad (7.1)$$

where Q_{22} is the intrinsic quadrupole moment around the rotation 3 axis. Assuming a uniformly charged ellipsoid we find

$$Q_{22} = \sqrt{\frac{3}{2}}\,\frac{Z}{5}\left(R_1^2 - R_2^2\right) , \qquad (7.2)$$

where R_1 and R_2 are the semi-axes lengths along the 1 and 2 axes.

If the equilibrium deformation is used in (7.2) we obtain for ^{154}Dy values of B(E2) which, along the most probable γ-decay path of Fig. 19, vary between 0 (in the non-collective oblate regime) and 180 W.u. However, if fluctuations in β,γ are taken into account according to the general formula (4.5), we obtained values of 240-300 W.u., in agreement with the experiment. It is interesting to note that the dysprosium surfaces may have a local superdeformed minimum. As a result the shape fluctuations are enhanced and lead to the above large values of B(E2).

The fourth order Landau theory of Sect. 3 does not predict a superdeformed minimum. In order to treat the above transitional nuclei in the Landau framework we have to carry this expansion to sixth order. This is however beyond the scope of the present review.

8. Conclusions

We have reviewed the main ideas and theoretical techniques used in the description of hot nuclei. In the temperature range $1 \lesssim T \lesssim 3$ MeV, a macroscopic approach to hot rotating nuclei, based on the Landau theory combined with a uniform fluctuation theory in all five quadrupole degrees of freedom, can reproduce well measured physical observables. Such observables are the GDR cross-section and angular anisotropy of the emitted γ-rays, and the B(E2) quasi-continuum transitions. Time-dependent fluctuations play an important role in non-adiabatic situations and comparison with the GDR data can be used to determine a dissipation parameter. Several interesting theoretical issues are still open:

(i) In our treatment of fluctuations we have assumed that the shape parameters ($\alpha_{2\mu}$) are classical (for $T \gtrsim 1$ MeV). It is not clear what is the role played by quantal fluctuations. They are certainly important at low temperatures.

(ii) The role played by higher-order shape multipoles (such as octupole).

(iii) A theoretical estimate of the damping of quadrupole motion at finite temperature.

(iv) The importance of temperature fluctuations in the transformation from the canonical to the physical microcanonical ensemble.

On the experimental side it is hoped that with the improvement of the present generation of experiments it will become possible to control more accurately the nuclear phase space variables (temperature and spin). It is also necessary to increase the precision with which quantities such as angular anisotropy are measured.

Acknowledgements

I would like to thank S. Levit, J. Zingman and J. Manoyan for their collaboration on the Landau theory of the nuclear shapes transitions, and B. Bush for his collaboration on shape fluctuations and the GDR in hot nuclei.

References

[1] B. Borderie et al., Z. Phys. **Z316**, 243 (1984).
[2] D. Jacquet et al., Phys. Rev. Lett. **53**, 2226 (1984).
[3] R.J. Charity et al., Phys. Rev. Lett. **56**, 13 (1986).
[4] A.L. Goodman, Nucl. Phys. **A352**, 30 (1981);
K. Tanabe, K. Sugawara-Tanabe and M.J. Mang, Nucl. Phys. **A357**,20,45(1981);
M. Faber, J.L. Egdio and P. Ring, Phys. Lett. **127B**, 5 (1983);
J.L. Egido and P. Ring,Nucl. Phys. **A388**, 19 (1982).
[5] L.G. Moretto, Nucl. Phys. **A182** (1972) 641.
[6] M. Brack and P. Quentin, Phys. Lett. **52B**, 159 (1974); Phys. Scripta **10A**, 163 (1974);
P. Quentin and H. Flocard, Ann. Rev. Nucl. Sci **28**, 523 (1978).
[7] A.K. Ignatiuk, I.N. Mikhailov, L.H. Molina, R.G. Nazmitdinov and K. Pomorsky, Nucl. Phys. **A346**, 191 (1980);
J. Diebel, K. Albrecht and R.W. Hasse, Nucl. Phys. **A355**, 66 (1981).
[8] P. Bonche, S. Levit and D. Vautherin, Nucl. Phys. **A427**, 278 (1984); **A436**, 265 (1985);
S. Levit and P. Bonche, Nucl. Phys. **A437**,426 1985.
[9] J.O. Newton et al., Phys. Rev. Lett. **46**, 46 (1981).
[10] J.J. Gaardhoje, C. Ellegaard, B. Herskind and S.G. Steadman, Phys. Rev. Lett. **53**, 148 (1984).
[11] W. Henneriei et al., Nucl. Phys. **A396**, 329c (1983).
[12] B. Haas et al., Phys. Lett **120B**, 79 (1983).
[13] A.M. Sandorfi et al., Phys. Lett. **130B**, 19 (1983).
[14] G.A. Gossett, K.A. Snover, J.A. Behr, G. Feldman and J.L. Osborne, Phys. Rev. Lett. **54**, 1456 (1985).
[15] D. Vautherin and N.V. Vinh Mau, Phys. Lett. **120B**, 271 (1983).

[16] P. Ring, L.M. Robledo, J.L. Egido and M. Faber, Nucl. Phys. **A415**, 261 (1984).
[17] H. Sagawa and G.F. Bertsch, Phys. Lett. **146B**, 138 (1984);
K. Sugawara-Tanabe and K. Tanabe, Phys. Lett. **192B**, 268 (1987); Prog. Theor. Phys. **76**, 1272 (1986).
[18] For a review see K.A. Snover, Ann. Rev. Nucl. Part. Sci. **36**, 545 (1986).
[19] J.J. Gaardhoje, C. Ellengaard, B. Herskind, R.M. Diamond, M.A. Deleplanque, G. Dines, A.O. Macchiavelli and F.S. Stephens, Phys. Rev. Lett. **56**, 1783 (1986).
[20] D.R. Chakrabarty, M. Thoennenssen, S. Sen, P. Paul, B. Butsch and M.G. Herman, Phys. Rev. **C37**, 1437, (1988).
[21] J.J. Gaardhoje, A.M. Bruce and B. Herskind, Nucl. Phys. **A482**, 121c (1988).
[22] P. Thirolf, D. Habs, D. Schwalm, R.D. Fisher and V. Metug, Nucl. Phys. **A482**, 93c (1988).
[23] K.A. Snover, Nucl. Phys. **A482**, 13c (1988).
[24] D.R. Chakrabarty, S. Sen, M. Thoennessen, N. Alamanos, P. Paul, R. Schicker, J. Stachel and J.J. Gaardhoje, Phys. Rev. **C36**, 1886 (1988).
[25] Y. Alhassid, S. Levit and J. Zingman, Phys. Rev. Lett. **57**, 536 (1986).
[26] Y. Alhassid, J. Zingman and S. Levit, Nucl. Phys. **A469**, 205 (1987).
[27] S. Levit and Y. Alhassid, Nucl. Phys. **A413**, 439 (1984).
[28] M. Gallardo, M. Diebel, T. Dossing and R. A. Broglia, Nucl. Phys. **A443**, 415 (1985).
[29] Y. Alhassid, B. Bush and S. Levit, Phys. Rev. Lett. **61**, 1926 (1988);
Nucl. Phys. **A482**, 57c (1988).
[30] M. Gallardo, F.J. Luis and R.A. Broglia, Phys. Lett. **191B**, 222 (1987); J.M. Pacheco et al., Phys. Rev. Lett. **61**, 294 (1988).
[31] A. Goodman, Phys. Rev. **C37**, 2162 (1988).
[32] Y. Alhassid and B. Bush, Nucl. Phys. **A509**, 461 (1990).
[33] Y. Alhassid and B. Bush, Phys. Rev. Lett. **65**, 2527 (1990).
[34] J. Gundlach, K.A. Snover, J.A. Behr, C.A. Gossett, M.K.-Habior and K.T. Lesko, Phys. Rev. Lett. **65**, 2523 (1990).

[35] Y. Alhassid and B. Bush, Phys. Rev. Lett. 63, 2452 (1989).
[36] Y. Alhassid and B. Bush, Nucl. Phys. **A514**, 434 (1990).
[37] R. Broglia, T. Dossing, B. Lauritzen and R.B. Mottelson, Phys. Rev. Lett. **53**, 326 (1987).
[38] B. Lauritzen, R.A. Broglia, W.E. Ormand and T. Dossing, Phys. Lett. **207B**, 2381 (1988).
[39] R. Holzmann,T.L. Khoo et al., Phys. Rev. Lett. **62**, 520 (1989).
[40] A. Bohr and B. Mottelson, Nucl. Structure, Vol. 1 (Benjamin, New York, 1969), Ch. 2. App. 2.
[41] L.D. Landau and E.M. Lifshitz, Statistical Physics, (Pergamon,
Oxford, 1950), Part 1, Ch. III.
[42] A.K. Kerman and S. Levit, Phys. Rev. **C24**, 1029 (1981).
[43] Y. Alhassid and J. Zingman, Phys. Rev. **C30**, 684 (1984).
[44] B. Lauritzen, P. Arve and G.F. Bertsch, Phys. Rev. Lett. **65**, 2835 (1988).
[45] Ref. 41, Ch. IVX.
[46] A. Aharony, Multicritical points, in: Critical Phenomena, Lecture Notes in Physics **186** (Springer, New York, 1983) p. 207.
[47] Y. Alhassid, J. Manoyan and S. Levit, Phys. Rev. Lett. **63**, 31 (1989).
[48] D.L. Hill and J.A. Wheeler, Phys. Rev. **89**, 1182 (1953).
[49] Ref. 41, Ch. XII.
[50] See in Ref. 40, Vol. 2, Ch. 6.
[51] J.M. Eisenberg and W. Greiner "Nuclear Theory" (North Holland,
Amsterdam, 1975), Vol. 1, Chs. 10-11.
[52] M. Danos and W. Greiner, Phys. Lett. **8**, 113 (1964); M.G. Huber, M. Danos, H.J. Weber and W. Greiner, Phys. Rev. **155**, 1073 (1967).
[53] B. Bush and Y. Alhassid, Yale preprint YCTP-N12-90.
[54] See, for example, C.P.Slichter, "Principles of Magnetic Resonance,"
Springer Verlag, New York 1978; P.W. Anderson and P.R. Weiss, Rev. Mod. Phys. **25**, 269 (1953);
R. Kubo, "Fluctuations Relaxation and Resonance in Magnetic Systems," D. ter Haa ed., Oliver and Boyd, Edinburgh, 1961.

[55] For a discussion of general non-linear Langevin equations see in R. Balian, Y. Alhassid and H. Reinhardt, Physics Reports 131, 1 (1986).
[56] Y. Alhassid and B. Bush, Yale preprint YCTP-N11-90.
[57] See, for example, in N.G. Van Kampen, "Stochastic Processes in Physics and Chemistry" (Amsterdam, North Holland, 1981).
[58] See for example in "The Maximum Entropy Formalism," ed. R.D. Levine and M. Tribus, (MIT Press, Cambridge, 1978).
[59] Ref. 40, Vol. 2, Ch. 4.

Dissipation and Thermal Fluctuations in Heavy-Ion Collisions

P. Fröbrich

Hahn-Meitner-Institut and Freie Universität Berlin,
Glienicker Strasse 100, W-1000 Berlin 39, Fed. Rep. of Germany

1. Introduction

The concept of friction has turned out to be a useful one not only in solid state physics but also in the description of heavy-ion collisions and fission. In the following I shall concentrate on applications to low energy ($E \leq 10$ MeV/nucleon) heavy-ion collisions. I shall put emphasis on the phenomenological side in showing that by using frictional forces (and the associated fluctuating forces) in a semi-phenomenological model one is able to put some order into a large variety of experimental data. These concern above- and below-barrier fusion, spin distributions, deep-inelastic scattering and the emission of δ electrons in deep-inelastic collisions.

Because of lack of time I cannot go into the many (different) theoretical approaches toward a microscopic understanding of dissipation in nuclear physics, such as perturbation theory, linear response theory, one- and two body dissipation mechanisms, wall and window friction, random matrix models, or diabatic dissipative dynamics. I give only a very limited and incomplete list of more recent articles, in which further literature can be found [1-6]. In spite of these efforts, in my opinion, a satisfactory quantitative microscopic understanding of the dissipation mechanism in nuclear physics has not yet been reached. None of the microscopic models until now yields friction coefficients quantitatively compatible with those that are necessary to describe the data in the phenomenological surface friction model discussed below. This, however, could also mean that the latter is a too simplified approach. In order to make progress in answering this question we confront the surface friction model systematically with as many data as possible. In this way we hope to detect shortcomings which might motivate more refined investigations.

This lecture is supposed to be a pedagogical one. Therefore I do not begin with one of the microscopic theories. I do not show, for example, how a Fokker Planck equation with microscopic drift and diffusion coefficients can be obtained from the Wigner transformed von Neumann equation by eliminating the intrinsic (bath) variables using projection operator techniques. Instead I start in Sect. 2 with a simple but very instructive model in which dissipation and fluctuations are produced by many oscillators coupled to the relative motion. We mainly discuss the classical version of this model [7], which has many of the essential features which enter a theoretical description of collisions with frictional and fluctuating forces. We show how this model leads to a Langevin equation of the kind used in Sect. 3 to describe heavy-ion scattering data. Some remarks on the quantal version of the model with respect to the problem of dissipative tunneling are also made.

Section 3 deals with applications of Langevin equations to heavy-ion collisions in the framework of the surface friction model. Calculated fusion barriers are compared to the data, the competition of fusion and deep-inelastic collisions close to the barrier is discussed, examples for spin distributions are shown, results for deep-inelastic collisions are displayed, and the effects of thermal fluctuations on the energy spectra of δ electrons emitted in coincidence with deep-inelastic collisions are studied. We investigate to what extent δ electrons can be used as an atomic clock for the timescale of deep-inelastic collisions.

In Sect. 4 we draw some conclusions.

2. An Instructive Model

In the following we show how in a simple model, whose main features can be found in [7], many of the essential characteristics of dissipative motion can be discussed. Equations of motion (Langevin equations) with conservative, dissipative and fluctuating forces are derived. A microscopic derivation of the friction and fluctuating force is given within the framework of the model. The relation between the dissipative and fluctuating forces is expressed by the fluctuation-dissipation theorem. The connection between Langevin and Fokker-Planck descriptions is exemplified.

2.1. The Model Lagrangian and its Equations of Motion

We start with the classical version of the model. To be specific one may think in terms of heavy-ion collisions. The relative motion of two heavy ions and its coupling to intrinsic degrees of freedom is described by the following Lagrangian

$$L = \frac{1}{2} M \dot{R}^2 - V(R) + \sum_i \frac{m_i}{2} (\dot{q}_i^2 - \omega_i^2 q_i^2) - \sum_i q_i f_i(R). \qquad (1)$$

Here, R is the relative motion between the nuclei, M its reduced mass and V(R) its conservative interaction potential. The intrinsic degrees of freedom are modelled by harmonic oscillators. The coupling between relative motion and the intrinsic degrees of freedom is assumed to be separable and linear in the intrinsic coordinates q_i (linearizing also with respect to R : $f_i(R) = c_i R$ leads to "Ohmic" friction). For the moment we neglect additional degrees of freedom relevant for heavy-ion scattering like the orbital motion, deformation modes, neck formation, and particle transfer.

From equation (1) we obtain the classical equations of motion

$$M\ddot{R} = -\frac{dV(R)}{dR} - \sum_i q_i \frac{df_i(R)}{dR}, \tag{2}$$

$$m_i\ddot{q}_i = -m_i\omega_i^2 q_i - f_i(R). \tag{3}$$

2.2. Elimination of the Intrinsic Variables

In the simple classical model example it is easy to eliminate the intrinsic coordinates in eq.(2) by an explicit solution of eq.(3). This is not so easy in the more general case, where one can e.g. start from the Wigner transformed von Neumann equation and where the intrinsic (bath) variables have to be eliminated by complicated projection operator techniques. Alternatively they would have to be integrated out when applying Feynman's influence functional technique.

The solution of eq.(3) is

$$q_i = q_i(0)cos\omega_i t + \frac{p_i(0)}{m_i\omega_i} sin\omega_i t + \int_0^t ds \frac{f_i(R(s))}{m_i\omega_i} sin(\omega_i(t-s)). \tag{4}$$

The initial conditions of the intrinsic degress of freedom $q_i(0)$ and $p_i(0)$ are not known. Therefore statistical assumptions on their behaviour are made. They are assumed to be randomly distributed with $< q_i(0) >= 0$ and $< p_i(0) >= 0$, i.e. their mean values vanish. The uncertainty in the initial conditions for the intrinsic variables, i.e. the quantity $q_i^o(t) = q_i(0)cos\omega_i t + \frac{p_i(0)}{m_i\omega_i} sin\omega_i t$ is the origin of the fluctuating force, the Langevin force. Defining the momentum of the relative motion as

$$P = M\dot{R} \tag{5}$$

and substituting eq.(4) into eq.(2) yields the Langevin equation

$$\dot{P} = -\frac{dV(R)}{dR} - \sum_i \int_o^t ds \frac{f_i(R(s))}{m_i\omega_i} sin(\omega_i(t-s)) \frac{df_i(R)}{dR} - \sum_i q_i^o(t) \frac{df_i(R)}{dR} \tag{6}$$

The terms on the right hand side represent a conservative force, a retarded dissipative force and a fluctuating force.

2.3. The Friction Force

In order to extract from the above expression a friction force proportional to the velocity we perform a partial integration of the second term. One obtains

$$\begin{aligned}\dot{P} = & - \frac{dV}{dR} + \sum_i \frac{f_i(R)}{m_i\omega_i^2}\frac{df_i(R)}{dR} \\ & - \sum_i \frac{1}{m_i\omega_i^2}\int_o^t ds \frac{df_i(R(s))}{dR} cos(\omega_i(t-s))\frac{df_i(R(t))}{dR}\dot{R}(s) \\ & - \sum_i q_i^o(t)\frac{df_i}{dR}. \end{aligned} \tag{7}$$

The surface term of the partial integration $\sum_i \frac{f_i(R(s))}{m_i\omega_i^2} cos(\omega_i(t-s))|_o^t \frac{df_i(R)}{dR} = \sum_i \frac{f_i(R)}{m_i\omega_i^2}\frac{df_i(R)}{dR}$ (in a scattering situation $f_i(R(0)) = 0$) gives rise to a term that renormalizes the conservative potential. In applications this term is usually not calculated but assumed to be contained in the phenomenological potential which is used in trajectory calculations. An alternative possibility is to prepare the model Lagrangian in such a way that the renormalization term does not occur in the equations of motion [26]. In this case the coupling term in the Lagrangian has to be written as $-\sum_i \frac{m_i\omega_i^2}{2}(q_i - \frac{f_i(R)}{m_i\omega_i^2})^2$.

The third term in eq.(7) is a non-Markovian friction force proportional to the velocity with the structure of a force-force correlation function

$$\int_o^t ds \sum_i \frac{df_i(R(s))}{dR}\frac{cos(\omega_i(t-s))}{m_i\omega_i^2}\frac{df_i(R(t))}{dR}\dot{R}(s) = \int_o^t ds\gamma(t-s)\dot{R}(s). \tag{8}$$

In solid state physics "Ohmic" dissipation is often studied, where $f_i(R) = c_i R$ is linearized; then $\gamma(t-s) = \sum_i \frac{c_i^2}{m_i\omega_i^2} cos(\omega_i(t-s))$, which still represents a non-Markovian friction term. In the scattering situation of a heavy-ion collision one has to take into account the complete R-dependence in the form factor f(R), which has to vanish for large distances. On the other hand one tries, for reasons of simplicity, to obtain Markovian friction. This can be strictly achieved in our model example only if the intrinsic spectrum is assumed to be of quadratic form

$g(\omega) = g_o\omega^2$. Replacing $\sum_i$ by $\int_o^\infty d\omega g(\omega)$ and with $f_i \approx f$ and $\omega_i \approx \omega$ a friction term local in time is obtained

$$\int_o^t ds \sum_i \frac{df_i(R(s))}{dR} \frac{cos(\omega_i(t-s))}{m_i\omega_i^2} \frac{df_i(R(t))}{dR} \dot{R}(s)$$

$$\approx \int_o^t ds \frac{df(R(s))}{dR} \int_o^\infty d\omega \frac{cos(\omega(t-s))}{m\omega^2} g_o\omega^2 \frac{df(R(t))}{dR} \dot{R}(s)$$

$$= \int_o^t ds \frac{df(R(s))}{dR} \frac{g_o}{m} \pi\delta(t-s) \frac{df}{dR} \dot{R}(s) = \frac{g_o\pi}{m2} \frac{df}{dR} \frac{df}{dR} \dot{R}(t). \qquad (9)$$

Thus the equation of motion contains besides a renormalized conservative force a time-local (Markovian) friction force and a random force

$$\dot{P} = -\frac{d\tilde{V}(R)}{dR} - \gamma(R)\dot{R} - F(t). \qquad (10)$$

The tilde is an indication for the renormalization of the potential ($\tilde{V} = V + \int d\omega \frac{g(\omega)}{2m\omega^2} f^2(R)$). For a quadratic spectrum one has to introduce a cut-off frequency in order to avoid a divergence of the integral; using a Drude spectrum leads to a convergent integral but to non-Markovian friction. The friction form factor is given by

$$\gamma(R) = \frac{g_o\pi}{2m}(\frac{df(R)}{dR})^2. \qquad (11)$$

In the derivation above, no argument on the strength of the interaction had to be made, only separability had to be assumed, which allowed the analytical calculation. The radial dependence of the friction form factor in (11) will be used in the applications of the surface friction model in Sect. 3. The derivation above involves of course a certain approximation, because the relative coordinate R itself is a stochastic variable. Therefore the R dependence of the form factor and also of the conservative potential has to be understood as the dependence on the mean value of R: The mean value of a function is assumed to be the function of the mean value. This is certainly the better fulfilled the weaker the interaction. This assumption is only not necessary if f_i depends only linearly on R (Ohmic dissipation). But for a scattering situation a non-linear dependence of the form factor on R is necessary in order to account for the correct boundary conditions.

2.4. The Fluctuating Force

Now the fluctuating force

$$F(t) = \sum_i (q_i(0) cos\omega_i t + \frac{p_i(0)}{m_i\omega_i} sin\omega_i t)\frac{df_i}{dR} \qquad (12)$$

is investigated.

In general the fluctuating force is not $\delta-$correlated in time. In order to be able to deal with a Markovian process (no time retardation of the frictional kernel) we investigate under which conditions a $\delta-$correlated random force within the model can be constructed and how a relation between its strength and the friction force can be obtained (fluctuation-dissipation theorem).

A δ-correlated fluctuating force has the structure

$$F(t) = g\xi(t), \qquad (13)$$

where $\xi(t)$ is a stochastic variable with the properties

$$\begin{aligned} < \xi(t) > &= 0 \\ < \xi(t)\xi(t') > &= 2\delta(t-t'). \end{aligned} \qquad (14)$$

The strength g will be shown below to be related to the diffusion coefficient D occurring in a Fokker-Planck equation by $g^2 = D$, so that

$$< F(t)F(t') >= 2D\delta(t-t'). \qquad (15)$$

Considering eq.(12) we write

$$\begin{aligned} < F(t)F(t') > &= \\ \sum_i\sum_j \frac{df_i}{dR}\frac{df_j}{dR} < &(q_i^o cos\omega_i t + \frac{p_i^o}{m_i\omega_i} sin\omega_i t)(q_j^o cos\omega_j t' + \frac{p_j^o}{m_j\omega_j} sin\omega_j t') > \\ = \sum_i\sum_j \frac{df_i}{dR}\frac{df_j}{dR} &[< q_i^o q_j^o > cos\omega_i t cos\omega_j t' + < p_i^o p_j^o > \frac{1}{m_i m_j \omega_i\omega_j} sin\omega_i t\omega_j t' \\ + < q_i^o p_j^o &> \frac{1}{m_j\omega_j} cos\omega_i t sin\omega_j t' + < q_j^o p_i^o > \frac{1}{m_i\omega_i} sin\omega_i t cos\omega_j t']. \end{aligned} \qquad (16)$$

Again the radial dependence of $df_i(R)/dR$ has to be considered as the dependence on the mean value of $< R >$. Assuming q_i^o and p_i^o to be Gaussian distributed random variables we have

$$< q_i^o q_j^o > = \delta_{ij}\overline{q_i^o}^2$$

$$
\begin{aligned}
< p_i^o p_j^o > &= \delta_{ij}\bar{p}_i^{o^2} \\
< q_i^o p_j^o > &= 0
\end{aligned}
\tag{17}
$$

and

$$
< F(t)F(t') >= \sum_i (\frac{df_i}{dR})^2 [\bar{q}_i^{o^2} cos\omega_i t cos\omega_i t' + \frac{1}{m_i^2\omega_i^2}\bar{p}_i^{o^2} sin\omega_i t sin\omega_i t'] \tag{18}
$$

Using the addition theorems for the trigonometric functions one finds that the main contribution comes from $t \approx t'$. With a quadratic spectrum ($g(\omega) = g_o\omega^2$) and extending the frequency integration to infinity one obtains with the mean energy of the oscillator $< \epsilon_i >= \frac{1}{2}m_i\omega_i^2\bar{q}_i^{o2} + \bar{p}_i^{o^2}/2m_i$

$$
\begin{aligned}
< F(t)F(t') > &\simeq \sum_i \frac{< \epsilon_i >}{m_i\omega_i^2}(\frac{df_i}{dR})^2 cos(\omega_i(t-t')) \\
&\simeq (\frac{df}{dR})^2 < \epsilon > \frac{g_o}{m}\int_o^\infty d\omega cos\omega(t-t') \\
&= \frac{g_o\pi}{m}(\frac{df}{dR})^2 < \epsilon > \delta(t-t').
\end{aligned}
\tag{19}
$$

Comparing with eq.(15) we can read off the diffusion coefficient

$$
D = \frac{1}{2}\frac{g_o\pi}{m}(\frac{df}{dR})^2 < \epsilon > . \tag{20}
$$

Assuming thermodynamic equilibrium we apply the equipartition theorem for the oscillator energy $< \epsilon >$

$$
< \epsilon >= kT, \tag{21}
$$

where k is Boltzmann's constant and T is the bath temperature. Comparing eq.(11) with eq.(20) we establish the fluctuation-dissipation theorem which connects the friction form factor with the diffusion coefficient,

$$
D = \gamma kT. \tag{22}
$$

This relation reflects the fact that dissipation and fluctuation are not independent of each other, which is not surprising because they have the same physical origin, namely the collisions of the particles with each other.

2.5. The Langevin Equation

Using the fluctuation-dissipation theorem the final Langevin equation reads

$$\dot{R} = \frac{P}{M}$$
$$\dot{P} = -\frac{d\tilde{V}}{dR} - \gamma(R)\frac{P}{M} + \sqrt{\gamma(R)kT}\xi(t). \quad (23)$$

The generalization of this equation to many dimensions will be used in the applications to heavy-ion collisions in Sect. 3. The classical distribution function is calculated by sampling many Langevin trajectories. Cross sections are then obtained by a Monte Carlo integration of the distribution function over the impact parameter.

In practice the Langevin equation has to be discretized

$$P(t+\tau) - P(t) = \int_t^{t+\tau} dt'[-\frac{d\tilde{V}}{dR} - \gamma\frac{P}{M}] + \sqrt{D}\int_t^{t+\tau} dt'\xi(t')$$
$$\simeq [-\frac{d\tilde{V}}{dR} - \gamma\frac{P}{M}]\tau + \sqrt{D}\Gamma(t). \quad (24)$$

Here τ is a finite time step and $\Gamma(t) = \int_t^{t+\tau} dt'\xi(t')$ is a superposition of Gaussian distributed random numbers which again is of Gaussian form: $\Gamma(t) = a\omega(t)$ with $<\omega>= 0$ and $<\omega^2>= 2$. The coefficient a is obtained by comparing

$$<\Gamma\Gamma> = a^2 <\omega^2>= 2a^2$$
$$<\Gamma\Gamma> = \int_t^{t+\tau} dt' \int_t^{t+\tau} dt'' <\xi(t')\xi(t'')>$$
$$= 2\int dt' \int dt''\delta(t'-t'') = 2\tau \quad (25)$$

as $a = \sqrt{\tau}$.

The final form of the Langevin equation suitable for numerical integration is

$$P_{n+1} - P_n = [-\frac{d\tilde{V}}{dR} - \gamma\frac{P}{M}]_n\tau + \sqrt{D_n}\sqrt{\tau}\omega(t). \quad (26)$$

2.6. Equivalence of Langevin and Fokker-Planck Equations

For Markovian processes the Langevin equations can be transformed into a corresponding Fokker-Planck equation for the classical distribution function d(R,P,t) of the form ($x_i = R, P$)

$$\frac{d}{dt}d(x_i,t) = -\sum_i \frac{d}{dx_i}(D_i^{(1)}d(x_i,t)) + \sum_{ij}\frac{d^2}{dx_i dx_j}(D_{ij}^{(2)}d(x_i,t)). \quad (27)$$

This can be shown as follows. We use the Kramers-Moyal expansion for the drift $D^{(1)}$ and diffusion $D^{(2)}$ terms

$$\begin{aligned} D_i^{(1)} &= \lim_{\tau\to 0}\frac{1}{\tau} < x_i(t+\tau) - x_i(t) > \\ D_{ij}^{(2)} &= \frac{1}{2}\lim_{\tau\to 0}\frac{1}{\tau} < (x_i(t+\tau) - x_i(t))(x_j(t+\tau) - x_j(t)) > . \end{aligned} \quad (28)$$

Writing the Langevin equation as an integral equation and using the properties of the random force (eq.14) we identify

$$\begin{aligned} D_P^{(1)} &= \lim_{\tau\to 0}\frac{1}{\tau} < P(t+\tau) - P(t) > \\ &= \lim_{\tau\to 0}\frac{1}{\tau}\int_t^{t+\tau} dt' < -\frac{d\tilde{V}(r(t'))}{dR} - \gamma(R(t'))\frac{P(t')}{M} + \sqrt{D}\xi(t') > \\ &\simeq \lim_{\tau\to 0}\frac{1}{\tau}(-\frac{d\tilde{V}}{dR} - \gamma\frac{P}{M})\tau = -\frac{d\tilde{V}}{dR} - \gamma\frac{P}{M} \\ D_R^{(1)} &= \lim_{\tau\to 0}\frac{1}{\tau} < R(t+\tau) - R(t) > \\ &= \lim_{\tau\to 0}\frac{1}{\tau}\int_t^{t+\tau} dt'\frac{P}{M} = \frac{P}{M} \\ D_{PP}^{(2)} &= \lim_{\tau\to 0}\frac{1}{2}\frac{1}{\tau} < (P(t+\tau) - P(t))^2 > \\ &= \lim_{\tau\to 0}\frac{1}{2\tau}[\int_t^{t+\tau} dt'(-\frac{d\tilde{V}}{dR} - \gamma\frac{P}{M})(\int_t^{t'}(-\frac{d\tilde{V}}{dR} - \gamma\frac{P}{M})dt'' \\ &+ 2\int_t^{t+\tau} dt'(-\frac{d\tilde{V}}{dR} - \gamma\frac{P}{M})\int_t^{t'} dt'\sqrt{D} < \xi(t') > \\ &+ \int_t^{t+\tau} dt'\sqrt{D} < \xi(t')\int_t^{t'+\tau} dt''\sqrt{D}\xi(t'') >] = D \\ D_{RP}^{(2)} &= 0 \\ D_{RR}^{(2)} &= 0. \end{aligned} \quad (29)$$

Thus the Fokker-Planck equation physically equivalent to the Langevin equation (23) reads

$$\begin{aligned} \frac{d}{dt}d(R,P,t) &= \frac{d}{dP}[(\frac{d\tilde{V}}{dR} + \gamma\frac{P}{M})d(R,P,t)] \\ &- \frac{d}{dR}[\frac{P}{M}d(R,P,t)] + \frac{d^2}{dP^2}[Dd(R,P,t)]. \end{aligned} \quad (30)$$

In calculating heavy-ion collisions one has more degrees of freedom: radial and tangential motion, deformation degrees of freedom, charge and mass transfer. In this multi-dimensional case it is technically not possible to solve the corresponding multi-dimensional Fokker-Planck equation numerically with grid methods. Therefore many authors have used the so called quasi-linear method [8–14], which consists in the assumption that the distribution function is of multi-dimensional Gaussian form and is determined by calculating its first and second moments. This is, however, only accurate if the drift terms are linear in the relevant variables, which is not the case in heavy-ion collisions. In particular a situation close to fusion where the distribution function, when approaching a barrier, is split into two parts (one that surmounts the barrier and gives the probability for fusion and one that is reflected from the barrier giving the probability for deep inelastic collisions) cannot be described by a single Gaussian. Therefore one needs a more accurate method which consists in solving the corresponding Langevin equations. These have been suggested for use in nuclear collisions for quite some time [15–17] and in fission by [18], and have been elaborated for fusion and deep-inelastic collisions in [19–23]. Further examples will be given in Sect. 3.

2.7. Quantal Features of the Model

In the above-discussed model one can try to add quantal features. These can consist in adding features of zero-point motion to the diffusion process or in adding tunneling processes in the presence of friction, in particular when going down in energy.

2.7.1. Quantal Diffusion

If quantal zero-point fluctuation processes are assumed to be important the most simple generalization is to consider an oscillator in a heat bath. The corresponding energy is

$$< \epsilon >= \frac{\hbar\omega}{2} ctgh \frac{\hbar\omega}{2kT}. \tag{31}$$

This equation gives the previous result in the high temperature limit, whereas it gives the zero-point energy for $T \to 0$.

For a non-Markovian friction, whose Fourier transform is frequency dependent, the following form for the force-force correlation has been suggested in [24], [25]

$$< F(t)F(t') >= \frac{1}{2\pi} \int_o^{\Omega} d\omega e^{-i\omega(t-t')} \gamma(\omega) \frac{\hbar\omega}{2} ctgh(\frac{\hbar\omega}{2kT}). \tag{32}$$

This equation approaches the old result in the high temperature limit and $\Omega \to \infty$ and for frequency-independent (Markovian) friction: $\gamma(\omega) = \gamma$.

2.7.2. Dissipative Tunneling

The oscillator model above was applied by Caldeira and Leggett [26] to calculate the decay of a metastable state in the presence of dissipation using the influence functional of Feynman [27], which can be calculated analytically for the linearly coupled oscillators above. Within the same formalism scattering can be treated. It was shown how classical equations of motion with friction can be obtained for energies above the barrier [28]. For energies below the barrier the formalism is also suitable for formulating a theory for dissipative tunneling [29]; e.g. for the situation of subbarrier fusion of heavy ions. The transmission probability for a particle at energy E to move from R_1, where the oscillators are in the ground state, to R_2, where the final oscillator states are summed over, can be written as a path integral [30]

$$P(R_1, R_2; E) = \dot{R}_1 \dot{R}_2 \int_o^\infty dT e^{\frac{i}{\hbar}ET} \int_{R_1}^{R_2} D[R] e^{\frac{i}{\hbar}\int_o^\infty dt L(R)}$$
$$\int_o^\infty d\tilde{T} e^{-\frac{i}{\hbar}E\tilde{T}} \int_{R_1}^{R_2} D[\tilde{R}] e^{-\frac{i}{\hbar}\int_o^{\tilde{T}} dt L(\tilde{R})} e^{\frac{i}{\hbar}\Phi(T,\tilde{T};[R,\tilde{R}])}. \tag{33}$$

Here $exp(\frac{i}{\hbar}\Phi)$ is the two-time influence functional with

$$\Phi(T,\tilde{T};[R,\tilde{R}]) = \sum_i [-\frac{\hbar\omega_i}{2}(T-\tilde{T}) + \frac{i}{2m_i\omega_i}$$
$$(-\int_o^T dt f_i(R(t)) e^{i\omega_i(t-s)} \int_o^{\tilde{T}} ds f_i(\tilde{R}(s)) e^{-i\omega_i(s-\tilde{T})}$$
$$+\int_o^T dt \int_o^t ds f_i(R(t)) f_i(R(s)) e^{-i\omega_i(t-s)}$$
$$+\int_o^{\tilde{T}} dt \int_o^t ds f_i(\tilde{R}(t)) f_i(\tilde{R}(s)) e^{i\omega_i(t-s)})]. \tag{34}$$

Performing a stationary phase approximation and replacing $\sum_i$ by an integral over the intrinsic spectrum of the system, $\int d\omega g(\omega)$, we find the following equation for the stationary trajectory:

$$\dot{P} = -\frac{dV}{dR} + \frac{df(R(t))}{dR} \int_o^t ds f(R(s)) \int d\omega \frac{g(\omega)}{m\omega} sin\omega(t-s)$$
$$-\frac{i}{2}\frac{df(R(t)}{dR}[\int_o^{\tilde{T}} ds f(\tilde{R}(s)) \int d\omega \frac{g(\omega)}{m\omega} e^{i\omega(t-s-T+\tilde{T})}$$

$$- \int_o^T ds f(R(s)) \int d\omega \frac{g(\omega)}{m\omega} e^{i\omega(t-s)}]. \qquad (35)$$

In the classically allowed region one is dealing with real times: $T = \tilde{T}$. In this case the last term in eq.(35) vanishes and one obtains the result of ref.[28]. When using a quadratic spectrum for the intrinsic excitations one gets back the equations of motion of Section 2.3 without the fluctuating term. Tunneling can be treated by allowing for complex times in eq.(35). We have not been able to solve the problem of dissipative subbarrier scattering in general. However, with the choice $\tilde{T} = T^*$ we investigated the case of coupling one oscillator to the relative motion in the sudden limit [29], which is defined as $\omega \to 0$ and $\sqrt{\frac{\hbar}{m\omega}} F_o = const$, where $f(R) = F_o F(R)$. This case sheds light on some problems which occur in the more general situation. This is because in the sudden limit the exact solution of the influence functional method is known [30],[31],[32], and we can compare it with the stationary phase solution (35) , which corresponds to a single transmission WKB result. The exact penetrability is given by the so called zero-point fluctuation formula, which is the solution of the corresponding quantal coupled-channel problem in the sudden limit. It is given by

$$P(E) = \int_{-\infty}^{\infty} dq \frac{1}{\sqrt{\pi}\bar{q}} exp(-\frac{q^2}{\bar{q}^2}) P_o(E, V(R) + qF(R)). \qquad (36)$$

Here $\bar{q} = \sqrt{\hbar/m\omega}$ and $P_o(E, V(R)+qF(R))$ is the penetrability through a potential with the coupling term added.

Using an Eckart potential and Eckart coupling we can compare the result for single transmission obtained from eq.(35) with the exact solution (36). We find that the result is only good for very small penetrabilities ($< 10^{-5}$) if a coupling strength relevant for heavy-ion fusion is used [29]. In the case with coupling we also find that it is (contrary to the usual no coupling WKB) not at all sufficient to use the Hill-Wheeler formula: $P^{HW} = P^{WKB}/(1 + P^{WKB})$, which takes into account the proper summing-up of a multiple reflection series in the no-coupling situation [33]. This result should be a warning against using a Hill-Wheeler formula in the non-sudden ($\omega \neq 0$) decay of a metastable state , as proposed in [34], [35]. As long as the proper uniform approximation for summing up multiple reflections in the case with coupling is not found, one might avoid all WKB problems by a direct numerical calculation of the path integrals by using Monte Carlo techniques, as is attempted in [36].

A proper theory for dissipative tunneling in the non-sudden limit applicable in scattering is lacking. It is not a simple task to develop the influence functional method discussed above up to a stage where e.g. subbarrier fusion of heavy ions can be treated.

3. Treating Heavy-Ion Collisions by Langevin Equations

In this Sect. we present applications of Langevin equations of the type derived in Sect. 2 to low energy ($<$ 10 MeV/nucleon) heavy-ion scattering. This is done in the framework of the surface friction model [12],[37]. We use the Langevin Monte Carlo method [20] because it yields more accurate solutions for multi-dimensional problems than solving the corresponding Fokker Planck equation by a moment expansion under the assumption of a Gaussian distribution function.

In Section 3.1 we treat the problem of the competition of fusion and deep inelastic collisions (DIC) close to the barrier, which is an ideal application of the Langevin method because it demonstrates the bifurcation of the distribution function into the two corresponding branches. In connection with this problem we apply the surface friction model to the calculation of the relevant fusion barriers. These are dynamical barriers which are (due to the effects of friction and deformations) higher than the static barrier of the conservative potential. The model explains the extra energy (above the static barrier energy) necessary for the systems to fuse: the extra-push effect. In Section 3.2 we are dealing with the calculation of fusion excitation functions and spin distributions, which an up-to-date model for fusion has to describe simultaneously. An important finding will be that at subbarrier energies one has to use enhanced diffusion coefficients (as compared to those from Einstein relations) in order to obtain a consistent description of fusion cross sections and spin distributions in a systematic way. Section 3.3 gives a few examples of the state of the art to describe differential cross sections of deep inelastic collisions. In Section 3.4 we investigate the role which thermal fluctuations play in describing the energy spectra of δ electrons emitted in DIC. We also show in which sense δ-electrons can serve as an atomic clock for determining the time scale of deep-inelastic collisions.

3.1. The Competition of Fusion and Deep-Inelastic Collisions Close to the Barrier

The Langevin method is an ideal tool to calculate the bifurcation of the distribution function into a fusion and a DIC branch when approaching

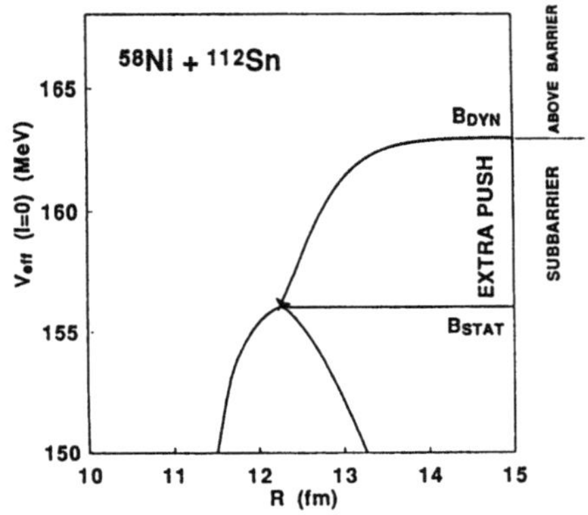

Figure 1: Static (B_{STAT}) and dynamical barrier (B_{DYN}) in the surface friction model for $^{58}Ni +^{112} Sn$ are illustrated. A frictional trajectory (without fluctuations) at barrier energy ($B_{DYN} = 163.0MeV$) is shown, demonstrating the extra-push effect.

the barrier [38 , 39]. We exemplify the situation in Fig.1 by the Ni + Sn system, for which fusion and DIC cross sections have been measured by Wolfs et al. [40 , 41] down to subbarrier energies. The static potential V_{eff} at angular momentum l=0 is plotted versus the radial distance

The lowest energy at which a classical trajectory without friction can fuse is given by the static barrier height B_{STAT}. A trajectory with friction starting at the same energy cannot overcome this barrier because it loses energy on its way to the barrier. Therefore a frictional trajectory (see Fig.1) needs an extra energy- the so called extra push- to overcome the original barrier and to fuse. This new energy defines the dynamical barrier B_{DYN} as the new limit for fusion. In the following the terms above barrier or subbarrier refer to energies above or below the dynamical barrier, which is the barrier actually measured in experiment. In order to calculate the bifurcation of fusion and deep inelastic processes at (or below or above) the barrier as in Fig.1 one has to follow many Langevin trajectories which create the classical distribution function. It is intuitively clear that the latter splits into two parts when approaching the barrier: one part that surmounts the barrier giving the probability for fusion, and one part that is reflected from the barrier giving the probability for inelastic processes, which one may subdivide into quasi-elastic and deep inelastic .

The surface friction model has been used in ref.[42] to calculate the extra push energies which are measured for many systems. As an example we compare in Fig.2 surface friction extra push energies for asymmetric systems with proximity model predictions [43] and the data [43],[42];

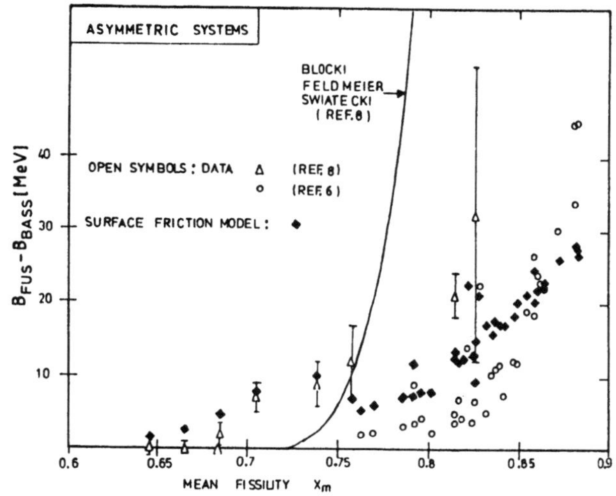

Figure 2: Measured extra energies (open symbols) over the Bass barrier are plotted versus the mean fissility and compared to the model of Blocki et al.[43] and to the surface friction model ([42], filled diamonds).

the data with the open circles are those of Iljinow et al. [44]. The average trend of the measured barriers is reproduced by the surface friction model, i.e. it essentially explains the extra push effect. However, there exist individual deviations from the data for the explanation of which individual nuclear structure effects are certainly important [4], but which are not included in the present version of the model.

In the following calculations for the Ni+Sn system we had to increase the radius of the potential by 10% as compared to the global parametrization [37] (in agreement with the systematics of ref.[45]) in order to obtain the correct onset of fusion and DIC. Otherwise we use the surface friction model in its standard form [37].

As we have shown in detail in ref.[20] and also in Section 2 how Langevin equations can be obtained from previously used Fokker Planck equations we give here only the final Langevin equations of the surface friction model and display them in discretized form suitable for numerical calculation:

$$\begin{aligned} R_{n+1} &= R_n + \frac{P_n}{\mu}\tau \\ P_{n+1} &= P_n - [\frac{dV}{dR} + K_R(R)\frac{P}{\mu} + \sum_i K_{R\alpha_i}\frac{\pi_i}{B_i}]_n\tau + \sqrt{D_R(R_n)\tau}w_R(t) \\ \phi_{n+1} &= \phi_n + \frac{l_n}{\mu R_n^2}\tau \end{aligned}$$

$$
\begin{aligned}
l_{n+1} &= l_n - K_\phi(R_n)l_n\tau + \sqrt{D_\phi(R_n)\tau}w_\phi(t) \qquad (37)\\
\alpha_{i(n+1)} &= \alpha_{i(n)} + \frac{\pi_{i(n)}}{B_i}\tau
\end{aligned}
$$

$$\pi_{i(n+1)} = \pi_{i(n)} - [\tfrac{dV}{d\alpha_i} + \textstyle\sum_j K_{\alpha_i\alpha_j}\tfrac{\pi_j}{B_j} + K_{\alpha_i R}\tfrac{P}{\mu} + C_i\alpha_i]_n\tau + \sqrt{D_{\alpha_i}\tau}w_{\alpha_i}(t).$$

Here R is the relative distance between the nuclei, P the radial momentum,l the angular momentum, ϕ its conjugate variable, μ the reduced mass, C_i and B_i are the liquid drop stiffness and mass parameters of the quadrupole modes of projectile and target and α_i and π_i their deformation variables and momenta. The potential consists of nuclear, centrifugal and Coulomb parts. For the nuclear potential we use a single folding potential with curvature corrections for the deformations; for details see ref.[37]. The radial and tangential friction form factors are proportional to the gradient of the nuclear potential squared (see eq.11), in which the system dependence is contained; the mass dependence is also contained in the form factor for the intrinsic damping of the deformation modes:

$$
\begin{aligned}
K_R &= K_R^o(\frac{dV_N}{dR})^2\\
K_\phi &= K_\phi^o(\frac{dV_N}{dR})^2 \qquad (38)\\
K_{\alpha_i} &= K_{\alpha_i}^o(C_iB_i)^{1/2}.
\end{aligned}
$$

The non-diagonal damping terms are related to the friction terms above by the assumption that the friction is proportional to the velocity of the surface distance $d = R - \sum_i R_i(1 + \alpha_i Y_{20})$ between the two colliding nuclei (for details, see ref.[37]):

$$
\begin{aligned}
K_{R\alpha_i} &= -R_iY_{20}K_R\\
K_{\alpha_i\alpha_j} &= R_iR_jY_{20}^2K_R + K_{\alpha_i}\delta_{ij} \qquad (39)
\end{aligned}
$$

Thus we describe all systems with three universal parameters which have been fixed in ref.[12]:

$$
\begin{aligned}
K_R^o &= 3.5 * 10^{-23}[sec/MeV]\\
K_\phi^o &= 0.01 * 10^{-23}[sec/MeV] \qquad (40)\\
K_{\alpha_i} &= 20.
\end{aligned}
$$

In the calculations we assume that the diffusion coefficients are related to the friction coefficients above by Einstein relations: $D_R = K_RT, D_\phi = K_\phi T, D_{\alpha_i} = K_{\alpha_i}T$. The temperature T is calculated from the dissipated energy E^* along each trajectory by $T = \sqrt{8E^*/A}$. The quantities

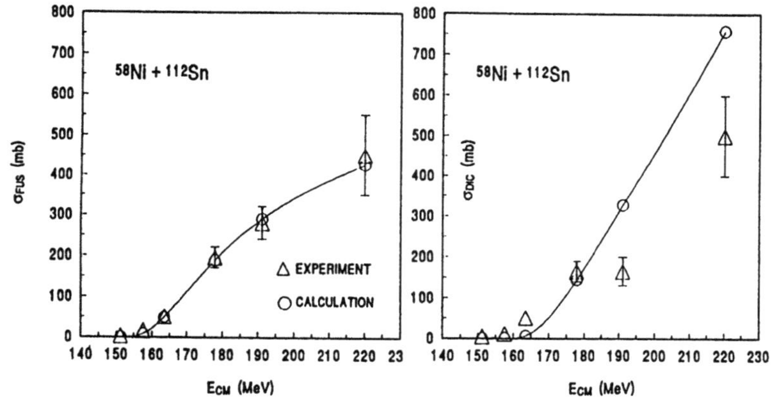

Figure 3: The calculated total fusion and deep-inelastic cross sections (open circles, the solid lines are to guide the eye) for the $^{58}Ni +^{112} Sn$ system are shown in comparison to experiment(open triangles with error bars). The dynamical barrier is at $E_{cm} = 163.0 MeV$.

$w_R, w_\phi, w_{\alpha_i}$ are Gaussian distributed random variables for the corresponding degrees of freedom.

Solving eqs.(37) with fixed initial conditions generates distributions of trajectories due to the action of the fluctuating force. Cross sections are calculated by integrating the distribution function over the l-values. The total fusion cross section e.g. is calculated as a Monte Carlo integral

$$\begin{aligned} \sigma_{FUS} &= \frac{\pi}{k^2}\sum_l (2l+1)T_l \simeq \frac{\pi}{k^2}\int_o^{l_>} 2lT_l dl \\ &= \frac{\pi}{k^2}l_>^2 \int_o^1 dx T_{l(x)} = \frac{\pi}{k^2}l_>^2 \sum_i \frac{T_i}{N} = \frac{\pi}{k^2}l_>^2 \frac{N_F}{N}. \end{aligned} \tag{41}$$

The initial angular momenta are chosen according to $l = l_>\sqrt{x}$(x is a random number, $0 \leq x \leq 1$) and N_F is the number of fusing and N the total number of trajectories; $l_>$ can be taken as the grazing angular momentum.

Simultaneously with the fusion cross section, one obtains the total deep-inelastic cross section by counting the number N_{DIC} of reflected (DIC) trajectories. The formula for the total deep inelastic cross section is obtained in complete analogy to eq.(41),

$$\sigma_{DIC} = \frac{\pi}{k^2}l_>^2 \frac{N_{DIC}}{N}. \tag{42}$$

As an example we calculate the fusion and deep inelastic cross sections for the reaction $^{58}Ni+^{112}Sn$ and compare in Fig.3 with the experiments of Wolfs et al.[41],[40]. The data for the highest energy are from ref.[46]. The agreement between theory and experiment is quite reasonable for the fusion cross section, whereas the calculated deep-inelastic cross section is somewhat too small. In the calculation and in the experiment deep-inelastic events are defined as those which have an energy loss of more than 20 MeV.

The application of the Langevin method certainly reaches its limit if the energy of the system goes below the energy of the static barrier. Then a classical trajectory can no longer fuse and the application of a theory for dissipative tunneling (see Section 2.7.2) would be most adequate.

We mention that a similar agreement with the above data has been obtained by Dasso and Pollarolo [47] using a friction model in which the fluctuations are due to quantal zero-point motion.

The Langevin method allows also the calculation of the spin distributions for fusion and DIC simultaneously with the total cross sections. The spin distributions are the partial differential (with respect to the angular momentum) cross sections calculated by the formulas

$$(\frac{d\sigma^{FUS}}{dl})_i = \frac{2\pi}{k^2} l_i \frac{N_i^F}{N_i} \tag{43}$$

$$(\frac{d\sigma^{DIC}}{dl})_i = \frac{2\pi}{k^2} l_i \frac{N_i^{DIC}}{N_i}, \tag{44}$$

where l_i is the angular momentum, N_i^F the number of fusing, N_i^{DIC} the number of DIC, and N_i the total number of trajectories in the ith angular momentum bin.

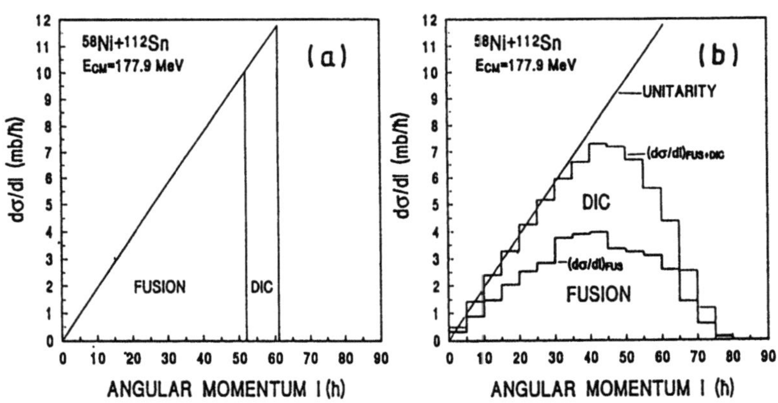

Figure 4: Calculated spin distributions for fusion and DIC without (Fig.4a) and with thermal fluctuations (Fig.4b) for the system $^{58}Ni +^{112} Sn$ at $E_{cm} = 177.9 MeV$.

Considering spin distributions it has for a long time been the general belief that at least above the barrier, deep-inelastic and fusion processes are well separated in l-space, as e.g. shown in Fig.4a, which is a result of the surface friction model if the thermal fluctuations are switched off. The example is $^{58}Ni +^{112} Sn$ at $E_{cm} = 177.6$ MeV, which is an energy well above the barrier energy of 163.0 MeV. Taking into account thermal fluctuations we obtain the spin distributions in Fig.4b. We predict that even at energies above the barrier, fusion and DIC are not well separated in l-space. The fusion spin distribution does not reach the unitarity limit. Only when DIC is added, will unitarity be satisfied.

3.2. Fusion Spin Distributions

Whereas the spin distributions for fusion and DIC of the last section are predictions, there exist a variety of experimental data on fusion spin distributions for other systems; for a recent review on the data see [48]. We have made a systematic theoretical analysis of the data for heavier systems within a somewhat simplified version of the surface friction model (neglect of deformations) in ref.[45].

In Fig.5 we give an example of our calculations for a situation above the barrier. We compare the theory with experimental [51] spin distributions for an asymmetric system ($^{16}O +^{144} Nd$) and a symmetric system ($^{80}Se +^{80} Se$) leading to the <u>same</u> compound nucleus at the <u>same</u> excitation energy. Although the energies of the systems are identical their spin distributions are very different. The calculations satisfactorily reproduce the fusion cross sections, the flat and broad spin distributions

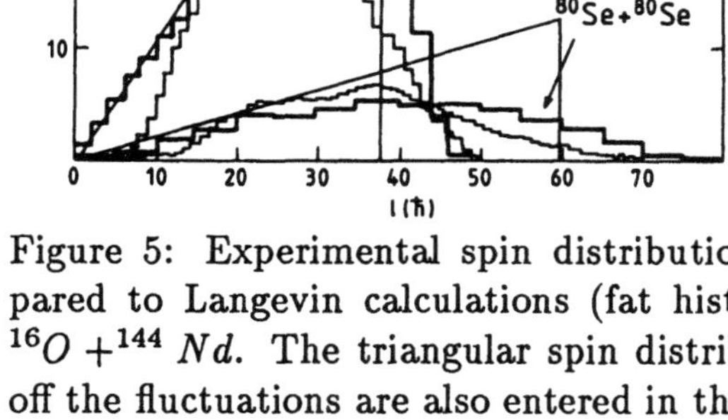

Figure 5: Experimental spin distributions (thin histograms)are compared to Langevin calculations (fat histograms) for $^{80}Se +^{80} Se$ and $^{16}O +^{144} Nd$. The triangular spin distributions obtained by switching off the fluctuations are also entered in the figure.

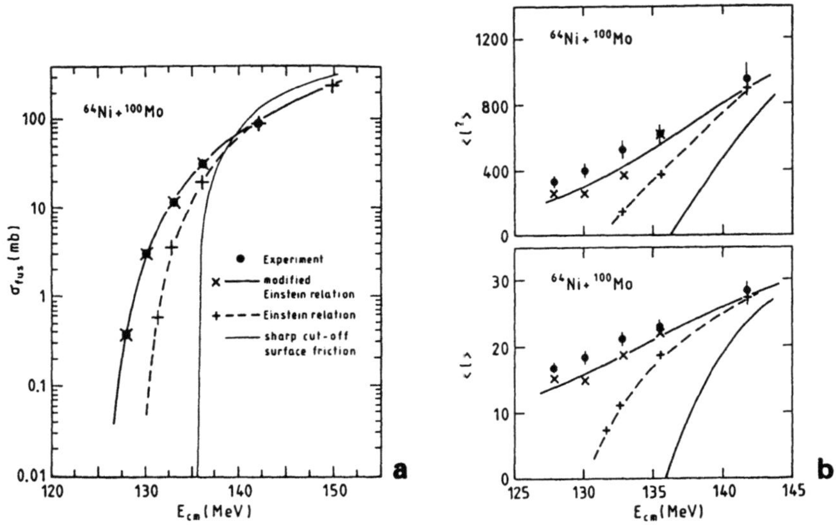

Figure 6: The experimental fusion cross sections (a) and $< l >$ and $< l^2 >$ (b) for the $^{64}Ni +^{100} Mo$ system are compared with surface friction model calculations without(sharp cut-off) and with fluctuations using the Einstein relation and the modified Einstein relation (enhanced diffusion).

of the symmetric system, as well as the higher and narrower distribution of the asymmetric system.

Up-to-date theories for fusion have to reproduce simultaneously fusion excitation functions and spin distributions down to subbarrier energies. For reviews on theoretical models, see refs.[52],[53]. In order to obtain also a satisfactory description below the barrier within the Langevin model we had to modify the Einstein relation in this energy range. For this modification there exist theoretical arguments [3]. It is intuitively clear that the diffusion is not allowed to go to zero with zero temperature (as is the case with the Einstein relation) because the quantal zero-point motion survives.

As an example we show in fig.6 the $^{64}Ni +^{100} Mo$ system, where we compare the data [54] with calculations without fluctuations(sharp cut-off), and with fluctuations using the Einstein relation and using enhanced diffusion. Above the barrier fusion cross section(Fig.6a) and the moments $< l >$ and $< l^2 >$ (Fig.6b) are reproduced using the Einstein relation, whereas only enhanced diffusion coefficients give a consistent description also at subbarrier energy. The same enhancement of the diffusion coefficients yields also systematically reasonable results for other fusion-evaporation systems [45].

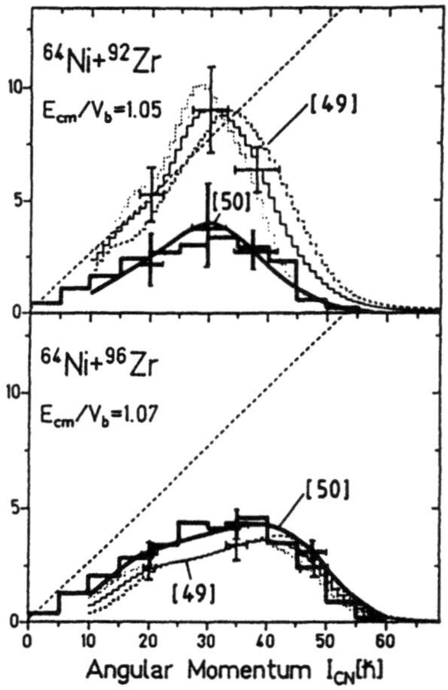

Figure 7: Original [49] and corrected [50] experimental spin distributions for the $^{64}Ni+^{92,96}Zr$ systems are compared with Langevin calculations (fat histograms) which agree with the corrected data.

The enhanced diffusion does not, however, help to reproduce the large $< l^2 >$ measured at subbarrier energies for fusion-fission systems. To understand these measurements is an open problem because other models also fail to explain these data [55],[53].

Finally, we look at the systems $^{64}Ni+^{92,96}Zr$, for which a strong isotope dependence of the spin distributions is claimed to be discovered [49]. This, however, does not seem to be true since the cross sections used for the normalization of the spin distributions have been remeasured [50]. With the corrected normalization both systems behave similarly, and Langevin calculations with eqs.(37) (including deformations) reproduce the corrected spin distributions, see Fig.7.

3.3. Deep-Inelastic Collisions

The Langevin method also allows one to calculate the differential cross sections for deep-inelastic collisions. These are obtained by integrating the distribution function over impact parameters, e.g.

$$\begin{aligned} \frac{d^2\sigma}{dEd\theta} &= \frac{\pi}{k^2}\int_o^{l_>} 2l\,dl\,d(E,\theta,t_\infty) = \frac{\pi}{k^2}l_>^2\frac{N(E,\theta)}{\Delta E\Delta\theta}\frac{1}{N}, \\ \frac{d\sigma}{d\theta} &= \frac{\pi}{k^2}l_>2\frac{N(\theta)}{\Delta\theta}\frac{1}{N}. \end{aligned} \tag{45}$$

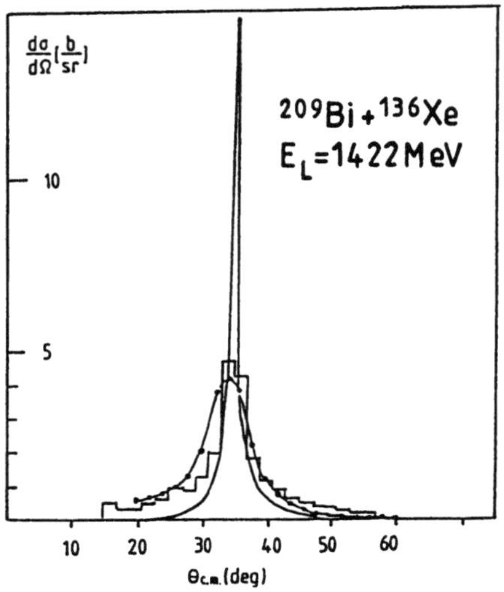

Figure 8: Comparing the moment expansion of the Fokker Planck equation (solid line) with a Langevin solution (histogram) and the data (dots) for the angular distribution of Xe+Bi at 1422 MeV.

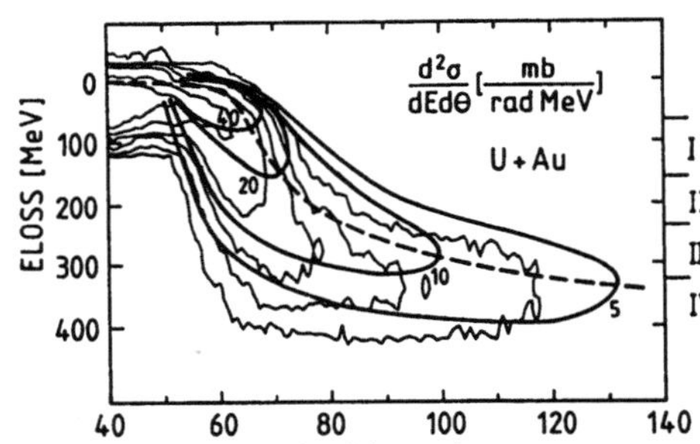

Figure 9: The experimental Wilczynski diagram (thin lines) for U+Au at 8.65 MeV/u is compared to a Langevin calculation (thick lines). The dashed line is the theoretical $E_{LOSS} - \theta$ correlation without fluctuations. The numbers I to IV indicate the energy loss windows for which the δ electrons are measured in coincidence with DIC (see Section 3.4).

Here Langevin trajectories are sampled in energy-loss and angle bins of bin widths ΔE and $\Delta\theta$, and $l_>$ is the grazing angular momentum, N(E,θ) is the number of trajectories in the corresponding E and θ bins and N is the total number of trajectories.

For a first example we compare in Fig.8 the measured [56] angular distribution for the Xe+Bi collision at $E_{Lab} = 1422 MeV$ with surface friction model calculations using the moment expansion of a Fokker Planck equation [12] and by sampling Langevin trajectories from eq.(37)

[21]. This example clearly demonstrates the superiority of the Langevin method over the approximate solution of the Fokker Planck equation.

As a second example we display in Fig.9 an experimental Wilczynski diagram [57] for the U+Au collision at 8.65 MeV/u in comparison to a Langevin calculation [58]. We use this example because it is the system for which δ electrons (see Section 3.4) are measured in coincidence with DIC for the energy-loss windows indicated in the figure.

3.4. δ Electrons as an Atomic Clock for Deep-Inelastic Collisions

In the last section we have already mentioned that energy spectra of δ electrons which are emitted from the quasimolecule formed in a deep-inelastic collision have been measured in coincidence with various energy loss bins of DIC [57]. The interest in measuring δ electrons comes from the idea of using them as an atomic clock for the time scale of DIC [59]. For Coulomb trajectories (nuclear contact time T=0) the energy spectra of δ rays are monotonically decreasing functions of the electron energies. In case of prolonged nuclear contact, however, oscillations in the δ-ray spectra are expected with a period related to the nuclear contact time. These oscillations are predicted to be very pronounced if nuclear sticking is assumed [59]. There are, however, a number of effects that to a large extent wipe out the oscillatory pattern. We have calculated the influence of these effects on the energy spectra of δ electrons [58]: (i) In the calculation we use surface friction model trajectories from which we see that nuclear sticking is not at all fulfilled. (ii) It is not sufficient to use a single trajectory with a mean impact parameter corresponding to each energy loss window; rather one has to integrate over the impact parameter distribution. (iii) In comparing to data one has to fold in the energy-loss resolution of the experimental set up for DIC. (iv) Finally, and this is the most important physical effect, one has to take into account statistical fluctuations by using Langevin trajectories. This quite naturally leads to interaction time distributions instead of a single interaction time.

The δ-electron emission probability is calculated from

$$\begin{aligned} dP_{e^-}/dE_{e^-} &= |a(\omega)|^2, \\ a(\omega) &= \frac{N}{\omega}\int_{-\infty}^{\infty} dt \frac{\dot{R}(t)}{R(t)} exp(i\omega t). \end{aligned} \tag{46}$$

Here N is an overall normalization factor, $\hbar\omega = E_{e^-} + E_b$, where E_{e^-} is the kinetic energy of the electron and E_b its mean binding energy.

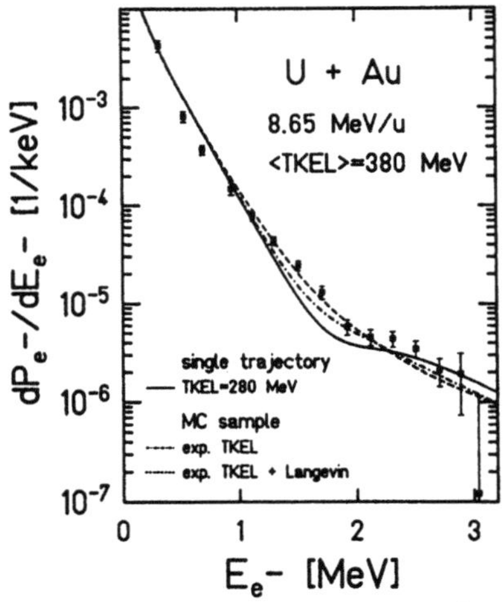

Figure 10: The experimental energy-differential emission probability of δ electrons as a function of the kinetic energy of the emitted electrons is compared to the calculations discussed in the text.

The quantities N and E_b are fixed by a fit of the electron spectra in coincidence with the elastic scattering and are then also used for the events with energy loss. The time variation of $\dot{R}(t)/R(t)$ determines the δ-ray spectra, where R(t) is the distance between the centers of mass of the nuclei. In Fig.10 we give as an example the energy differential emission probability for the energy window III (280± 45 MeV).

The calculations allow to disentangle the different effects discussed above. The dashed line represents the complete calculation, which involves integration over the impact parameter distribution, folding over the experimental energy resolution, and sampling of fluctuating trajectories. Included is also a calculation (solid line) with a single non-fluctuating trajectory corresponding to an energy loss of 280 MeV, and a calculation (dashed-dotted line) without fluctuations but with integrations over the impact parameters and the experimental energy resolution.

Although the original hope of measuring directly the interaction times for DIC has not been vindicated, because in agreement with the calculations no pronounced oscillations are found in the energy spectra of the δ electrons, one can nevertheless extract not only mean interaction times but interaction time distributions from surface friction model calculations. This should be rather reliable because the model does not only reproduce the δ ray spectra but also, as we have seen, many features of DIC and fusion. We give as a result the mean interaction times for DIC for the Au+U system for the four energy loss windows in Fig.9 in units of 10^{-21}sec:I (100±45MeV):1.13; II(190±45 MeV): 1.22; III(280±45 MeV): 1.81; IV(>325 MeV): 2.50.

4. Conclusions

We have discussed the role of dissipation and thermal fluctuations in heavy-ion collisions with a Langevin Monte Carlo method within the framework of the surface friction model.

We have started with a derivation of Langevin equations from a simple but very instructive model in which frictional and fluctuating forces arise from coupling the relative motion to many oscillators which represent the bath system. Only under very restrictive assumptions (separability of the interaction, quadratic spectrum for the intrinsic excitations, frequency integration to infinity, mean value of the form factor is equal to the form factor at the mean value of its argument, thermal equilibrium for the intrinsic system) could we obtain a Markovian friction with a $\delta-$correlated fluctuating force. This is certainly a too simplified model for nuclear collisions, and can hopefully only be a reasonably good simulation of the actual situation. If more realistic microscopic models yield non-Markovian expressions with more complicated time correlations for the fluctuating force, Langevin methods would still be suitable to treat such situations.

In order to decide whether it is necessary to proceed to such more complicated descriptions and to detect where non-Markovian features show up in measured quantities (or alternatively to give arguments for a Markovian description by its success in reproducing experimental results) we have used Markovian Langevin equations with white noise for a systematical analysis of heavy-ion data. We review the results and list a number of open questions:

(i) The surface friction model explains essentially the extra push effect by reproducing the general trend of measured fusion (or capture) barriers for heavy systems. Individual deviations are certainly due to more detailed nuclear structure effects which are not contained in the global version of the model discussed here. It is also not settled whether statistical fluctuations alone can explain the barrier fluctuations observed e.g. in measured fusion probabilities, or whether quantal fluctuations are of importance.

(ii) As an ideal application for the Langevin approach we described the measured bifurcation of the distribution function into fusion and deep-inelastic scattering for large-mass heavy ions down to subbarrier energies. Here certainly a limit of the classical Langevin description will be reached at energies below the static barrier, where a theory for dissipative tunneling would be most adequate. But - as discussed in Section 2 - until now such a theory has not been developed sufficiently to be applicable to subbarrier fusion of heavy ions.

(iii) For heavy systems the Langevin description predicts that spin distributions for fusion and DIC are not well separated in l-space even at energies above the barrier. It is desirable to measure not only spin distributions for fusion but also those for DIC, in order to test the model predictions.

(iv) Spin distributions have been measured for fusion and up-to-date theories have to explain them simultaneously with the fusion excitation functions. The Langevin description reproduces measured spin distributions above the barrier, when using the Einstein relation. At subbarrier (below the dynamical barrier) energies, however, one has to enhance the diffusion coefficient as compared to the Einstein relation in order to account systematically for fusion cross sections as well as for spin distributions of fusion-evaporation systems. More detailed theoretical investigations on modifications of the Einstein relation are called for.

(v)The Langevin approach cannot account for the measured large $< l^2 >$ values for fusion-fission systems at subbarrier energies; nor are alternative models able to do so. This is an open problem.

(vi)For differential cross sections of DIC the multi-dimensional Langevin description gives improved results as compared to the quasi-linear approach (calculating the assumed Gaussian distribution function by the first two moments) to the corresponding Fokker-Planck equation.

(vii)It was shown to be necessary to take into account thermal fluctuations of nuclear trajectories in order to describe quantitatively energy spectra of δ electrons measured in coincidence with DIC. Interaction times for DIC cannot be obtained directly by looking at the δ ray energy spectra, but indirectly by calculating interaction time distributions from Langevin trajectories.

In summary we can say that calculations with the Langevin Monte Carlo method based on the surface friction model can describe the general trend of a large variety of different data. By comparing the calculations with experiment we are not really forced to abandon the simple Markovian description with white noise. In order to be able to describe the finer details of the data (e.g. the scatter of the measured barriers around the average trend, different widths of measured fusion probabilities) one has certainly to take into account nuclear shell effects.

This can be done to a certain extent within the surface friction Langevin description , e.g. by calculating the nuclear folding potential (and also the friction form factor: it is the gradient of the potential squared) with densities from Hartree Fock calculations performed to describe electron scattering data. Besides improving the nuclear structure input it is of course an important problem to clarify to what extent quantal effects have to be added to the Langevin description, which involves only classical statistical mechanics. In particular in the low energy regime quantal tunneling and quantal fluctuations should play a role.

Further investigations on the theoretical understanding of dissipation in nuclear physics are necessary. If it turns out that one has to proceed to a non-Markovian description with coloured noise, Langevin methods remain a useful tool .

The surface friction model has been extended to higher energies to account for projectile break-up processes [60],[61]. Thermal fluctuations should also play a role in these processes and could be treated by a Langevin procedure. A Langevin description could also turn out to be a helpful tool for treating the dynamics of induced fission (including particle emission) [62], [63], in particular if fission is treated as a multidimensional problem [64].

I am indebted to R.Lipperheide, J.Marten, K.Möhring, J.Richert , J.Stroth and S.Y.Xu for collaboration.

References

[1] H.A.Weidenmüller, Nucl.Phys. **A 502** (1989) 387c

[2] H.J.Krappe, Nucl.Phys. **A 505** (1989) 417

[3] H.Hofmann, R.Samhammer, and G.Ockenfuss, Nucl. Phys. **A 496** (1989) 269

[4] D.Berdichevsky, A.Lukasiak, W.Nörenberg, and P.Rozmeij, Nucl. Phys. **A 499** (1989) 609

[5] H.Feldmeier, Rep.Progr. in Phys. **50** (1987) 915

[6] S.Pal and D.H.E.Gross, Z.Phys. **A 329** (1988) 349

[7] D.H.E.Gross, Lecture Notes in Physics **117** (1980) 81

[8] C.Ngo and H.Hofmann, Z.Phys. **A 282** (1977) 83

[9] M.Berlanger, P.Grange, H.Hofmann, C.Ngo, and J.Richert, Z.Phys.A-Atomic Nuclei **286** (1978) 207

[10] D.Agassi, C.M.Ko, and H.A.Weidenmüller, Phys.Rev. **C 18** (1978) 22

[11] H.L.Yadav and W.Nörenberg, Phys.Lett. **B 115** (1982) 179

[12] P.Fröbrich, B.Strack, and M.Durand, Nucl.Phys. **A 406** (1983) 557

[13] H.Feldmeier and H.Spangenberger, Nucl.Phys. **A 435** (1985) 229

[14] Ch.Leclercq-Willain, Nucl.Phys. **A 459** (1986) 227

[15] K.E.Rehm, Phys.Lett. **B 86** (1979) 260

[16] J.Randrup, Nucl.Phys. **A 327** (1979) 490

[17] V.C.Barbosa, P.C.Soares, E.C.de Oliveira, and L.C.Gomes, Rev.Bras.de Fizika **15** (1985) 211

[18] Y.Abe, C.Gregoire, and H.Delagrange, J.de Phys. **C4, vol.47** (1986) 329

[19] P.Fröbrich, in *Gross Properties of Nuclei and Nuclear Excitations*, ed. H.Feldmeier, Hirschegg, Austria 1987, 230

[20] P.Fröbrich and S.Y.Xu, Nucl.Phys. **A 477** (1988) 143

[21] P.Fröbrich, in *3rd Int. Conf. on Nucleus-Nucleus Collisions*, Contributed Papers, Saint-Malo,France 1988, 124

[22] P.Fröbrich, Lecture Notes in Physics **317** (1988) 61

[23] C.E.Aguiar, V.C.Barbosa, R.Donangelo, and S.R.Souza, Nucl.Phys. **A 491** (1989) 301

[24] R.H.Koch, D.J.Van Harlingen, and J.Clarke, Phys.Rev.Lett. **45** (1980) 2132

[25] R.Benguria and M.Kac, Phys.Rev.Lett. **46** (1981) 1

[26] A.O.Caldeira and A.J.Leggett, Ann.Phys. **149** (1983) 374

[27] R.P.Feynman and A.R.Hibbs, *Quantum mechanics and path integrals* (McGraw Hill, New York 1965)

[28] K.Möhring and U.Smilansky, Nucl. Phys. **A 338** (1980) 227

[29] P.Fröbrich, R.Lipperheide, and K.Möhring, Z.Phys.B-Condensed Matter **78** (1990) 325

[30] A.B.Balantekin and N.Takigawa, Ann.Phys. **160** (1985) 441

[31] P.M.Jacobs and U.Smilansky, Phys.Lett. **B 127** (1983) 313

[32] H.Esbensen, Nucl.Phys. **A 352** (1981) 147

[33] D.M.Brink and U.Smilansky, Nucl.Phys. **A 405** (1983) 301

[34] G.W.Ford, J.T.Lewis, and R.F.O'Connel, Phys.Lett. **A 128** (1988) 29

[35] H.Dekker, Phys.Rev. **A 38** (1988) 6351

[36] K.Yamashita and W.H.Miller, J.Chem.Phys. **82** (1985) 5475

[37] P.Fröbrich, Phys.Rep. **116** (1984) 337

[38] P.Fröbrich and J.Marten, in *Workshop on the Interface between Nuclear Structure and Heavy-Ion Reaction Dynamics*, Notre Dame, USA 1990, in print

[39] P.Fröbrich and J.Marten, Z.Phys. **A** (submitted)

[40] F.H.L.Wolfs, W.Henning, K.E.Rehm, and J.P.Schiffer, Phys.Lett. **B 196** (1987) 113

[41] F.H.L.Wolfs, Phys.Rev. **C 36** (1987) 1379

[42] P.Fröbrich, Phys.Lett. **B 215** (1988) 36

[43] J.Blocki, H.Feldmeier, and W.J.Swiatecki, Nucl.Phys. **A 459** (1986) 145

[44] A.S.Iljinow et al., Il Nuovo Cimento **101 A** (1989) 225

[45] P.Fröbrich and J.Richert, Phys.Lett **B 237** (1990) 328

[46] A.M.van den Berg, W.Henning, L.L.Lee, K.T.Lesko, K.E.Rehm, J.P.Schiffer, G.S.F.Stephans, and F.L.H.Wolfs, Phys.Rev. **C 37** (1988) 178

[47] C.H.Dasso and G.Pollarolo, Phys.Rev. **C 39** (1989) 2073

[48] D.E.DiGregorio, K.T.Lesko, B.A.Harmon, E.B.Norman, J.Pouliot, B.Sur, Y.D.Chan, and R.G.Stokstad, in *Workshop on the Interface between Nuclear Structure and Heavy-Ion Reaction Dynamics*, Notre Dame, USA 1990, in print

[49] W.Kühn, A.Ruckelshausen, R.D.Fischer, G.Breitbach, H.J.Hennrich, V.Metag, R.Novotny, R.V.F.Janssens, T.L.Khoo, D.Habs, D.Schwalm, B.Haas, and R.S.Simon, Phys.Rev.Lett. **62** (1989) 1103

[50] A.M.Stefanini, in *Workshop on Heavy Ion Collisions at Energies near the Coulomb Barrier*, Daresbury, England 1990, in print

[51] B.Haas and G.Duchene, private communication

[52] G.R.Satchler, in *Proc. Int. Nucl. Phys. Conf., vol.II*, World Scientific, Sao Paulo Brasil 1989, 541

[53] P.Fröbrich, in *Int. School Seminar on Heavy-Ion Physics*, Dubna, USSR 1989, in print

[54] M.L.Halbert et al., Phys.Rev. **C 40** (1990) 2558

[55] P.Fröbrich, in *Int. Conf. Fiftieth Anniversary of Nuclear Fission*, Leningrad, USSR 1989, in print

[56] H.J.Wollersheim, W.W.Wilcke, J.R.Birkelund, J.R.Huizenga, W.U.Schröder, H.Freiesleben, and D.Hilscher, Phys.Rev. **C 24** (1981) 2114

[57] H.Backe et al., (1990) to be published

[58] P.Fröbrich and J.Stroth, Phys.Rev.Lett. **64** (1990) 629

[59] J.Reinhardt, B.Müller, W.Greiner, and G.Soff, Z.Phys. **A 292** (1979) 211

[60] K.Möhring, T.Srokowski, D.H.E.Gross, and H.Homeyer, Phys.Lett. **B 203** (1988) 210

[61] K.Möhring, T.Srokowski, and D.H.E.Gross, preprint HMI-90/P1-Mö (1990)

[62] I.Gonschar, G.I.Kosenko, and N.D. Mavlitov, in *Int. Conf. Fiftieth Anniversary of Nuclear Fission,* Leningrad, USSR 1989, in print

[63] E.Strumberger, K.Dietrich, and K.Pomorski, preprint (1990)

[64] Y.Abe, N.Carjan, M.Ohta, and T.Wada, in *Proc. Second IN2P3-RIKEN Symposium on Heavy-Ion Collisions,* Obernai, Japan 1990, in print

Multi-Dimensional Tunneling and Nuclear Fission

A. Iwamoto

Japan Atomic Energy Research Institute,
Tokai, Naka-gun, Ibaraki 319–11, Japan

Abstract. We discuss the method for calculating tunneling probability in multi–dimensional space and apply it to calculate the decay width of nuclear fission. If the problem is one–dimensional, we have the simple semiclassical WKB method, which is widely used. In many cases, however, it is not possible to reduce the problem to one dimension, and thus we are faced with tunneling in multi–dimensional space. In nuclear fission calculations, we need several kinds of shape parameters to specify the shape of the fissioning nucleus in order to reproduce the correct fission saddle configuration. So the problem is to find the most probable fission path in multi–dimensional space if we proceed semi–classically. Two methods are proposed in this paper for the calculation of the quantum decay–width, a new technique to obtain the minimum action trajectory and the adiabatic approximation. Applying these two methods to nuclear fission–width calculations, we found that the former method works quite well and the latter method, which is much easier to use numerically, is a good approximation to the former method. We will explain these two methods and discuss related problems. Numerical calculations are performed for the decay width of actinide nuclei as an application of these methods. We will also investigate under what conditions the adiabatic method becomes a good approximation to the minimum action trajectory method.

1. Introduction

The fission process involves large distortions of nuclear matter and thus many shape parameters are required to calculate the potential surface. Spontaneous fission is considered as a multi–dimensional penetration problem in this respect. We do not, however, have standard methods to treat the multi–dimensional penetration problem. In contrast, we have a well established WKB method for the one–dimensional problem. Therefore, many traditional calculations for fission employ so called "adiabatic" approximation to reduce multi–dimensional problems to one–dimensional ones. In this model, the fission path is defined in the valley of the fission potential energy surface. Some groups [1–4] pointed out, however, that the action integral along the adiabatic path does not yield the minimum action. In addition, whereas the physical quantity should be invariant with respect to the coordinate transformation, the adiabatic path has the serious defect that the path and the value of action calculated along it depend on the choice of the coordinate system [5].

The methods used in [1–4] are based on numerical calculations of the action integral and are free from the adiabatic assumption. But these methods require considerable amounts of numerical calculation to get good results. Furthermore, it is hard to get a good physical insight into the problem.

We consider two methods of fission half–life calculation[6], which are invariant with respect to coordinate transformation. The first method is an application of the Schmid model[7], which is proposed as a method to determine multi–dimensional wave functions for metastable states. The fission path in multi–dimensional parameter space which minimizes the WKB action is determined in this method. The second method we propose is a new adiabatic model in which the path and the WKB action are invariant with respect to general coordinate transformations. In these two methods, the fission path is obtained by solving simple differential equations with suitable initial conditions. The methods are simpler in their structure and easier to handle numerically than the models of [1–4].

Our concern is to get insight into general properties of the multi–dimensional penetration problem and not necessarily to reproduce the data in detail. Therefore we will apply these methods to simple problems, namely the calculation of decay widths for symmetric fission in a two–dimensional space, neglecting shell and pairing effects on potential and mass tensors. The results of these methods are then compared and their features discussed, especially the validity of the adiabatic approximation.

In Sect.2, we explain a new method to obtain the least action trajectory in multi–dimensional space. In Sect.3, definition and properties of the new adiabatic path are given. Sect.4 is devoted to the numerical calculations for symmetric fission of actinide nuclei based on these two methods. The calculations show that our adiabatic approximation is quite good, and we will investigate why this is so in Sect.5. A summary will be given in Sect.6,

2. A New Method to Obtain the Least Action Trajectory

The first method of the life time calculation is based on the model of Schmid[7]. We generalize the formalism to the case where mass tensors are functions of position, which is necessary in fission life time calculation[6]. We start from the Schrödinger equation,

$$\mathcal{H}\psi(\vec{q}) = E\psi(\vec{q}), \tag{1}$$

where $\vec{q}$ represents shape parameters and E is the energy eigenvalue. The Hamiltonian $\mathcal{H}$ is assumed to have the form

$$\mathcal{H} = -\frac{\hbar^2}{2}\frac{\partial}{\partial q^i}(M^{-1})^{ij}(\vec{q})\frac{\partial}{\partial q^j} + V(\vec{q}), \tag{2}$$

where q^i represents a component of shape parameter $\vec{q}$ and M is the mass tensor. We will study a decay problem and thus the energy E is put

$$E = \hbar\cdot(\omega_t - i\Gamma) / 2 , \qquad (3)$$

where ω_t represents zero–point energy in the case of decay from the ground state and Γ is the decay width. We put the wave function in the form,

$$\psi(\vec{q}) = exp\{ -W(\vec{q})/\hbar \} . \qquad (4)$$

The exponent W is expanded in power of $\hbar$ in the form,

$$W(\vec{q}) = W_0(\vec{q}) + \hbar\, W_1(\vec{q}) + \hbar^2\, W_2(\vec{q}) + \cdots . \qquad (5)$$

This expansion is thought to be valid under quasi–classical conditions whose one–dimensional form is

$$\hbar\cdot\left| \frac{\partial^2 W_0}{\partial q^2} \right| << \left|\frac{\partial W_0}{\partial q}\right|^2 \qquad (6)$$

and is a familiar form in the WKB approximation.

Under the quasi–stationary condition,

$$\Gamma << \omega_t , \qquad (7)$$

we insert eq.(5) into eq.(1) and equate separately the terms of equal power of $\hbar$. In leading order, we obtain the eikonal equation,

$$H_0 = \frac{1}{2}\, p_i\, (M^{-1})^{ij}\, p_j - V(\vec{q}) = 0 , \qquad (8)$$

where

$$p_i = \frac{\partial W_0}{\partial q^i} , \qquad (9)$$

plays the role of momentum. In obtaining eq.(8), the energy E in eq.(1) was assumed to be much smaller than barrier height V_b,

$$\hbar\omega_t/2 << V_b . \qquad (10)$$

This energy is of first order in $\hbar$ and is treated in the next higher order equation called transport equation,

$$H_1 = \frac{1}{2}\left(\frac{\partial}{\partial q^i} M_{ij}^{-1}\right) p_j + \frac{1}{2} M_{ij}^{-1} \frac{\partial^2 W_0}{\partial q^i\, \partial q^j} - M_{ij}^{-1} p_i p^{(1)}{}_j - \frac{1}{2}\omega_t = 0 , \qquad (11)$$

where $p^{(1)}{}_i$ is given by

$$p^{(1)}{}_i = \frac{\partial W_1}{\partial q_i} . \qquad (12)$$

The eikonal equation (8) is identical to the energy conservation condition for inverted potential, that is, $V \to -V$, and can be solved by the method of characteristics. The characteristics can be represented in parametric form as trajectories in the 2N–dimensional $\{\vec{q},\vec{p}\}$–space,

$$q^i = q^i(\tau), \qquad p_i = p_i(\tau), \tag{13}$$

such that they obey the canonical equations of motion

$$\frac{dq_i}{d\tau} = \frac{\partial H_0}{\partial p_i}, \tag{14}$$

and

$$\frac{dp_i}{d\tau} = -\frac{\partial H_0}{\partial q_i}, \tag{15}$$

with the initial condition $H_0 = 0$ at $\tau=0$. We can introduce the concept of path which is defined as projections of the trajectories on N–dimensional $\{\vec{q}\}$–space. The path obeys the equations of motion

$$M_{ij}\ddot{q}^j = \frac{\partial V}{\partial q^i} - \dot{q}^k \frac{\partial M_{ij}}{\partial q^k} \dot{q}^j + \frac{1}{2} \dot{q}^j \frac{\partial M_{jk}}{\partial q^i} \dot{q}^k, \tag{16}$$

where the dot means a differentiation with respect to τ. Equation (16) is an extension of eq.(4.7) of [7] to the coordinate–dependent mass tensors. It represents a classical equation of motion in the inverted potential.

The boundary condition used to solve the eikonal equation is $\vec{p} = 0$ at the origin [7]. The origin corresponds to the ground state deformation of the fissioning compound nucleus which is taken as $\vec{q}=0$. We show in Fig.1 an illustration showing the situation for the two–dimensional problem. Because the origin is a maximum point in our inverted potential, we need to boost the system infinitesimally to get meaningful solutions to eq.(16). If we choose a suitable initial direction of $\vec{p}$ for the boosting, we are able to get a special path which can touch the outer equipotential surface given by $V(\vec{q})=0$, which is a line denoted by O in Fig.1., with zero velocity. This special path is called the escape path and the point at which this path touches the outer equipotential surface is the escape point [7] , which is denoted by E in Fig.1. All other paths can not touch this outer equipotential surface. By trying all 2π directions for the initial boost, we can find the escape path by solving eq.(16). We put an object at the maximal point denoted by M and boost it infinitesimally. The energies of M and of the line denoted by O are set to be equal and there is one saddle point S between them in this example. By changing the direction of initial boosting, we get a group of trajectories denoted by solid curves with arrows in this figure. Among them, only one trajectory can touch the curve O and it is called the escape path. The path touches the line O at point E with zero velocity. As is clear from this figure, there occurs a region, shown by a hatched area in this figure, into which no trajectory

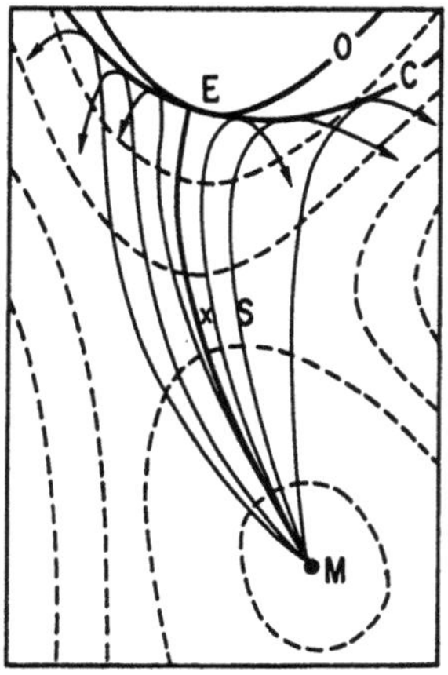

Fig.1. An illustration which shows how to solve the eikonal equation and to obtain the minimum action trajectory. Arbitrary two-dimensional space is treated here. Dotted lines represent equipotential surfaces. Point M represents the maximum point in the inverted potential and point at × (S) represents the saddle point. The line with O is the equipotential surface whose energy is the same as the point M and the line with C represents the caustic. The lines O and C touch each other at point E which is the escape point. The bold line which connects M and E is called escape path which gives the minimum action trajectory. To obtain it, we change the initial condition and generate trajectories represented by lines with arrows.

can enter. It means that, in the level of approximation we are using, there is no solution to the differential equation, that is, no real solution exists for W_0 of Eq.(5) in this area. The envelope of trajectory denoted by C is called the caustic. Therefore, beyond the caustic, there is no solution at this level of approximation.

Inside the caustic, we get solutions of eq.(16) from which we can construct the wave function (4) in the following way. Using eqs.(9) and (14), we get the relation,

$$\frac{dW_0}{d\tau} = \frac{dq^i}{d\tau} \cdot \frac{\partial W_0}{\partial q^i} = \dot{q}^i \cdot M_{ij} \cdot \dot{q}^j \,. \tag{17}$$

From this equation, the value of $W_0(\vec{q}(\tau))$ is obtained by integration as

$$W_0 = \int \dot{q}^i \cdot M_{ij} \cdot \dot{q}^j \, d\tau, \tag{18}$$

where the integration is performed along a solution of eq.(16), that is, along a trajectory given in Fig.1. The integration is performed starting from point M (original point) in Fig.1 till a point whose wave function we are seeking. The most important trajectory is the escape path and we denote by S_{es} the value of W_0 which is obtained when we integrate from the origin to the escape point along the escape path,

$$S_{es} = \int_{es} \dot{q}^i \cdot M_{ij} \cdot \dot{q}^j \, d\tau \, . \tag{19}$$

The quantity S_{es} is the action integral. It is essentially equivalent to the popular form of the action integral S given as

$$S = \int_{\tau_i}^{\tau_f} [2|E-V| \cdot M_{ij} \cdot \dot{q}^i \cdot \dot{q}^j]^{1/2} \, d\tau \, . \tag{20}$$

In the lowest order, the fission half life $T_{\frac{1}{2}}$ is expressed as

$$T_{\frac{1}{2}} = \left(\frac{2\pi}{\omega} \right) exp \left(-2S_{es} / \hbar \right) , \tag{21}$$

where ω is the typical frequency of the nuclear surface vibration and which is often taken as a value near

$$\hbar\omega/2 = 0.5\text{MeV}. \tag{22}$$

The higher order terms like W_1, W_2,.. in eq.(5) play an important role for the matching of the wave function and to determine the pre–factor of the fission half life expression. The contribution from the pre–factor, however, is not as decisive as the exponential contribution from W_0 when the potential satisfies the WKB conditions along the fission path. Lastly, we should notice that the escape action S_{es} is invariant with respect to the coordinate transformation, as is clear from the transformation property of mass tensor and the form of action integral (19) or (20). This point will be more discussed in the next section.

3. Invariant Adiabatic Path

In the previous section, we gave a method to calculate the least action trajectory in a much easier and more transparent way than the numerical trial–and–error method[1–4]. But obtaining the escape path numerically is not easy because any small change of the initial boost direction can cause a big change of the trajectory near the escape point. Also, it is not so simple to understand the relation between the escape path and the geometry of potential and mass tensor. Therefore, we propose here an alternative model which is much easier in its structure and for numerical calculation but, at the same time, is invariant with respect to coordinate transformation. This model has the feature of the adiabatic model as will be shown below.

To define it, we start from solving the equation,

$$M_{ij} \frac{dq_i}{ds} = \text{const} \cdot \frac{\partial V}{\partial q^i} , \tag{23}$$

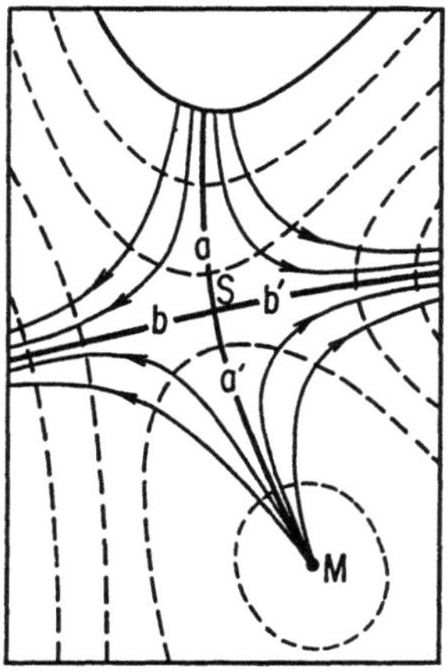

Fig.2. Definition of the invariant adiabatic path. Arbitrary two–dimensional space is treated as in Fig.1 and dotted lines represent equipotential surfaces. Point M represents the maximum point in the inverted potential and S represents the saddle point of the potential. A bold line in the upper part is the equipotential surface whose energy is the same as point M. Thin solid lines with arrows are solutions of eq.(23). The lines a–a' and b–b' are two asymptotes for the solution of eq.(23), and the line a–a' defines the invariant adiabatic path.

in the region where the following inequality holds,

$$(\frac{\partial V}{\partial q^i}) \cdot (\frac{\partial V}{\partial q^i}) \neq 0 \,. \tag{24}$$

Here const. can be chosen in any way because our concern is only the form of the trajectory. In this respect, also s can be any parameter. As will be discussed below, the WKB action does not depend on s, but depends only on the shape of the trajectory determined from eq.(23). We show in Fig.2, the solution of this equation for a simple potential which has one minimum and one saddle point. The dashed lines in this figure are the equipotential energy surfaces and the solid lines with arrows are the solution of eq.(24). Paths obeying this equation have two asymptotes, as shown in this figure by a–a' and b–b'. We define the former asymptote a–a' as the new adiabatic path.

The important feature of this adiabatic path is that it is invariant with respect to the coordinate transformation,

$$x^i \to q^i \,, \tag{25}$$

where old and new coordinates are related to each other by

$$q^i = q^i(x^1,..,x^n) \,, \tag{26}$$

or

$$x^i = x^i(q^1,..,q^n) \,. \tag{27}$$

The invariance is easily checked by remembering the transformation property of mass tensor,

$$M_{ij} \rightarrow \mathcal{M}_{kl} = \frac{\partial x^i}{\partial q^k} \cdot M_{ij} \cdot \frac{\partial x^j}{\partial q^k} \cdot \tag{28}$$

Or, in other words, we can say that it is because eq.(23) is a tensor equation, that is, the left and the right hand sides are both covariant vectors. Thus our new adiabatic action is invariant with respect to the coordinate transformation, and we call this path the invariant adiabatic path for that reason .

This invariance is important since the adiabatic path is determined by the incorrect manner which is not invariant with respect to coordinate transformation in many cases. A typical case is to define a valley line trajectory as a line of gradient vector which connects the ground–state minimum and saddle point. This way of fixing a line is equivalent to generating the velocity field by using the equation

$$\frac{dq_i}{ds} = \text{const} \cdot \frac{\partial V}{\partial q^i}, \tag{29}$$

which has no physical meaning, because it is not a tensor equation.

The WKB action for this adiabatic path can be calculated by using eq.(20). In the numerical calculation which will be performed in the next section for the fission half life, it turns out that the use of this invariant adiabatic path is a good approximation to the minimum action trajectory. In addition, to solve eq(23) and to determine the adiabatic path is much easier than to solve eq.(26). To solve eq.(23), we should only start from the point before and also beyond infinitesimally near to the saddle point and then solve eq.(23) assuming the value of const. is negative. We can get two lines by solving first order differential equation twice, one of which connects the ground state and saddle point and the other saddle point and an exit point on the zero energy contour. These two lines are connected at the saddle point and thus they are considered as parts of one line. This line is defined as the invariant adiabatic path.

4. Numerical Calculation for Spontaneous Fission Half Life

In this section, we will apply the methods given in Sects.2 and 3 to the spontaneous fission of actinide nuclei. Our concern is to check the applicability of the methods to nuclear fission; accordingly, we will not seek the quantitative solutions. We neglect shell and paring effects in the potential and mass tensor calculation, that is, we adopt the classical liquid drop model. Our aim is to calculate multi–dimensional tunneling, and we choose the 2–dimensional barrier penetration in a spontaneous fission process. The calculation is done only in the lowest order, that is, we solve eikonal equation (8) and calculate action integral. We will neglect higher order terms, e.g., eq.(11). Higher order terms affect the pre–exponential factor for the decay width (21) and also are necessary when we connect

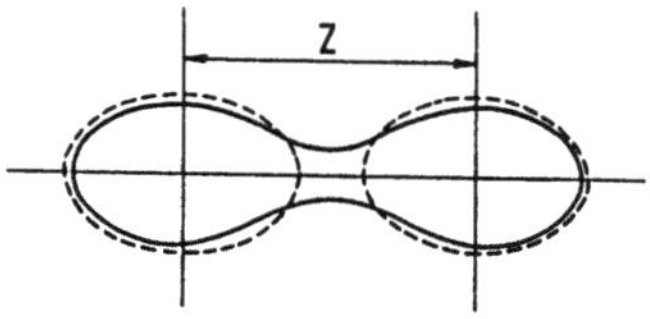

Fig.3. Shape of fissioning nucleus generated by the two–center parameterization. Two parameters are potential center separation z and deformation of fragments δ. Fixing neck parameter ϵ=0.5, the shape is given by the solid line. Dotted lines are for no neck configuration.

the wave function in negative and positive energy regions. In these respects, the higher order terms are important and their study should be done in the near future.

For the two–dimensional fission calculation, we choose two shape parameters. We start from the two–center parameterization of [8] in which there were originally four shape parameters. If we restrict ourselves to symmetric fission, the asymmetric parameter drops and three parameters remain: potential center separation z, spheroidal deformation parameter of fragments δ and neck correction parameter ϵ. In our calculations, ϵ is fixed at 0.5. Thus we treat a two–dimensional $\{z,\delta\}$ problem. The two center model is defined for a single–particle potential. From it, we calculate the shape of the equipotential surface corresponding to the Fermi energy and identify it with the shape of the fissioning nucleus. The potential energy is calculated by the Yukawa–plus–exponential model [9] and the mass tensors by the Werner–Wheeler method [10].

Our Werner–Wheeler mass [10] is an approximation to the incompressible and irrotational flow mass. The latter is known to give an order of magnitude smaller mass for low lying quadrupole vibration than those deduced from experimental data [11]. In fission study, it is necessary to introduce the factor k ≈ 7 [12] to renormalize the mass. In our two–dimensional model, we adopt the following procedure for this renormalization.

We transform the coordinate system from $\{z,\delta\}$ to $\{R,\delta\}$ according to eq.(28), where R is the distance between the center–of–mass of nascent fragments. After calculating the diagonal mass M_{RR}, we modify it to the new $\bar{M}_{RR}$ in the following manner

$$\bar{M}_{RR} = k\cdot[M_{RR}(R,\delta) - \mu] + \mu\,, \tag{30}$$

where k is a fitting parameter and μ is the reduced mass. Other masses $M_{R\delta}$ and $M_{\delta\delta}$ are modified to new ones $\bar{M}_{R\delta}$ and $\bar{M}_{\delta\delta}$ as

$$\bar{M}_{R\delta} = k\cdot M_{R\delta} \tag{31}$$

and

$$\bar{M}_{\delta\delta} = k\cdot M_{\delta\delta}\,. \tag{32}$$

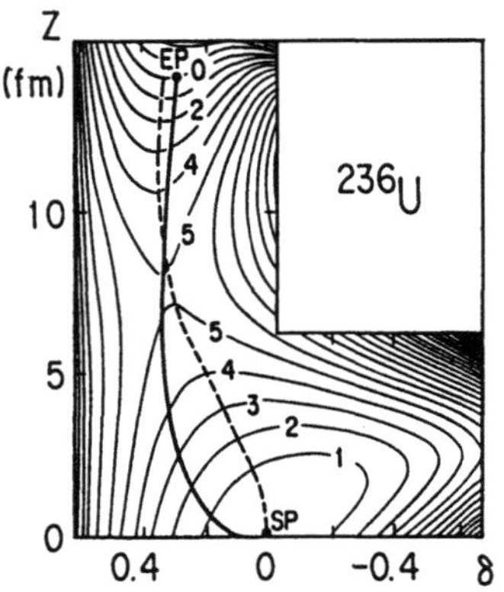

Fig.4. Potential energy surface for fissioning ^{236}U nucleus. The abscissa is fragment's deformation δ and the ordinate is the potential center separation z. The equipotential surface shown by thin solid lines are plotted in units of MeV. Solid line is the escape path and dashed line is the usual adiabatic path for the spontaneous fission. Point denoted by EP is the escape point and SP is the starting point which corresponds to the spherical ground–state of the compound nucleus. The invariant adiabatic path is not shown because it almost coincides with the escape path for this nucleus.

After performing these modifications, we transform the new mass tensors into those in the original coordinate $\{z,\delta\}$ and equations of motion are solved there by using these modified masses. This double–fold transformation was used as a technique to solve a problem which occurs in the numerical calculation.

In Fig.(4), potential energy and fission paths for the spontaneous fission of ^{236}U are shown. The solid line is the escape path which gives the minimum action and the dashed line is the "traditional" adiabatic path which is defined as the curve which is always orthogonal to the equipotential surface and, at the same time, passes through the saddle point. This latter curve is the one criticized in the discussion of eq.(29). From this figure, we see a large difference between two fission paths. The origin of this difference is due to the kinetic energy which is neglected in the adiabatic treatment. The invariant adiabatic path which was determined in the method given in Sect.3 is not shown because it almost coincides with the minimum action trajectory within the width of the line in the figure. It is a surprise that these two almost coincide with each other, because they are defined quite independently.

To see differences and similarities among various methods of path determination more quantitatively, we show in table 1 the WKB actions calculated along the minimum action trajectory, along the invariant adiabatic path and along the usual adiabatic path for the spontaneous fission of ^{236}U and ^{212}Po. For ^{236}U, the "experimental" action, which is calculated from eq.(13) by using the fission half life data, is also shown. The value of k=23 for ^{236}U is chosen for the invariant actions to agree with the experimental data but k=15 is needed if we postulate that the usual adiabatic action agrees with the data. The value of k=20 used for

^{212}Po is based on a rough estimate obtained by comparing the k=23 for ^{236}U and Fig.6–25 of Ref.[11].

The most prominent result is that the value of invariant adiabatic action deviates little from the minimum value of action determined by the method of Sect.2. These two paths give smaller value of WKB actions than the usual adiabatic one for a fixed k. Another point is that the coefficient of the mass renormalization for ^{236}U takes the value k = 23 for the invariant actions to reproduce the experimental half life. This value is consistent with the one shown in fig.6–25 of Ref.[11], which is obtained from the data of the low–energy quadrupole vibrations for actinide nuclei. The value of k≈7 used in [12] is too small in this respect.

The conclusion drawn from the numerical calculation is that the method of Sect.2 works well for the calculation of fission half life. The method gives us the fission path that minimizes the action integral and is invariant with respect to the coordinate transformation. An alternative model we proposed in Sect.3, that is, the invariant adiabatic path, which is also invariant with respect to the coordinate transformation is a good approximation to the minimum action path. We insist that this definition of the adiabatic path should be used in place of the usual adiabatic path which is coordinate dependent. Two methods we proposed are applicable for the decay process in strongly coupled multi–dimensional space which can not be reduced into the product of one–dimensional subspaces.

Although we showed that the invariant adiabatic path is a good approximation to the minimum action trajectory numerically, it is not clear at this stage why it is so. The adiabatic path does not contain any effect of kinetic energy or inertia force, while these are important to determine the minimum action trajectory. Thus the next section is devoted to investigating more details of the invariant adiabatic path in order to get an insight as to why it is a good approximation, at least for the problems we calculated.

Table 1. The values of action integral for spontaneous fission of ^{236}U and ^{212}Po. "Invariant dynamical" corresponds to the method of Sect.2, "invariant adiabatic" to the method of Sect.3 and "usual adiabatic" is discussed in the text. k is the multiplication factor defined in eqs.(30), (31) and (32). Experimental value of action integral is determined from the life time data and eqs.(21) and (22). k=23 is required to fit the invariant dynamical method to the data and k=15 is required to fit usual adiabatic result to the data for the case of ^{236}U.

System	^{236}U $(k=23)$	$(k=15)$	^{212}Po $(k=20)$
invariant dynamical	$53\hbar$	$45\hbar$	$88\hbar$
invariant adiabatic	$53\hbar$	$45\hbar$	$90\hbar$
usual adiabatic	$63\hbar$	$53\hbar$	$114\hbar$
experimental	$53\hbar$		*unknown*

5. Validity of Adiabatic Approximation

In order to investigate the features of the invariant adiabatic path and its applicability, we will start from more popular formulation for the minimum action trajectory, which was obtained in Sect.2 in a different way. It is based on the standard variational method of action integral.

Let the energy be denoted by E and let the potential V be normalized to zero at the ground state minimum. When we treat the decay from the ground state, E can be set to zero in the following calculation, which corresponds to our eikonal equation (8). Classical action S is defined in the form,

$$S = \int_{\tau_i}^{\tau_f} F(\vec{q},\dot{\vec{q}})\, d\tau \,, \tag{33}$$

and

$$F(\vec{q},\dot{\vec{q}}) = [2|E-V| \cdot M_{ij} \cdot \dot{q}^i \cdot \dot{q}^j]^{1/2} \,, \tag{34}$$

where $\dot{q}$ means derivative of q with respect to τ. We are mainly concerned with the decay–width calculation, and in this case action S is calculated in negative energy region. Boundary conditions have a special feature in that the initial point is fixed at the ground state minimum but the final point is not fixed but lies somewhere on the equipotential energy surface which has the same energy as the ground state. Therefore, the variation of the action S leads to the Euler equation [1],

$$\frac{\partial F}{\partial q^i} - \frac{d}{d\tau} \cdot \frac{\partial F}{\partial \dot{q}^i} = 0 \tag{35}$$

with boundary conditions

$$\frac{\partial F}{\partial \dot{q}^i} \cdot \delta q^i \Big|_{\tau=\tau_f} = 0 \,. \tag{36}$$

In our problem, the final point is on the equipotential surface and thus we get

$$\delta V = \frac{\partial V}{\partial q^i} \cdot \delta q^i \Big|_{\tau=\tau_f} = 0 \,. \tag{37}$$

From (36) and (37), we get the relation

$$\frac{\partial F}{\partial \dot{q}^i} \Big/ \frac{\partial V}{\partial q^i} \Big|_{\tau=\tau_f} = \text{const. for each i} \tag{38}$$

Using the form of F given in eq.(34), the relation (38) is rewritten as

$$M_{ij} \cdot \dot{q}^j = \text{const} \cdot \frac{\partial V}{\partial q^i} \Big|_{\tau=\tau_f} \,, \tag{39}$$

which is called as transversality condition [1]. Therefore, our problem is to solve (35) under the condition (39). We see that the transversality condition (39) has the same form as the relation (23), which was used to define the adiabatic path. It means that near the end point, the minimum action trajectory should follow one of the solutions of (23).

In order to discuss the trajectory near the initial point (ground state), we return to the equation which determines the dynamical (minimum action) trajectory (16). It is rewritten as

$$\frac{d}{d\tau}\vec{p} = \vec{\nabla}V + \dot{q}^i(\vec{\nabla}M_{ij})\dot{q}^j / 2 , \tag{40}$$

where $\vec{p}$ denotes the momentum vector defined in (9), which corresponds to pseudo momentum in imaginary time formalism. The first term of the right–hand–side of eq.(40) is larger that the second term near two end points if the fluctuation of mass tensors with respect to the position q^i is not large compared to the typical values of the mass tensors. This will be shown as follows.

The fluctuation of mass within the region of $\Delta\vec{q}$ is written as

$$\Delta M_{ij} = \vec{\nabla}M_{ij}\cdot\Delta\vec{q} . \tag{41}$$

In the region of $\Delta\vec{q}$ where the inequality

$$M_{ij} >> \Delta M_{ij} \tag{42}$$

holds, we see the following relation

$$\dot{q}^i M_{ij}\dot{q}^j >> [\dot{q}^i(\vec{\nabla}M_{ij})\dot{q}^j]\cdot\Delta\vec{q} . \tag{43}$$

The left–hand–side of this inequality is twice the kinetic energy. In so far as the decay of ground state is concerned, we can put the energy E as zero. We thus have the following relation

$$\dot{q}^i M_{ij}\dot{q}^j = 2\cdot\vec{\nabla}V\cdot\Delta\vec{q} , \tag{44}$$

from the eikonal equation (8). From (40),(43) and (44), we see that the first term of the left hand side of (40) is much larger than the second term if the condition (42) holds. The condition (42) is satisfied for small $\Delta\vec{q}$ if we use the classical mass tensors, as was done in the previous section. We could therefore prove that near the two end points where the momentum is small, the dynamical trajectory is along the gradient field $\vec{\nabla}V$ provided the linearity of $\vec{\nabla}V$ is good. Thus, we see that the minimum action trajectory deviates little from one of the trajectories of (16) near these two end points.

As (33) and (34) show, the major contribution to the action integral comes from the region near the saddle point. In this region, the trajectory determined from eq.(23) changes very rapidly as seen in Fig.2, and we need another approach to investigate this region. We will restrict ourselves hereafter to the two-dimensional case and change the notation (q^1,q^2) to (x,y) for simplicity. To obtain the shape of the trajectory, we need to solve eq.(35). There the quantity τ is an arbitrary parameter, and thus we can choose it within the coordinate x. Kinetic energy operator T in this case is written as

$$T = a + 2b(\frac{dx}{dy}) + c(\frac{dy}{dx})^2 , \tag{45}$$

where a,b,c correspond to M_{xx}, M_{xy} and M_{yy}. Equation (35) is written after some calculation as [1]

$$\begin{aligned} &K\cdot[2\ddot{y}(b^2-ac)+a_x(b+c\dot{y})+a_y(a+3b\dot{y}+2c\dot{y}^2)-2b_x(a+b\dot{y})+ \\ &2b_y\dot{y}^2(b+c\dot{y})-c_x\dot{y}(2a+3b\dot{y}+c\dot{y}^2)-c_y\dot{y}^2(a+b\dot{y})]+ \\ &T[(a+b\dot{y})\frac{\partial K}{\partial y}-(b+c\dot{y})\frac{\partial K}{\partial x}] = 0 , \end{aligned} \tag{46}$$

where $\dot{y}$ and $\ddot{y}$ are the first and the second derivative of y with respect to x and K is given by

$$K = | E-V | , \tag{47}$$

and a_x, for example, represents partial derivative of a with respect to x. If we assume a constant mass near the saddle point, eq.(46) reduces to

$$2K(b^2-ac)\ddot{y} + T [(a+b\dot{y})\frac{\partial K}{\partial y} - (b+c\dot{y})\frac{\partial K}{\partial x}] = 0 . \tag{48}$$

Thus the term in the square bracket in (48) is zero for the solution of eq.(23).

If we can assume that the mass tensors are changing smoothly near the saddle point, which is the case if we calculate the mass in the classical model, as was done in the previous section, the equation to be solved is (48). It is natural to assume the potential function V is represented by

$$V = \alpha_0 x^2 + \alpha_1 xy + \alpha_2 y^2 , \tag{49}$$

near the saddle point, where $\alpha_0 \times \alpha_1 < 0$. In this case, it is possible to show that there are two solutions of (48), which are represented by

$$y_i = c_i \cdot x , \quad i=1,2 , \tag{50}$$

where c_i are two real constants. The solution (50) should coincide with the two separatrices shown in Fig.2, because comparing (23) and (48), we notice that (50) is the solution of (23). In other words, two separatrices satisfy the Euler equation (35) near the saddle point.

Up to now, we showed that two separatrices are solutions of the Euler equation near the saddle point and thus the adiabatic path satisfies the minimum action requirement there, because we define the adiabatic path as one of the separatrices. When we follow the path to both directions until the potential becomes small enough, it satisfies the Euler equation again as we showed before. The path satisfies the correct boundary conditions as well. Between saddle and two end points, no definite discussion is possible. We can say, however, that the adiabatic path we defined will approximately satisfy the minimum action requirement.

The above discussion is a qualitative argument for the similarity of the dynamical and invariant adiabatic fission paths. To estimate the difference in more detail, we need to know the whole feature of potential and mass tensors, which is in general too complicated and no general discussion seems to be possible. If, on the other hand, we settle on a simple model which is manageable, it is then hard to make correspondence to the real physical system. We know that even in the classical liquid drop model, potential and mass tensors are very complicated.

Lastly, we would like to point out that the subsidiary equation (23) to generate the invariant adiabatic trajectory has a special feature: we can diagonalize the mass tensors for the two–dimensional case with it. It is also possible at the same time to let the potential depend only on one dimension, i.e. the potential is flat in one direction. Because we need a lot of space to give the complete formulation, interested readers are recommended to refer to [13].

6. Summary

Although the problem of quantum tunneling in multi–dimensional space is an attractive and important item, there is not much work done on this problem. Nuclear fission is a typical problem in which we encounter multi–dimensional tunneling. In this field, multi–dimensional tunneling was attacked by some groups by calculating action integrals for all different paths within some numerical precision. Such a calculation is quite time consuming and, at the same time, scarcely provides clear physical insight into the problem. For example, the structure of caustic is hard to see in these calculations. Another problem is that it is a zeroth order approximation to quantum tunneling and does not tell us how to calculate the higher order effects. In the case of quantum decay, the pre–exponential factor can not be calculated in this method. The higher–order effects are needed to investigate and obtain the wave function in the whole multi–dimensional space.

We showed that there is a method that could overcome such difficulties. This method was developed to calculate the multi–dimensional wave function in negative energy region for the decay problem, and is based on the proposal of Schmid [7] and was generalized to be applied to nuclear fission problem [6]. In this method, we could get the value of the minimum action by integrating a kind of equation of motion with initial conditions. When the geometry of the potential is simple, this procedure works quite well. During this investigation, we also proposed a simpler method to get fission paths. This method is similar to the method of steepest descent trajectory, which is often used in the practical fission life time calculations. The problem associated with this latter method is that it does not satisfy a physically necessary condition that it is invariant with respect to the general coordinate transformation. But, the new adiabatic method is free from this difficulty and the calculation is as easy as the steepest descent method. In addition, we found that this adiabatic path is a good approximation to the minimum action trajectory. Because we found that it is an unexpectedly good approximation numerically, we studied the reason for it and found that, if the potential and mass tensors change gradually as functions of position, we could understand qualitatively why it is a good approximation to the minimum action trajectory. Since the calculation of invariant adiabatic trajectory is much easier and is more transparent than the minimum action trajectory, it will serve as a good starting point for the search of the minimum action trajectory in many fields of physics.

It is of importance to extend the method we used here to obtain the minimum action trajectory to higher order. The lowest–order equation we used is eikonal equation (8), while the next order is transport equation (11). To get the pre–factor for the decay width expression, it is necessary to solve the transport equation. Also it is used to join the wave function for the positive and negative energy region. How easily we can apply this model to three or more dimensional problems may be another interesting feature of this method. Formally, it is clear that the method can be applied to any dimensions, but the degree of difficulty to do the calculation should be examined.

Restricting ourselves to nuclear fission problem, we still have the following open problems. In this work, fission potential and mass tensors were calculated by the classical model. Shell and pairing effects play an important role in the nuclear fission problem and, if we include them, potential and mass tensors become more complicated. The methods proposed here should be checked beforehand if we wish to use them in such complicated geometry. It is clear that more than two trajectories which give minimum actions exist in such a case. Also, it is possible that there occur more than two invariant adiabatic trajectories. In any case, two methods should be checked for their applicability to this more realistic situation in nuclear fission. Our methods are of interest to not only for the double–humped barrier, which occurs quite generally in actinide nuclei, but also for bimodal fissions. This follows because it is expected [14] that there are two or more trajectories which give minimum actions in bimodal fissions. When we speak of quantum tunneling, we should also recognize another quite troublesome problem of quantum mass tensors which might show a larger fluctuation than the potential itself. Such a large fluctuation occurs in the cranking formula or in the adiabatic TDHF model. The large fluctuation may question the validity of using the semiclassical approximation. Thus the problem of mass tensors should be studied more extensively.

I would like to thank Dr.Y.Tomita for valuable discussions and Dr.H.J. Kim and Dr.P. Möller for careful reading of the original manuscript.

References

[1] M. Brack et.al., Rev.Mod.Phys.**44**, 320 (1972)
[2] T. Ledergerber and H.C. Pauli, Nucl.Phys.**A207**, 1 (1973)
[3] A. Baran, Phys.Lett.**B76**, 8 (1978)
[4] A. Baran et al., Nucl.Pnys.**A361**, 83 (1981)
[5] L. Wilets, Theories of Nuclear Fission (Clarendon Press, Oxford 1964).
[6] T. Kindo and A. Iwamoto, Phys.Lett.**B225**, 203 (1989)
[7] A. Schmid, Ann.Phys.**170**(1986)333.
[8] J. Maruhn and W. Greiner, Z.Phys.**251**, 431 (1972).
[9] H.J. Krappe, J.R. Nix andd A.J. Sierk, Phys.Rev.**C20**, 992 (1979).
[10] K.T.R.Davies, A.J. Sierk and J.R. Nix, Phys.Rev.**C13**, 2385 (1976).
[11] A. Bohr and B.R. Mottelson, Nuclear Structure vol.II, p531 (W.A.Benjamin, Massachusettss 1975).
[12] J. Rundrup et.al., Nucl.Phys.**A217**(1973)221.
[13] A. Iwamoto and Y. Tomita, preprint.
[14] P. Möller, J.R. Nix and W.J. Swiatecki, Nucl.Phys.**A469**, 1 (1987).

Phase-Space Dynamics of Heavy-Ion Reactions†

W. Cassing, S. Hannuschke, U. Mosel, K. Niita, and G. Wolf

Institut für Theoretische Physik,
Heinrich-Buff-Ring 16, W-6300 Giessen, Fed. Rep. of Germany

Abstract: The description of heavy–ion collisions up to bombarding energies of a few hundred MeV per nucleon requires nonperturbative many–body approaches far from equilibrium which include the selfconsistent mean–field dynamics as well as the residual nucleon–nucleon collisional history. In this contribution we present such approaches within the framework of quantum correlation dynamics in the nonrelativistic case and derive selfconsistent transport equations for the baryonic phase–space distribution. This quantity allows one to follow the space–time evolution of the reaction, to determine the degree of thermalization and to calculate one–body as well as collective observables. We explicitly study the approximations performed in the derivation of the transport theory. As an application we evaluate the differential hard photon yield for heavy–ion reactions on the basis of the incoherent on–shell nucleon–nucleon production mechanism and compare it to the available experimental data. Furthermore, we analyze deeply inelastic binary collisions of Mo+Mo from 14.7 MeV/u to 25 MeV/u and evaluate double differential cross sections with respect to the total kinetic energy (TKE) and mass or scattering angle.

1. Introduction

Heavy–ion collisions in the energy regime from roughly 15 MeV/u to a few hundred MeV/u offer a unique possibility to study nonequilibrium dynamics of strongly interacting fermion systems as well as the properties of dense and hot nuclear systems [1–5]. Whereas at low bombarding energy the time–dependent Hartree–Fock (TDHF) theory provides a proper understanding of the reaction dynamics [6], residual nucleon–nucleon collisions become more important with increasing laboratory energy [1–5]. In the past semiclassical phase–space simulations have been widely used for the description of heavy–ion reactions [1–5,7–21] and successfully applied to the production of energetic particles too [4,5,19–21]. In this paper we show the relation of such transport models to the underlying quantum many–body dynamics (section 2) and discuss explicitly the approximations performed in the context of the derivation (section 3). We, furthermore, briefly review the application of the transport theory to hard photon production and compare with the experimental data (section 4). Finally, we test the applicability of the fully microscopic approach for deeply–inelastic heavy–ion reactions at low bombarding energies from 14.7 to 25 MeV/u (section 5).

---

†This work has been supported by BMFT, GSI Darmstadt, the PROCOPE program and a NATO science grant.

Eds: Y. Abe, H. Horiuchi, K. Matsuyanagi

2. General Derivation of Nonrelativistic Transport Theories

The most reliable theoretical framework nowadays for the description of intermediate energy heavy–ion reactions is based on the Vlasov–Uehling–Uhlenbeck (VUU) equation for the one–body phase–space distribution. This equation has originally been proposed by Nordheim [22] and later on by Uehling and Uhlenbeck [23] for weakly interacting Fermion systems as an extension of the Boltzmann equation. While its derivation has commonly been based on perturbative arguments [24,28] or 'plausibility' [1], it has been shown by Botermans and Malfliet [26,27,30] and Cassing et al. [29], that it may consistently be derived also in a nonperturbative framework. In particular, the time–dependent mean–field in the Vlasov term as well as the in–medium nucleon–nucleon cross section in the collision term, which determine the nuclear dynamics, are uniquely extracted from the bare nucleon–nucleon interaction within the framework of many–body theory. In section 2.1 we establish these relationships by investigating different truncation schemes in the BBGKY density–matrix hierarchy.

2.1 The BBGKY Hierarchy and its Truncation Schemes

The dynamical description of the nuclear many–body problem in the nonrelativistic limit is based on the von–Neumann equation for the A–body density matrix ρ_A, i.e. ($\hbar$=1)

$$i\partial/\partial t\ \rho_A = [H_A, \rho_A] \tag{2.1}$$

where H_A denotes the A–body hamiltonian in the restriction to mutual two–body interactions v(ij)

$$H_A = \sum_{i=1}^{A} t(i) + \sum_{i<j}^{A-1} v(ij)\ . \tag{2.2}$$

The quantity t(i) denotes the kinetic energy of particle i which in the case of nucleons is characterized by position $\mathbf{r}_i$, spin σ_i and isospin τ_i. Equation (2.1) is equivalent to a coupled set of equations for the reduced density matrices of rank n

$$\rho_n = 1/(A-n)!\ \ \mathrm{Tr}_{(n+1,\,..,A)}\{\rho_A\} \tag{2.3}$$

known as the BBGKY hierarchy [25], i.e.

$$i\partial/\partial t\ \rho_n = [\sum_{i=1}^{n} t(i), \rho_n] + [\sum_{i<j}^{n-1} v(ij), \rho_n] + \mathrm{Tr}_{(n+1)} [\sum_{i=1}^{n} v(i,n+1), \rho_{n+1}] \tag{2.4}$$

for $1 \leq n \leq A$ and ρ_{A+1}=0. We rewrite the density matrices ρ_n in terms of antisymmetrized products of density matrices of lower order [31,32], i.e.

$$\rho_1(11') = \rho(11') \tag{2.5a}$$

$$\rho_2(12,1'2') = (1-P_{1'2'})\rho(11')\rho(22')+c_2(12,1'2') \tag{2.5b}$$

$$\begin{aligned}\rho_3(123,1'2'3') &= (1-P_{1'2'}-P_{1'3'})(1-P_{2'3'})\rho(11')\rho(22')\rho(33') \\ &+(1-P_{1'2'}-P_{1'3'})\rho(11')c_2(23,2'3')+(1-P_{2'1'}-P_{2'3'})\rho(22')c_2(13,1'3') \\ &+(1-P_{3'1'}-P_{3'2'})\rho(33')c_2(12,1'2') + c_3(123,1'2'3') ,\end{aligned} \tag{2.5c}$$

where P_{ij} denotes the permutation operator between particle i and j. This yields coupled equations of motion for the one-body density matrix $\rho(11';t)$ [31]

$$\begin{aligned} i\partial/\partial t\, \rho(11';t) &= t(1)\rho(11';t) - \rho(11';t)t(1') + \\ &\quad \underset{(2=2')}{\mathrm{Tr}} [v(12)\rho_2(12,1'2';t)-\rho_2(12,1'2';t)v(1'2')] \\ &= [t(1)+U(1)]\rho(11';t) - \rho(11';t)[t(1')+U(1')] \\ &\quad + \underset{(2=2')}{\mathrm{Tr}} [v(12)c_2(12,1'2';t)-c_2(12,1'2';t)v(1'2')] \end{aligned} \tag{2.6}$$

and the two-body correlation function $c_2(12,1'2';t)$;

$$\begin{aligned} i\partial/\partial t\, c_2(12,1'2';t) &= [t(1)+t(2)+U(1;t)+U(2;t)]\, c_2(12,1'2';t) \\ &\quad - c_2(12,1'2';t)[t(1')+t(2')+U(1';t)+U(2';t)] \end{aligned}$$

$$+ [Q_{12}v(12)\rho_{20}(12,1'2';t)-\rho_{20}(12,1'2';t)v(1'2')Q_{1'2'}] \tag{2.7a}$$

$$+ [Q_{12}v(12)c_2(12,1'2';t) - c_2(12,1'2';t)v(1'2')Q_{1'2'}] \tag{2.7b}$$

$$\begin{aligned} + \underset{(3=3')}{\mathrm{Tr}} &[v(13)\{\mathcal{A}_{13}\mathcal{A}_{1'2'}\rho(11';t)c_2(23,2'3';t)\} \\ &- \{\mathcal{A}_{1'3'}\mathcal{A}_{12}\, c_2(23,2'3';t)\rho(11';t)\}v(1'3') \\ &+ v(23)\{\mathcal{A}_{23}\mathcal{A}_{1'2'}\rho(22';t)c_2(13,1'3';t)\} \\ &- \{\mathcal{A}_{2'3'}\mathcal{A}_{12}\, c_2(13,1'3';t)\rho(22';t)\}v(2'3')] \end{aligned} \tag{2.7c}$$

$$+ \underset{(3=3')}{\mathrm{Tr}} \{[v(13)+v(23)]\,c_3(123,1'2'3';t)-c_3(123,1'2'3';t)[v(1'3')+v(2'3')]\} . \tag{2.7d}$$

with $\rho_{20}(12,1'2') = \mathcal{A}_{12}\rho(11')\rho(22')$. In equations (2.6) and (2.7) $\mathrm{Tr}_{(i=i')}$ stands for a summation over spin $\sigma(i)$ and isospin $\tau(i)$ as well as an integration over d^3r_i with $i=i'$, while $\mathcal{A}_{ij}$ is the antisymmetrization operator for fermions defined by

$$\mathcal{A}_{ij} = 1 - P_{ij} . \tag{2.8}$$

For the sake of convenience we have introduced the bare mean-field potential

$$U(i;t) = \underset{(n=n')}{\mathrm{Tr}} \{v(in)\mathcal{A}_{in}\rho(nn';t)\} \tag{2.9}$$

and the Pauli-blocking operator

$$Q_{ij} = 1 - \underset{(n=n')}{\mathrm{Tr}} (P_{in}+P_{jn})\rho(nn';t) \qquad (2.10)$$

In order to close (2.6) and (2.7) the three–body correlation term c_3 (2.7d) is neglected and it is assumed that the nuclear many–body problem can be adequately described by the preceding set of equations in the nonrelativistic regime.

We note that for the bare interaction v(ij) in the hamiltonian (2.2) the TDHF limit does not lead to nuclear saturation and we have to go beyond this limit in the case of nuclear physics where we deal with strong two–body correlations. The limit retaining all terms up to (2.7a) corresponds to the Born approximation (BA) for nucleon–nucleon scattering of two nucleons moving in the mean field U(i), while the additional terms in (2.7b) lead to time–dependent G–matrix theory (TDGMT) [29]. The limit (2.7a+b+c) finally corresponds to nuclear quantum correlation dynamics (NQCD) [32].

Introducing additional operators H(i) that describe particle–hole correlations,

$$\begin{aligned} H(i;1'2') &= \underset{(3=3')}{\mathrm{Tr}} \; v(i3)\mathcal{A}_{i3}\mathcal{A}_{1'2'}\rho(ii';t)P_{i3}P_{i'3'}; \\ H^{\dagger}(i';12) &= \underset{(3=3')}{\mathrm{Tr}} \; P_{i3}P_{i'3'}\rho(ii';t)\mathcal{A}_{i'3'}\mathcal{A}_{12}v(i'3'), \end{aligned} \qquad (2.11)$$

where the operators P and $\mathcal{A}$ in $H^{\dagger}(i')$ act to the left, we can rewrite (2.7a+b+c) in shorthand operator form, i.e.

$$\begin{aligned} i\partial/\partial t \; c_2(t) = \; & [\{t(1)+t(2)+U(1)+U(2)\}c_2(t) \\ & - c_2(t)\{t(1')+t(2')+U^{\dagger}(1')+U^{\dagger}(2')\}] \\ & + [Qv\rho_{20}(t) - \rho_{20}(t)v^{\dagger}Q^{\dagger}] \qquad (2.12a) \\ & + [Qvc_2(t) - c_2(t)v^{\dagger}Q^{\dagger}] \qquad (2.12b) \\ & + [(H(1)+H(2))c_2(t) - c_2(t)(H^{\dagger}(1')+H^{\dagger}(2'))] \qquad (2.12c) \end{aligned}$$

where the particle coordinates have been partly contracted. The explicit matrix elements of the operators t(1),t(2),U(1),U(2),Q,H(1),H(2) and their adjoint operators are given in [33] for an arbitrary s.p. basis. Equation (2.12) includes one–body terms as well two–body interaction terms; H(i;1'2'), furthermore, is a three–body operator which for $\mathcal{A}_{1'2'} = 1-P_{1'2'} \approx 1$ reduces to a one–body operator, too. Adding up the 'one–body' terms by

$$\begin{aligned} h(i) &= t(i)+U(i)+H(i); \qquad h^{\dagger}(i') = t(i')+U^{\dagger}(i')+H^{\dagger}(i'), \\ h_{12} &= h(1)+h(2); \qquad h_{12}^{\dagger} = h^{\dagger}(1')+h^{\dagger}(2') \end{aligned} \qquad (2.13)$$

we arrive at

$$\begin{aligned} & i\partial/\partial t \; c_2(t) - [h_{12} + Qv]c_2(t) + c_2(t)[v^{\dagger}Q^{\dagger} + h_{12}^{\dagger}] = \\ & \quad [Qv\rho_{20}(t) - \rho_{20}(t)v^{\dagger}Q^{\dagger}]. \end{aligned} \qquad (2.14)$$

The formal solution of (2.14) – which includes all terms up to (2.7c) – is given by

$$c_2(t) = -\frac{i}{8\pi^2}\iiint d\omega d\omega' dt' \exp\Big[-i(\omega-\omega')(t-t')\Big] \cdot G_{12}(\omega)\{Qv\rho_{20}(t')-\rho_{20}(t')v^\dagger Q^\dagger\} G^\dagger_{12}(\omega') \quad (2.15)$$

with the full two–body propagator

$$G_{12}(\omega) = [\omega - \{h_{12}+ Qv\} + i\epsilon]^{-1}. \quad (2.16)$$

Introducing the generalized Moeller operator Ω_{12} via the relation

$$G_{12}(\omega) = \sum_{n=0}^{\infty}(g_{12}(\omega)Qv)^n g_{12}(\omega) = \Omega_{12}(\omega)g_{12}(\omega) \quad (2.17)$$

with

$$g_{12}(\omega) = [\omega - h_{12} + i\epsilon]^{-1}, \quad (2.18)$$

eq. (2.15) can be written as

$$c_2(t) = -\frac{i}{8\pi^2}\iiint d\omega d\omega' dt' \exp\Big[-i(\omega-\omega')(t-t')\Big] \cdot\Big\{[\Omega_{12}(\omega)\rho_{20}(t')g^\dagger_{12}(\omega')\Omega^\dagger_{12}(\omega') - \Omega_{12}(\omega)g_{12}(\omega)\rho_{20}(t')\Omega^\dagger_{12}(\omega')] - [\rho_{20}(t')G^\dagger_{12}(\omega') - G_{12}(\omega)\rho_{20}(t')]\Big\} . \quad (2.19)$$

We expand the operators $\Omega(\omega)$ and $\Omega^\dagger(\omega')$ in (2.19) around a fixed mean energy E and neglect the correction, i.e.

$$\Omega_{12}(\omega) \approx \Omega_{12}(E) = 1/(1-g_{12}(E)Qv) = \sum_{n=0}^{\infty}(g_{12}(E)Qv)^n = \Omega_{12}. \quad (2.20)$$

Now the integrations over $d\omega$ (or $d\omega'$) can be carried out and yield a δ–function in (t–t'). We note that this markovian limit corresponds to an on–shell approximation for collisions (cf. section 3.3). The residual integrations lead to

$$\rho_2 = \Omega_{12}\rho_{20}\Omega^\dagger_{12} . \quad (2.21)$$

Operating with v and $v^\dagger$ from the left and the right of (2.21) we have ($G := v\,\Omega$)

$$v\rho_2 v^\dagger = G\rho_{20}G^\dagger \quad (2.22)$$

which relates the matrix elements of the bare interaction v with respect to the full two–body density–matrix ρ_2 to the matrix elements of the effective interaction G with respect to the unperturbed two–body density–matrix ρ_{20}. The interaction G according to (2.20) follows

$$G(E) = v + v\, g_{12}(E)\, Q\, G(E) , \quad (2.23)$$

which is the stationary integral equation for the G–matrix. The propagator $g_{12}(E)$ (2.18), furthermore, which still includes particle–hole type correlations, obeys the integral equation

$$g_{12}(E) = \mathcal{g}_{12}(E) + \mathcal{g}_{12}(E)\{H(1)+H(2)\}g_{12}(E) \tag{2.24}$$

with the bare two–body mean–field propagator

$$\mathcal{g}_{12}(E) = (E - \{t(1)+t(2)+U(1)+U(2)\} + i\epsilon)^{-1} . \tag{2.25}$$

The integral equation (2.23) corresponds to a resummation of ladder diagrams for the effective interaction G while (2.24) includes an infinite summation of medium polarizations for the mean–field propagator. In lowest order we have $g_{12}(E) = \mathcal{g}_{12}(E)$ and the integral equation (2.23) is equivalent to that for the Brueckner G–matrix [34] when neglecting hole–hole scattering [27], i.e. (in a diagonal representation of $\rho_{\alpha\alpha'} = \delta_{\alpha\alpha'} n_\alpha$)

$$Q = (1-n_\alpha-n_\beta) = (1-n_\alpha)(1-n_\beta)-n_\alpha n_\beta \approx (1-n_\alpha)(1-n_\beta) . \tag{2.26}$$

Thus (2.12a+b+c) shows up as an extension of Brueckner theory where ladder and ring diagrams are summed up on the same footing, while (2.12a+b) is equivalent to time–dependent G–matrix theory (TDGMT) which includes not only particle–particle (p–p) but also hole–hole (h–h) scattering. Furthermore, in the limit (2.12a) we have G = v, which shows up as the perturbative Born approximation for residual collisions of two nucleons moving in the bare mean field U(i).

In the further limit $Tr_3 = 0$, i.e. for a two–particle system, we have Q = 1, U(i)=H(i)=0 and equation (2.23) becomes identical to the integral equation for the free scattering amplitude, i.e.

$$G(E) = T(E) = v + v\, g_{12}(E)\, T(E) . \tag{2.27}$$

We now rewrite the term $Tr_2[v, \rho_2]$ in (2.6) as a function of G and ρ_{20}. Using (2.22) and (2.23) we obtain

$$\underset{(2)}{Tr}[v, \rho_2] = \underset{(2)}{Tr}[G, \rho_{20}] + \underset{(2)}{Tr}[G\rho_{20}G^\dagger Q^\dagger g_{12}^\dagger - g_{12}QG\rho_{20}G^\dagger] \tag{2.28}$$

which yields the commutator of G with ρ_{20} plus terms of second order in G. In order to extract the mean–field part of this expression one might be tempted to neglect the second order terms in G of (2.28). However, the resulting equation of motion for ρ no longer conserves particle number since G is a complex valued operator. It is therefore consistent to retain only the real part of G in the mean–field limit of (2.28) which defines a mean–field potential via the relation

$$[U_m, \rho] = \mathop{\mathrm{Tr}}_{(2)} [\mathrm{Re}(G), \rho_{20}] \; . \tag{2.29}$$

The remaining terms are attributed to a two-body collision term and given by

$$I_2(t) = -i \mathop{\mathrm{Tr}}_{(2)} \Big\{ i\, G^\dagger Q\, G\, \mathrm{Im}(g_{12}) \rho_{20} - i \rho_{20} \mathrm{Im}(g_{12}^\dagger) G^\dagger Q\, G + G\, \rho_{20} G^\dagger Q^\dagger g_{12}^\dagger - g_{12} Q\, G\, \rho_{20} G^\dagger \Big\} \tag{2.30}$$

using the identity

$$\mathrm{Im}(G) = G^\dagger Q\, G\, \mathrm{Im}(g_{12}) \; , \tag{2.31}$$

which yields the loss terms in $I_2(t)$ while the last two terms in (2.30) account for the gain terms [29]. Putting everything together we obtain

$$i\partial/\partial t\, \rho(11';t) - [t(1), \rho(11';t)] - [U_m(1;t), \rho(11';t)] = i\, I_2(11';t) \; . \tag{2.32}$$

We thus have closed the equation of motion for $\rho(11';t)$ on the one-body level by integrating out the two-body correlation function c_2. We note, however, that the respective many-body problem remains hidden in the integral equations (2.23) and (2.24) for the effective interaction G and the in-medium propagator $g_{12}(E)$.

2.2 Derivation of a Transport Equation from NQCD

In the following we assume that the single-particle basis is given by plane waves and we perform a Wigner transformation of (2.32), i.e.

$$f(\mathbf{r},\mathbf{p};t) = (2\pi)^{-3} \int d^3q\, \exp(i\mathbf{q}\mathbf{r}) \langle \mathbf{p}-\mathbf{q}/2 | \rho | \mathbf{p}+\mathbf{q}/2 \rangle \; . \tag{2.33}$$

Furthermore, by adopting the infinite nuclear matter limit for all quantities in the collision term $I_2(t)$,

$$\langle \mathbf{p}_3 | \rho | \mathbf{p}_1 \rangle \approx (2\pi)^3\, \delta(\mathbf{p}_1 - \mathbf{p}_3)\, f(\mathbf{p}_1;t) \tag{2.34a}$$

$$\langle \mathbf{p}_3\mathbf{p}_4 | Q | \mathbf{p}_2\mathbf{p}_1 \rangle \approx (2\pi)^6\, \delta(\mathbf{p}_1 - \mathbf{p}_3)\, \delta(\mathbf{p}_2 - \mathbf{p}_4)\, \{1 - f(\mathbf{p}_3;t) - f(\mathbf{p}_4;t)\} \tag{2.34b}$$

$$\langle \mathbf{p}_3\mathbf{p}_4 | G(E) | \mathbf{p}_2\mathbf{p}_1 \rangle \approx (2\pi)^3\, \delta(\mathbf{p}_1 + \mathbf{p}_2 - \mathbf{p}_3 - \mathbf{p}_4)\, G(\mathbf{p}_2 - \mathbf{p}_4;E) \tag{2.34c}$$

$$\langle \mathbf{p}_3\mathbf{p}_4 | g_{12}(E) | \mathbf{p}_2\mathbf{p}_1 \rangle \approx (2\pi)^6\, \delta(\mathbf{p}_1 - \mathbf{p}_3)\, \delta(\mathbf{p}_2 - \mathbf{p}_4) \cdot \{E - [(p_1^2 + p_2^2)/2m + U(\mathbf{p}_1) + U(\mathbf{p}_2) + H(\mathbf{p}_1) + H(\mathbf{p}_2)] + i\epsilon\}^{-1} \; , \tag{2.34d}$$

we obtain a transport equation of the VUU type [29], i.e.

$$\left\{\frac{\partial}{\partial t} + \{\frac{\mathbf{p}_1}{m} + \frac{\partial}{\partial \mathbf{p}_1} U_m(\mathbf{r},\mathbf{p}_1;t)\} \frac{\partial}{\partial \mathbf{r}} - \frac{\partial}{\partial \mathbf{r}} U_m(\mathbf{r},\mathbf{p}_1;t) \frac{\partial}{\partial \mathbf{p}_1}\right\} f(\mathbf{r},\mathbf{p}_1;t) =$$

$$\frac{4}{(2\pi)^3} \iint d^3p_2 d^3p_3 \int d\Omega \; |v_{12}| \; \frac{d\sigma}{d\Omega}(E;\mathbf{p}_2-\mathbf{p}_4) \cdot \delta(\mathbf{p}_1+\mathbf{p}_2-\mathbf{p}_3-\mathbf{p}_4)$$

$$\cdot \Big\{ f(\mathbf{r},\mathbf{p}_3;t) f(\mathbf{r},\mathbf{p}_4;t)(1-f(\mathbf{r},\mathbf{p}_1;t))(1-f(\mathbf{r},\mathbf{p}_2;t))$$
$$- f(\mathbf{r},\mathbf{p}_1;t) f(\mathbf{r},\mathbf{p}_2;t)(1-f(\mathbf{r},\mathbf{p}_3;t))(1-f(\mathbf{r},\mathbf{p}_4;t)) \Big\} \qquad (2.35)$$

where additionally an expansion of $U_m(\mathbf{r},\mathbf{p}_1;t)$ up to second order around $\mathbf{r}$ has been employed in the Vlasov term (l.h.s. of (2.32)) and the momentum dependence of $U(\mathbf{p})$ and $H(\mathbf{p})$ arising from (2.34d) has been neglected in the collision term. The approximations used in (2.34) are found to hold sufficently well for hard nucleon–nucleon scattering in the case of intermediate energy nucleus–nucleus reactions (cf. section 3).

In eq. (2.35) $|v_{12}|$ is the relative velocity of the colliding nucleons and the time–dependent mean field U_m and the in–medium NN cross section $d\sigma/d\Omega$ are given in terms of the G–matrix,

$$U_m(\mathbf{r},\mathbf{p};t) = 4/(2\pi)^3 \int d^3r' d^3p' \; \mathrm{Re}\left\{G_A(E;\mathbf{r}-\mathbf{r}',\mathbf{p}-\mathbf{p}')\right\} f(\mathbf{r}',\mathbf{p}';t) \qquad (2.36)$$

$$\frac{d\sigma}{d\Omega}(E;\mathbf{q}) = \frac{m^2}{16\pi^2} G(E;\mathbf{q}) G_A^{\dagger}(E;\mathbf{q}) \qquad (2.37)$$

where $f(\mathbf{r},\mathbf{p};t)$ denotes the local phase–space density averaged over spin and isospin and $\mathbf{q}$ the momentum transfer in the nucleon–nucleon collision. In eq. (2.36) and (2.37) the suffix A denotes the antisymmetrized matrix element of G. Whereas early studies [1] have used independent parametrizations for the mean field and $d\sigma/d\Omega$, the approach (2.35) with (2.36) and (2.37) is free from such ambiguities and avoids double counting of interactions.

2.3 Parametrizations of the G–Matrix at Low Energies

The selfconsistent solution of the one–body phase–space distribution via the VUU equation (2.35) and the iterative equation (2.23) for the G–matrix is presently out of scope even for the fastest computers since it involves the solution of a G–matrix equation (2.23) for each configuration in phase space. Recent efforts to solve the G–matrix equation for specific configurations appearing in the course of heavy–ion collisions have shed some light on in–medium effects on the NN cross section in the collision term [13]. However, in these attempts the mean–field part has not been related to the G–matrix. On the other hand, medium polarizations for the propagator (2.24) have not been evaluated so far for the relevant phase–space configurations nor have the hole–hole interaction terms. Some rather reliable information about the G–matrix is available for nuclear matter at finite temperature

[35–39]. There one finds that the explicit change of G with temperature or, for heavy–ion collisions, with moderate relative momentum is not very pronounced so that one may use a more universal parametrization of G fitted to nuclear groundstate properties as a convenient approximation for actual computations and dynamical simulations.

A simple parametrization of G for the groundstate configuration (neglecting the explicit energy dependence) has been proposed in [29], i.e.

$$\mathrm{Re}\left\{G(1,2)\right\} = t_0(1+P_\sigma)\delta(\mathbf{r}_1-\mathbf{r}_2) + t_3(1+P_\sigma)\delta(\mathbf{r}_1-\mathbf{r}_2)\rho((\mathbf{r}_1+\mathbf{r}_2)/2)^{1/3}$$
$$+ V_0 \frac{\exp(-\mu|\mathbf{r}_1-\mathbf{r}_2|)}{\mu|\mathbf{r}_1-\mathbf{r}_2|} + \delta_{pp}\frac{e^2}{|\mathbf{r}_1-\mathbf{r}_2|} \tag{2.38}$$

where P_σ denotes the spin exchange operator and δ_{pp} indicates that the Coulomb interaction is restricted to protons only. The four parameters are given by

$$t_0 = -1124 \text{ MeV fm}^3,\ t_3 = 2037 \text{ MeV fm}^4,\ V_0 = -378 \text{ MeV},\ \mu = 2.175 \text{ fm}^{-1}, \tag{2.39}$$

and are close to a previous Skyrme–type local approximation [6]. For infinite nuclear matter it yields a compressibility of K = 238 MeV, saturation at $\rho = \rho_0 = 0.17$ fm^{-3} and a binding energy of 16 MeV per nucleon. Furthermore, binding energies for closed–shell finite nuclei are reproduced within a few MeV [19]. The selfconsistent potential U_m for such nuclei is given by

$$U_m(\mathbf{r}) = \frac{3}{4}t_0\,\rho(\mathbf{r}) + \frac{7}{8}t_3\,\rho(\mathbf{r})^{4/3} + V_0\int d^3r'\,\frac{\exp(-\mu|\mathbf{r}-\mathbf{r}'|)}{\mu|\mathbf{r}-\mathbf{r}'|}\cdot\rho(\mathbf{r}') \tag{2.40}$$

where the Coulomb potential has to be added in case of protons and the exchange term of the finite range interaction has been dropped.

An important feature of the interaction (2.38) with (2.39) is that for $\rho=\rho_0$ only the finite range Yukawa term (except for Coulomb) survives so that at this density the effective interaction can be interpreted as being generated by scalar meson exchange only. Note, that for configurations close to the groundstate Im{G} is small due to large Pauli–blocking so that $GG^\dagger$ can be approximated by Re(G)Re(G). Therefore, the in–medium NN cross section (2.37) reads for p–n collisions at $\rho=\rho_0$,

$$\frac{d\sigma}{d\Omega}(q) \approx \frac{m^2}{16\pi^2}\left[\frac{4\pi V_0}{\mu(q^2+\mu^2)}\right]^2 . \tag{2.41}$$

The momentum transfer **q** and the angle θ are related by

$$q = p^{rel}\sin\frac{\theta}{2} \tag{2.42}$$

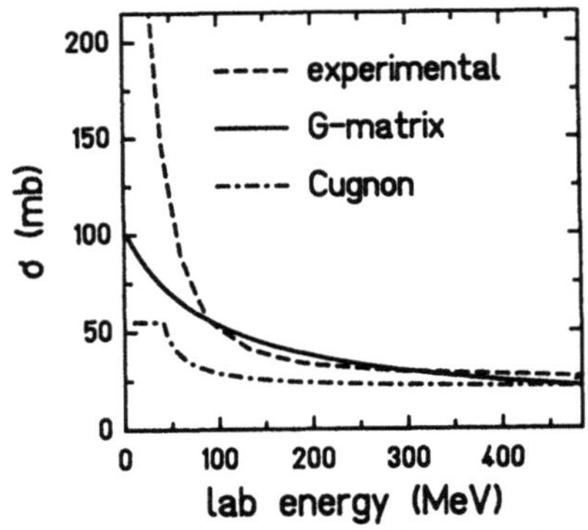

Fig. 2.1: Comparison of the effective in–medium cross section (2.41) (full line) with the corresponding free nucleon–nucleon cross section from Chen et al. [40] (dashed line) averaged over protons and neutrons and the Cugnon parametrization (dash–dotted line) used in cascade simulations so far [41].

where p^{rel} denotes the relative momentum of two nucleons. The elastic cross section averaged over isospin and integrated over $d\Omega$ is displayed in Fig. 2.1 (full line) for $\rho=\rho_0$ as a function of the laboratory energy of the nucleons in comparison with the free cross section from Chen et al. [40] (dashed line) averaged over isospin, too. This in–medium cross section is substantially smaller than the free one at low energy; however, it reaches the free value above 80 MeV. We, furthermore, show in Fig. 2.1 the parametrization of the NN cross section from Cugnon [41] used commonly in cascade or VUU simulations which is substantially smaller than the result of (2.41) since it has been fitted to pp elastic scattering above 45 MeV bombarding energy. Note, that the free np cross section is approximately three times the pp cross section at low energies up to 100 MeV so that the isospin averaged value is about twice the value of the Cugnon parametrization. Although the cross section (2.41) has been derived only for $\rho=\rho_0$, in the actual numerical simulations we will use eq. (2.41) as the in–medium NN cross section for all densities since at bombarding energies below 100 MeV/u the average density in a heavy–ion reaction remains close to ρ_0. For parametrizations of the G–matrix at higher energies as well as the treatment of inelastic NN scattering we refer the reader to refs. [1,4,5,43].

2.4 Numerical Techniques

We solve the VUU equation (2.35) by means of the testparticle method as proposed by Wong [44]. In this respect we use N testparticles per nucleon which propagate classically between collisions according to Hamilton's equation

$$\dot{\mathbf{p}}_i = -\nabla_{\mathbf{r}_i} U_m(\mathbf{r}), \qquad \dot{\mathbf{r}}_i = \frac{\mathbf{p}_i}{\sqrt{m^2+\mathbf{p}_i^2}} \tag{2.43}$$

where $\mathbf{r}_i$ and $\mathbf{p}_i$ are the coordinate and momentum of the individual testparticles. We, furthermore, introduce a Gaussian smearing for testparticles to obtain a smooth density distribution in coordinate space which we evaluate for fixed grid points $\mathbf{r}_g$, i.e.

$$\rho(\mathbf{r}_g) = \frac{1}{N} \sum_{i=1}^{N\cdot A} \frac{1}{(2\pi\Delta^2)^{3/2}} \exp\left[\frac{-(\mathbf{r}_g-\mathbf{r}_i)^2}{2\Delta^2}\right] \tag{2.44}$$

where N denotes the number of testparticles per nucleon, $\mathbf{r}_i$ is the coordinate of the individual testparticles and Δ is the width of the Gaussian, respectively. Since the Gaussian smearing has no physical meaning, the width Δ is only determined by the distance between grid points. We use a cubic grid of size $l = \Delta = 1$ fm in the numerical simulation for $\rho(\mathbf{r})$. The same technique is applied with respect to $f(\mathbf{r},\mathbf{p};t)$ for the Pauli–blocking factors in (2.35) which we also evaluate on the same grid in coordinate space. For each point $\mathbf{r}$ on the grid we choose a cube of sidelength 3 fm around $\mathbf{r}$ and a sphere of radius 0.45 fm^{-1} around $\mathbf{p}$ which corresponds to a phase–space volume of $\Delta V \approx 10.3$. In case of spin and isospin symmetric nuclear matter this phase space is completely filled with $4\Delta V/(2\pi)^3 \approx 1/6$ nucleons, i.e. with N/6 testparticles. The phase–space distribution $f(\mathbf{r},\mathbf{p};t)$ then is evaluated by

$$f(\mathbf{r},\mathbf{p};t) = \frac{6}{N}\cdot\frac{1}{N}\sum_{\mathbf{r}_g}\int d^3p \sum_{j=1}^{N\cdot A} \Theta(q_0-|\mathbf{p}-\mathbf{p}_j|)\,(2\pi\Delta^2)^{-3/2} \exp\frac{-(\mathbf{r}_g-\mathbf{r}_j)^2}{2\Delta^2} \tag{2.45}$$

with $\Theta(x) = \begin{cases} 1 \text{ for } x \geq 0 \\ 0 \text{ else} \end{cases}$

and $q_0 = 0.45\ fm^{-1}$ while the sum over $\mathbf{r}_g$ includes all grid points in a cube of volume 27 fm^3 around $\mathbf{r}$. With this method we achieve smooth density distributions and reliable Pauli–blocking factors $(1-f(\mathbf{r},\mathbf{p}_3;t))(1-f(\mathbf{r},\mathbf{p}_4;t))$ already for N>30 [19].

A second essential ingredient for the transport approach is the initial condition for the VUU equation (2.35). We first distribute the testparticles in coordinate space according to a distribution of Woods–Saxon shape with the radius parameter r_0 = 1.124 fm and surface parameter 'a' which is varied. Next, we determine the average density in each grid–cube. Using the local Thomas–Fermi (LTF) relation, the momenta of the testparticles are randomly distributed up to the local Fermi momentum in each cube. In order to obtain an approximately stationary solution of the Vlasov equation, we vary only the surface parameter of the initial Woods Saxon density distribution. For the optimum stability of the groundstate a suitable parametrization is given by $a = 0.024\ A^{1/3} + 0.29$ [fm] that is used in the numerical simulations.

With this procedure the stability of groundstates is quite good for all masses above A = 12 and binding energies are reproduced with the same quality as in HF calculations [45] for spherical nuclei [19]. Energy conservation for the groundstates, when propagated in time by (2.35) with timestep–size Δt = 0.5 fm/c, is quite accurately satisfied. We find for ^{40}Ca (with only 30 testparticles per nucleon), that a nucleus at rest gains only about $\Delta\epsilon \approx 0.2$ MeV/nucleon over a time period of

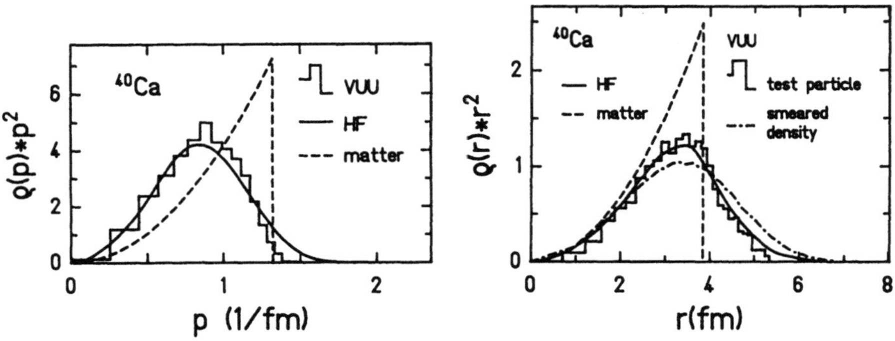

Fig. 2.2: Coordinate–space distribution $\rho(r)\cdot r^2$ and momentum distribution $\rho(p)\cdot p^2$ of the groundstate of ^{40}Ca obtained by the VUU simulation (histograms), a HF calculation (full line) and in the limit of infinite nuclear matter (dashed line). The dash–dotted line shows the density distribution after smearing of the testparticles as described in the text.

100 fm/c as compared to 0.8 MeV/nucleon in [1]. This value reduces to less than 0.1 MeV/nucleon for N>100. Furthermore, in nucleus–nucleus collisions, e.g. ^{92}Mo + ^{92}Mo at 19.5 MeV/nucleon, the deviation of the total energy is about 75 MeV (0.4 MeV/nucleon) for a time period of 400 fm/c in case of N=30 [4]. We can thus apply our VUU simulation to relatively low energy (< 20 MeV/u) heavy–ion reactions where the final excitation energy of the fragments is only a few MeV/nucleon but still large as compared to the numerical uncertainty $\Delta\epsilon$.

In Fig. 2.2 we show the density distribution in coordinate space $\rho(r)\cdot r^2$ for the groundstate of ^{40}Ca within the testparticle simulation (histogram) in comparison with a HF calculation (full line) and the conventional initialization by Bertsch and Das Gupta [1]. The smooth dash–dotted line is obtained after smearing out the testparticle distribution according to (2.44).

Another important quantity is the initial momentum distribution for the groundstates which affects very much the yield of energetic particle production in heavy–ion collisions, especially in the subthreshold region [46,47]. In Fig. 2.2 we additionally compare the momentum distribution for the groundstate of ^{40}Ca obtained by the VUU simulation (histogram) with that from a HF calculation (full line) adopting the same effective interaction (2.38) and the limit of infinite nuclear matter (dashed line). Since $\rho(p)\cdot p^2$ is shown in Fig. 2.2 one can directly extract the number of nucleons with momentum p available in the initial groundstate configuration. The VUU distribution follows the HF result very well except for the high momentum tail which can hardly be treated in a classical simulation like VUU. On the other hand the result of the nuclear matter limit (dashed line), which corresponds to the initialization quoted in [1], overestimates the number of nucleons in the high momentum region substantially.

In order to simulate nucleon–nucleon collisions we adopt the method of Bertsch and Das Gupta [1] and consider only collisions within the same ensemble. The collision criterion for testparticles within the same ensemble is close to that used in cascade simulations and is based on the minimum distance concept [41] which states that two nucleons can collide only if their distance fulfills the relation

$$|r_1 - r_2|_{min} \leq \left[\frac{\sigma_{NN}(\sqrt{s})}{\pi} \right]^{1/2} . \tag{2.46}$$

Furthermore, a collision actually takes place only if the final state phase–space factor

$$(1-f(r,p_3;t))(1-f(r,p_4;t)) > x \tag{2.47}$$

where x is a random number in the interval between 0 and 1.

A further modification has to be introduced in order to avoid double counting of interactions which happens numerically according to the criterion (2.46) if the testparticles are scattered towards each other after the first collision. To remedy this deficiency, which may yield up to 15 % higher collision numbers, all collisions between two testparticles are rejected if the previous collision partner has been the same. As shown by Niita et al. [19] this collision prescription reproduces quite accurately the collision rates for interpenetrating nuclear–matter problems over a wide range of bombarding energies.

3. Tests for the Validity of Semiclassical Limits

In the course of the derivation of the VUU transport equation (2.35) a number of approximations (2.34) have been introduced which do not necessarily apply in the nuclear physics context and have to be controlled explicitly for heavy–ion reactions in the energy regime of interest. In particular the energy–momentum relation for interacting nucleons is substantially different for quantum and classical systems in the groundstate and we will investigate these questions for nucleus–nucleus collisions in the following in the limit of TDHF, i.e. eq. (2.32) with vanishing collision term.

3.1 The Energy–Momentum Relation

In the time–dependent mean–field limit the one–body density matrix (averaged over spin and isospin) is conveniently expanded in terms of the time–dependent single–particle wavefunctions $\psi_\alpha(r;t)$, i.e.

$$\rho(r',r'';t',t'') = 1/4 \sum_{\alpha\ occ} \psi_\alpha(r',t')\psi^*_\alpha(r'',t'') \tag{3.1}$$

where $\psi_\alpha(r;t)$ follow from integrating the TDHF equations in time [48] with the in–medium interaction (2.38). Its Wigner transform is obtained by Fourier transformation with respect to $s = r'-r''$ and $\tau = t'-t''$,

$$f(\mathbf{r},\mathbf{k};t,\omega) = \int d\tau \int d^3s \exp\{-i\mathbf{k}\cdot\mathbf{s} + i\omega\tau\}\ \rho(r',r'';t',t'') \tag{3.2}$$

with $\mathbf{r} = (r'+r'')/2$ and $t = (t'+t'')/2$. The quantity (3.2) has the dimension [time] and goes over to the more conventional phase–space distribution $f(\mathbf{r},\mathbf{k};t)$ after integration over ω. Assuming $\hbar = 1$ (as before) we can identify momentum $\mathbf{p} = \hbar\mathbf{k}$ and energy $E = \hbar\omega$ as equivalent arguments of (3.2). When adopting the convention that integrations over energy and momentum carry additional phase–space factors of $(2\pi)^{-1} = (2\pi\hbar)^{-1} = h^{-1}$ (Planck's constant) and $(2\pi)^{-3} = h^{-3}$, respectively, quantities like density $\rho(\mathbf{r},t)$ or momentum distribution $\rho(\mathbf{k},t)$ are obtained from (3.2) by integration over the residual degrees of freedom.

In order to shed some light on the energy–momentum relation in violent heavy–ion collisions the Wigner function $f(\mathbf{r},\mathbf{k};t,E)$ (3.2) has to be studied with respect to energy E explicitly. In principle, the distribution

$$f(\mathbf{k};E) = \lim_{\Delta t\to\infty} (\Delta t)^{-1} \int d^3r \int_{t_0}^{t_0+\Delta t} dt\ f(\mathbf{r},\mathbf{k};t,E) \tag{3.3}$$

is the most favorable candidate to provide information about the relation between energy and momentum but its evaluation with present techniques is too involved numerically (see, however, [49] in the case of one–dimensional problems). We will thus proceed in a different way: we evaluate an average energy distribution $g^0(E)$ in the dinuclear reaction zone, i.e.

$$g^0(E) = (\Delta t\cdot\Delta V)^{-1} \int_{t_0}^{t_0+\Delta t} dt \int_{\Delta V} d^3r\ (2\pi)^{-3} \int d^3k\ f(\mathbf{r},\mathbf{k};E,t)\ , \tag{3.4}$$

where ΔV is a cylindrical volume of about 50 fm^3 around $\mathbf{r}=0$, and compare $g^0(E)$ with an average momentum distribution in the same volume,

$$g^1(\mathbf{k}) = (\Delta t\cdot\Delta V)^{-1} \int_{t_0}^{t_0+\Delta t} dt \int_{\Delta V} d^3r\ (2\pi)^{-1} \int dE\ f(\mathbf{r},\mathbf{k};E,t) \tag{3.5}$$

by assuming a quasi–free energy–momentum relation

$$E(|\mathbf{k}|) = |\mathbf{k}|^2/2m + \overline{U}\ ,$$

$$k(E) = (2m(E-\overline{U}))^{1/2}. \tag{3.6}$$

In eq. (3.7), m denotes the nucleons mass and $\overline{U}$ the average mean field in the dinuclear reaction zone

$$\overline{U} = (\Delta t\cdot\Delta V)^{-1} \int_{t_0}^{t_0+\Delta t} dt \int_{\Delta V} d^3r\ U_m(\mathbf{r};t) \tag{3.7}$$

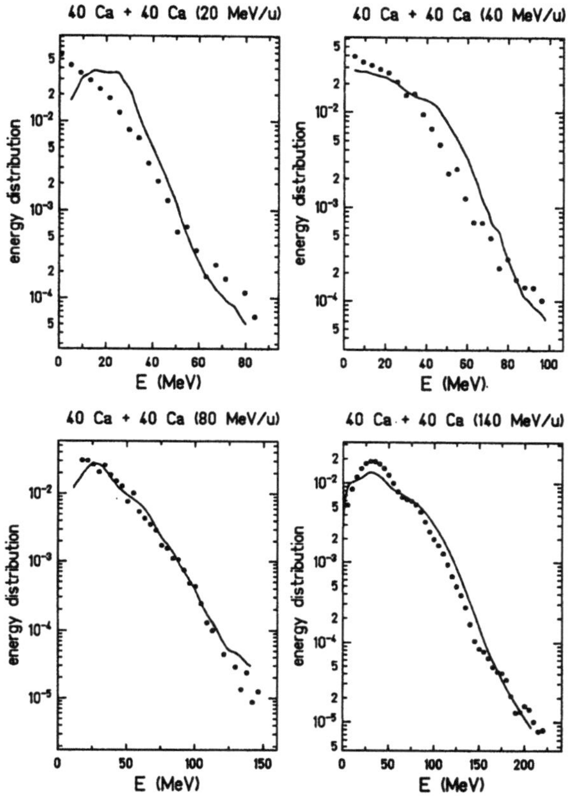

Fig. 3.1: The energy distribution $g^2(E)$ (3.8) (full dots) as determined by the average phase–space distribution $g^1(\mathbf{k})$ (3.5) in the limit of quasi–free particle propagation in comparison with $g^0(E)$ (3.4) (full lines) which is evaluated from the quantal phase–space distribution (3.2) by averaging over space and time and integrating over momentum. The reactions studied are head–on collisions of ^{40}Ca + ^{40}Ca at 20, 40, 80 and 140 MeV/u laboratory energy, respectively.

with ΔV and Δt as used in (3.4) and (3.5), while t_0 is given by the classical contact time of the two nuclei.

Rewriting $g^1(\mathbf{k})$ in terms of spherical coordinates in momentum we determine an energy distribution from $g^1(k,\Omega)$ (3.5) by

$$g^2(E) = m \cdot k/(2\pi)^2 \int d\Omega \; g(k,\Omega) \tag{3.8}$$

in analogy to (3.4) with $\overline{U}$ given by (3.7). The distribution $g^2(E)$ (3.8) is displayed in Fig. 3.1 in terms of the full dots for 20, 40, 80 and 140 MeV/u with $m \approx 938$ MeV/c^2 and compared to $g^0(E)$ from eq. (3.4) (full line). We note that the comparison in Fig. 3.1 does not depend on the time interval Δt for 20 fm/c $< \Delta t <$ 160 fm/c [50]. The apparent oscillations in $g^2(E)$ at higher energies are an artefact of the numerical transformation of $g^1(\mathbf{k})$ from cartesian to spherical coordinates and should be disregarded; however, they yield a measure of the numerical accuracy. The overall agreement of the two distributions within the restricted accuracy for 40, 80 and 140 MeV/u is striking and clearly indicates that the knowledge of the average phase-space distribution (3.5) is sufficient to determine also the average nucleonic energy distribution in the reaction zone over three order of magnitudes. The agreement is somewhat less reliable at 20 MeV/u. While in the higher energy domain for E>30 MeV the two distributions roughly coincide over two orders of magnitude, a significant discrepancy is found for the low energy regime where the free propagation limit (3.6) no longer holds well.

This trend becomes most apparent in the case of a single stationary nucleus where the quantity (3.3) can be evaluated analytically, i.e.

$$f(\mathbf{k},E) = (2\pi)^4 \sum_\alpha |\psi_\alpha(\mathbf{k})|^2 \, \delta(E-\epsilon_\alpha) \tag{3.9}$$

where the summation runs over occupied s.p. states with fixed spin and isospin. In (3.9) $\psi_\alpha(\mathbf{k})$ denote the Fourier transforms of the s.p. wavefunctions and ϵ_α their s.p. energies, respectively. In the limit (3.9) there is no unique energy for a given momentum component $\mathbf{k}$; especially high momentum components $|\mathbf{k}| > k_F \approx 1.36$ fm^{-1} are off-shell. On the other hand, for energies above 140 MeV/u, the free propagation limit is expected to hold when replacing the nonrelativistic energy-momentum relation (3.6) by its relativistic counterpart.

Thus TDHF as a quantal approach to the nuclear many-body problem allows for a simple and transparent picture of single-particle propagation in heavy-ion reactions at intermediate energies. When giving up resolution in space and time, i.e. averaging the quantal phase-space distribution over space and time (which are not experimental observables in heavy-ion collisions), smooth distributions in momentum and energy are obtained which follow closely the classical free relation (3.6) for higher momentum components. Thus quantal s.p. motion in nucleus-nucleus collisions in the one-body limit can approximately be understood in terms of classical mechanics provided that appropriate initial conditions or distributions are given. In contrast to classical mechanics high momentum components $|\mathbf{k}|>k_F$ appear in the quantal distributions which have to be associated with very energetic nucleons in the dinuclear reaction zone. We note that such high momentum components have been seen experimentally in neutron spectra from central heavy-ion collisions [51].

3.2 The Local Momentum Approximation

The problem addressed in the title of this subsection is intimately related to approximations for $\rho(\mathbf{q},\mathbf{q}';t)$, i.e. the double Fourier transform of (3.1) with respect to momentum. The infinite nuclear matter limit may serve as a guide;

$$\rho_\infty(\mathbf{k},\mathbf{k}';t) = (2\pi)^3 \, \delta_3(\mathbf{k}-\mathbf{k}') \, f_0((\mathbf{k}+\mathbf{k}')/2;t). \tag{3.10}$$

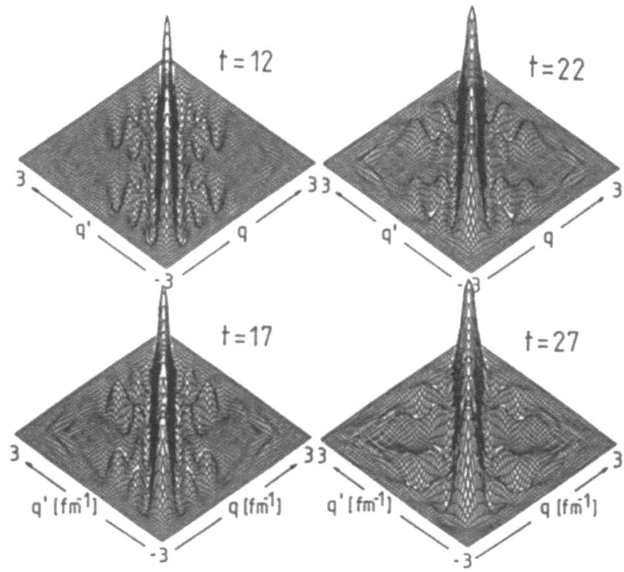

Fig. 3.2: The one–body density matrix $\rho(\mathbf{q},\mathbf{q}';t)$ in momentum space for a central collision of ^{40}Ca+^{40}Ca at 80 MeV/u as evaluated within TDHF. The momentum vectors correspond to $\mathbf{q}$ = (0,0,q) and $\mathbf{q}'$= (0,0,q′) in beam direction. The initial relative distance at t=0 is 14 fm while the time is indicated in units of fm/c.

In equation (3.10) $f_0(\mathbf{k};t)$ has the meaning of a 'local' momentum distribution since the inverse Wigner–transformation leads to

$$f(\mathbf{S},\mathbf{k};t) = \int d^3Q/(2\pi)^3 \exp(i\mathbf{Q}\cdot\mathbf{S})\, \rho_{\infty}(\mathbf{k}+\mathbf{Q}/2,\mathbf{k}-\mathbf{Q}/2;t) = f_0(\mathbf{k};t). \qquad (3.11)$$

In order to explore the properties of the density matrix in momentum space three–dimensional TDHF calculations have been performed for central collisions of ^{40}Ca + ^{40}Ca at various laboratory energies [28]. A typical picture is shown in Fig. 3.2 for ^{40}Ca + ^{40}Ca at 80 MeV/u at four different times in units of fm/c. The collision was assumed to proceed along the z–axis and $\rho(\mathbf{q},\mathbf{q}';t)$ is displayed in an isometric representation for $\mathbf{q}$=(0,0,q) and $\mathbf{q}'$=(0,0,q′). The density matrix is found to show long range oscillations in q and q′, however, it peaks considerably for q=q′ with a width between 0.15 to 0.3 fm^{-1}. This width is small compared to the momentum transfer in a hard nucleon–nucleon scattering process; we may thus adopt the diagonal approximation (2.34a) for the density matrix used in the derivation of the VUU equation (2.35).

3.3 The On–Shell Scattering Approximation

The effect of off–shell collisions is most easily discussed within the limit (2.7a) for the two–body correlation function c_2 which corresponds to the Born–approximation for nucleon–nucleon collisions. In this particular case the collision term (2.30) in momentum representation (using (3.10)) reads [28]

$$I_2(\mathbf{k};t) = g/(2\pi)^6 \iiint d^3q_2 d^3q_1' d^3q_2'$$
$$\cdot\, 2 \int_{-\infty}^{t} dt' \cos(1/(2m)(k^2+q_2^2-q_1^{2\prime}-q_2^{2\prime})(t-t'))$$
$$\cdot\, \delta_3(\mathbf{k}+\mathbf{q}_2-\mathbf{q}_1'-\mathbf{q}_2')\, v(\mathbf{q}_2-\mathbf{q}_2')\{v(\mathbf{q}_2'-\mathbf{q}_2)-v(\mathbf{q}_1'-\mathbf{q}_2)/g\}$$
$$\cdot\, \{f_0(\mathbf{q}_1';t')f_0(\mathbf{q}_2';t')\bar{f}(\mathbf{k};t')\bar{f}(\mathbf{q}_2;t') - f_0(\mathbf{k};t')f_0(\mathbf{q}_2;t')\bar{f}(\mathbf{q}_1';t')\bar{f}(\mathbf{q}_2';t')\} \tag{3.12}$$

with $\bar{f} = 1-f_0$ and spin–isospin degeneracy g=4. It is already of the Uehling–Uhlenbeck type apart from the additional time integration and second order terms in v instead of $d\sigma/d\Omega$. Using (2.37) with G replaced by v, i.e.

$$v(\mathbf{q}_2-\mathbf{q}_2')v_A^{\dagger}(\mathbf{q}_2-\mathbf{q}_2') = v(\mathbf{q}_2-\mathbf{q}_2')\{v(\mathbf{q}_2'-\mathbf{q}_2)-v(\mathbf{q}_1'-\mathbf{q}_2)/g\}$$
$$= 16\pi^2/m^2\, d\sigma/d\Omega(\mathbf{k}-\mathbf{q}_2,\mathbf{q}_1'-\mathbf{q}_2') \tag{3.13}$$

with $\mathbf{k}=\mathbf{q}_1'+\mathbf{q}_2'-\mathbf{q}_2$ due to momentum conservation, we obtain

$$I_2(\mathbf{k};t) = g/(4\pi^4m^2) \iiint d^3q_2 d^3q_1' d^3q_2'$$
$$\cdot\, 2 \int_{-\infty}^{t} dt' \cos\{1/2m(k^2+q_2^2-q_1^{2\prime}-q_2^{2\prime})(t-t')\}$$
$$\delta_3(\mathbf{k}+\mathbf{q}_2-\mathbf{q}_1'-\mathbf{q}_2')\, d\sigma/d\Omega(\mathbf{k}-\mathbf{q}_2,\mathbf{q}_1'-\mathbf{q}_2')$$
$$\{f_0(\mathbf{q}_1';t')f_0(\mathbf{q}_2';t')\bar{f}(\mathbf{k};t')\bar{f}(\mathbf{q}_2;t') - f_0(\mathbf{k};t')f_0(\mathbf{q}_2;t')\bar{f}(\mathbf{q}_1';t')\bar{f}(\mathbf{q}_2';t')\}. \tag{3.14}$$

The next step in approaching the VUU limit for the collision term is to assume that $f_0(\mathbf{k};t')$ is a smooth function of time and that it can be replaced by $f_0(\mathbf{k};t)$ in (3.14). The integral over dt′ then gives

$$I_2(\mathbf{k};t) = g/(2\pi^3m^2) \iiint d^3q_2 d^3q_1' d^3q_2'\; d\sigma/d\Omega(\mathbf{k}-\mathbf{q}_2,\mathbf{q}_1'-\mathbf{q}_2')$$
$$\delta(\mathbf{k}+\mathbf{q}_2-\mathbf{q}_1'-\mathbf{q}_2')\; \delta((k^2+q_2^2-q_1^{2\prime}-q_2^{2\prime})/2m)$$
$$\cdot\, \{f_0(\mathbf{q}_1';t)f_0(\mathbf{q}_2';t)\bar{f}(\mathbf{k};t)\bar{f}(\mathbf{q}_2;t) - f_0(\mathbf{k};t)f_0(\mathbf{q}_2;t)\bar{f}(\mathbf{q}_1';t)\bar{f}(\mathbf{q}_2';t)\} \tag{3.15}$$

which is identical to the r.h.s. of (2.35) (with g=4) except for the fact that the quantum mechanical on–shell version involves only the momentum distribution $f_0(\mathbf{k};t)$ and not the Wigner function itself because it holds strictly only in the infinite nuclear matter limit (3.10).

The assumption of an approximately frozen distribution $f_0(\mathbf{k};t')$, however, does not hold for colliding nuclei in the reaction zone; the on–shell scattering limit (3.15) is therefore expected to be a rather poor approximation. Indeed, Danielewicz [52]

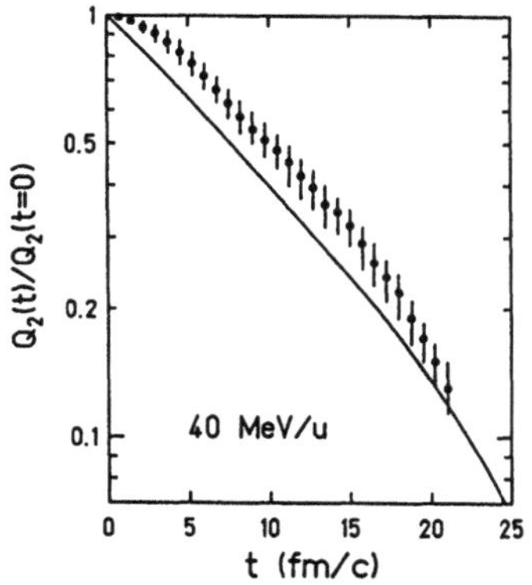

Fig. 3.3: The quadrupole moment in momentum space $Q_2(t)$ (3.16) as obtained from solutions of (3.14) (full dots) involving off-shell scattering in comparison with the result from the on-shell limit (3.15). The vertical error bars show the range of values obtained for $Q_2(t)$ due to different sets of random momenta used in the Monte Carlo integration of (3.14).

has shown in the relativistic energy regime that off-shell scattering might lead to a significant retardation for the nuclear approach to equilibrium. In order to obtain a clear picture on the validity of (3.15) we compare directly the solutions of (3.14) with those of (3.15) [28].

The actual comparison is performed for the quadrupole moments of the average phase-space distributions

$$Q_2(t) = \int d^3\mathbf{k}\ (2k_z^2 - k_x^2 - k_y^2)\ f_0(\mathbf{k};t) \tag{3.16}$$

and displayed in Fig. 3.3 for a heavy-ion reaction at 40 MeV/u. The solution corresponding to the on-shell collision limit (3.15) is given by the full line while the off-shell result from (3.14) is shown by the full dots. The vertical error bars show the range of values obtained for $Q_2(t)$ due to different sets of random momenta used in the Monte Carlo integration for (3.14). Similar to the relativistic case [52] we obtain a retardation for equilibration by off-shell scattering which, however, is only a minor effect on the 15% to 20% level at intermediate energies. The same picture arises at all bombarding energies from 20 MeV/u to 140 MeV/u indicating that memory effects play a minor role with respect to equilibration though $f_0(\mathbf{k};t)$ is a rapidly changing function of time.

We finally note that the on-shell expression (3.15) can be further simplified by integrating over the δ-functions, i.e.

$$\partial/\partial t\ f_0(\mathbf{k};t) = I_2(\mathbf{k};t) = g/(2\pi)^3 \int d^3q_2 \int d\Omega\ v_{12}\ d\sigma/d\Omega(\mathbf{k}-\mathbf{q}_2,\mathbf{q}_1'-\mathbf{q}_2')$$
$$\{f_0(\mathbf{q}_1';t)f_0(\mathbf{q}_2';t)\bar{f}(\mathbf{k};t)\bar{f}(\mathbf{q}_2;t) - f_0(\mathbf{k};t)f_0(\mathbf{q}_2;t)\bar{f}(\mathbf{q}_1';t)\bar{f}(\mathbf{q}_2';t)\} \tag{3.17}$$

with $v_{12} = |\mathbf{k}-\mathbf{q}_2|/m$, which is identical to the VUU collision term in (2.35) when replacing $f_0(\mathbf{k};t)$ by $f(\mathbf{r},\mathbf{k};t)$ in the local density approximation.

3.4 Quantum versus Classical One–Body Dynamics

In this subsection we proceed with dynamical tests of the VUU simulations and show the time evolution of the phase–space distribution

$$f(z,k;t) = \int d^2r_\perp \int d^2k_\perp / 4\pi^2 \; f(\mathbf{r},\mathbf{k};t) \tag{3.18}$$

for a central collision of ^{40}Ca + ^{40}Ca at 40 MeV/u (Fig. 3.4). Only the early stage of the reaction is presented since this region is essential for hard photon and meson production at this energy [20]. The left row is the result of a TDHF calculation with the effective interaction (2.38), while the middle and the right row are the results of the numerical simulation of the Vlasov–equation (without collision term) and full VUU equation (2.35), respectively. The results of the TDHF and the Vlasov equation, which corresponds to TDHF in the classical limit, show an excellent agreement. We find in both cases a considerable open phase space at the middle which is very important for the final blocking in case of particle production [21,46]. On the other hand, this open phase space is rapidly filled by NN collisions (see VUU result). This is one of the reasons why hard photon yields in heavy–ion reactions from secondary and further NN collisions are strongly suppressed. We note that similar results have been obtained even for the low bombarding energy of 20 MeV/u [53].

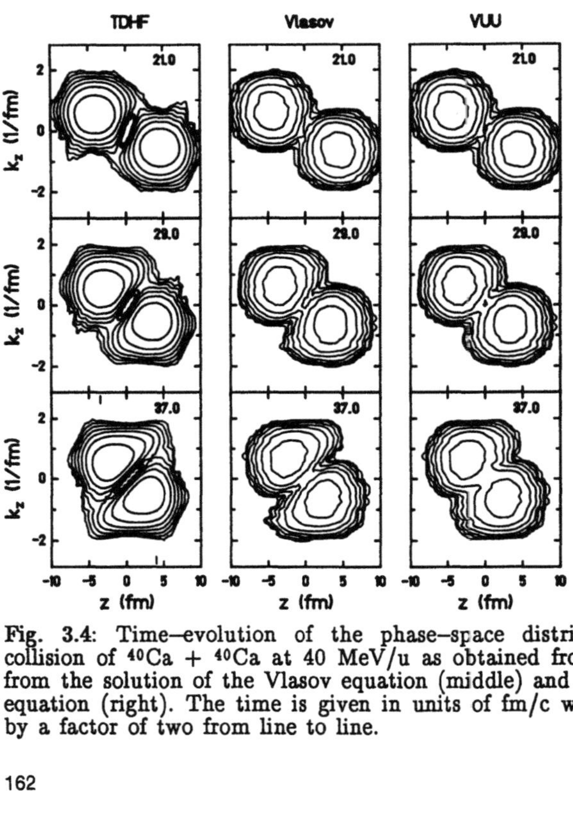

Fig. 3.4: Time–evolution of the phase–space distribution f(z,k;t) for a central collision of ^{40}Ca + ^{40}Ca at 40 MeV/u as obtained from a TDHF calculation (left), from the solution of the Vlasov equation (middle) and from the solution of the VUU equation (right). The time is given in units of fm/c while the contour lines increase by a factor of two from line to line.

Whereas the numerical treatment of the VUU equation is found in Fig. 3.4 to yield quite reasonable approximations to the full solution in the one–body limit, this does not necessarily apply to the solution of the collision term. However, as shown by Niita et al. [19], the nucleon–nucleon collision rate is also reasonably well reproduced for infinite–nuclear–matter problems where (2.35) can be solved directly (cf. section 3.3). We thus conclude that our testparticle simulation is close to the exact solution of the VUU equation (2.35) which, on the other hand, is well founded in nonperturbative nuclear many–body theory (cf. section 2) within reasonable approximations.

4. Production of Hard Photons

The dynamical evolution of heavy–ion reactions can not directly be controlled since, experimentally, we have no resolution in coordinate space and time and, especially, in phase space. Average properties like deflection angles or 'sidewards flow' provide only integral information about the reaction dynamics. The same holds for velocity or rapidity distributions which show some sensitivity to the nuclear stopping power [11]. In order to obtain a more precise clock of the reaction dynamics we need probes that are sensitive to the early stages of the reaction and not affected by the final (equilibrium) expansion phase. Such a clock may be provided by the emission of particles, that are not available as real particles in nuclei and that are strongly suppressed in the final stage of the reaction.

4.1 General Expressions for Production Cross Sections

One of the most promising candidates for such probes are hard photons since they couple perturbatively to the nuclear current, are emitted with the velocity of light and experience little rescattering due to the weak electromagnetic interaction with protons. Since photons couple to the electromagnetic current they are produced by accelerated or decelerated charges. It has been a longstanding suggestion that γ–rays might be good signals for collective nuclear deceleration [54,55]. Such a coherent bremsstrahlung yield should be proportional to the product of charges $Z_1 \cdot Z_2$ available in a central heavy–ion collision and thus be a unique signal for collective motion in these reactions. However, it was since found that collective bremsstrah– lung plays a minor role in the heavy–ion case for bombarding energies below 150 MeV/u [56] and that the radiation from nucleon–nucleon collisions clearly dominates [16]. Similar results have been also obtained for meson production where cooperative mechanisms [57] apparently play a minor role [4,5].

In the case of hard photon production mainly proton–neutron (pn) collisions have to be considered since their radiation is of dipole type and does not suffer from the destructive interference of the radiation amplitudes from two equal charges. The

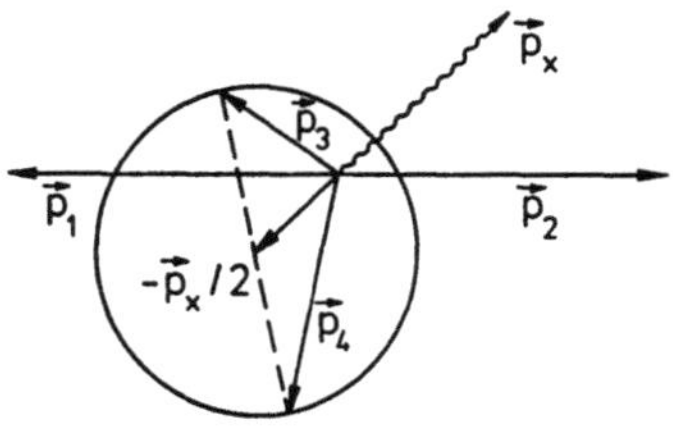

Fig. 4.1: Illustration of NN→NN+X kinematics in momentum space. The momenta of the colliding nucleons are denoted by p_1 and p_2, the final nucleon momenta by p_3 and p_4, respectively, while the momentum of the produced photon or meson is given by p_x.

latter is of quadrupole type at the lower bombarding energies and reduced by more than an order of magnitude [58]. The question remains whether the radiation from different pn collisions should be added coherently or incoherently. Detailed studies by Heuer et al. [16] for the energy–regime up to about 100 MeV/u have shown no significant effect of coherence for γ–radiation with energies above about 30 MeV. This implies that the total radiation in a heavy–ion collision can be obtained by incoherently summing over the radiation from the individual pn collisions. This result is due to the fact that the photon amplitudes from primary and secondary collisions are of different magnitudes and thus their interference becomes very small.

Within the picture of hard photon or meson production by incoherent on–shell nucleon–nucleon collisions the differential particle multiplicity emerging from a heavy–ion reaction at fixed impact parameter b is given by the sum over the individual probabilities of all possible nucleon–nucleon collisions with the Pauli–blocking for the final nucleon states taken into account:

$$E_x \frac{d^3N(b)}{d^3k_x} = \sum_{NN\ coll} \int \frac{d\Omega}{4\pi} E'_x \frac{d^3P_x(\sqrt{s})}{d^3k'_x} [1-f(r,p_3;t)]\cdot[1-f(r,p_4;t)] \ . \tag{4.1}$$

In this equation r and t indicate the space–time coordinates and $\sqrt{s}$ the invariant energy of each collision. The primes denote quantities in the individual NN–cms (center-of-mass system) which have to be transformed into the laboratory frame or the midrapidity frame, respectively. Ω denotes the solid angle of the relative momentum p_3-p_4 which is not fixed by energy and momentum conservation and has to be integrated over. The kinematical situation is illustrated in Fig. 4.1 where the index x stands for a photon (γ), pion (π), eta–meson (η) or kaon (K), respectively.

The invariant production probability per nucleon–nucleon collision $P_x(\sqrt{s})$ is given by the ratio of the invariant differential production cross section for the particle x and the total nucleon–nucleon cross section in the medium, i.e.

$$E_x \frac{d^3P_x(\sqrt{s})}{d^3k_x} \approx E_x \frac{d^3\sigma_x(\sqrt{s})}{d^3k_x} / \sigma_{NN}(\sqrt{s}) \ . \tag{4.2}$$

We note that due to energy and momentum conservation in each process NN $\to$ NN+x we treat all production vertices on–shell and thus neglect off–shell transitions which are expected to appear in the nuclear medium. In fact, a collision frequency of $\tau^{-1} \approx (10\mathrm{fm/c})^{-1}$ in the Fermi–energy regime would, according to the uncertainty relation, allow the nucleons to be off–shell by roughly 20 MeV between subsequent collisions. Significant deviations of our results from experimental data might indicate off–shell effects, but also other cooperative mechanisms (see below).

To obtain the double differential particle cross section in a nucleus–nucleus reaction in case of inclusive measurements, we have to integrate over impact parameter, i.e.

$$E_x\, d^3\sigma_x/d^3k_x = \int d^2b\, E_x \frac{d^3N_x(b)}{d^3k_x} . \tag{4.3}$$

Thus differential particle yields from proton or heavy–ion induced reactions can be calculated without introducing adjustable parameters provided that the differential production probability (4.2) is known and that the production process can be treated perturbatively in the heavy–ion reaction. The latter condition should be fulfilled for all 'subthreshold' processes since in this case P_x is very small compared to 1.

4.2 Photon Production

An essential ingredient for an understanding of the photon production yields in heavy–ion collisions is the microscopic nucleon–nucleon–γ cross section. The first temptation is naturally to take this cross section from experiment. In fact, the nucleon–nucleon bremsstrahlung was studied extensively in the late sixties [59–61] with the hope of learning from such studies about the off–shell behavior of the nucleon–nucleon interaction and in this way to be able to distinguish between different nucleon–nucleon potentials that all fitted the elastic scattering data equally well. However, because the γ–energies studied then were still fairly low, no convincing sensitivity of the observed γ–yield on the potentials used could be detected and consequently these studies were given up. We, therefore, have evaluated the differential γ–cross section in a relativistic OBE T–matrix approach in a broad kinematical range [58] based on a realistic T–matrix approximation that describes the elastic nucleon–nucleon cross sections very well [62,63].

For practical applications in VUU calculations we, furthermore, use an analytical fit to the angle–integrated cross sections obtained by Schäfer et al. [58] in the form of

$$\omega \frac{d\sigma}{d\omega} = \sqrt{1-x}\; f(\omega,T_{cm})\; [\mu b]$$

with

$$f(\omega,T_{cm}) = a + b\,\omega \exp(-c(\omega-d)^2);$$

$$x = \frac{\omega}{\omega_{max}}\;;\quad a = 9.74 + 0.037\; T_{cm}\;;$$

$$b = 0.91 - 0.14 \arctan(0.045\, T_{cm} - 7.66);$$
$$c = 12.365/(18.27 + T_{cm})^{2.38};\quad d = 1.5\, T_{cm} \tag{4.4}$$

where T_{cm} is the total kinetic energy of the incoming nucleons in the cm–system given in MeV and ω_{max} the maximum photon energy in an arbitrary NN collision.

4.3 Comparison with Double Differential Nucleus–Nucleus Data

We present our results with the OBE approximation (4.4) for the photon cross section in comparison with experimental data for the ^{86}Kr + ^{12}C reaction at 44 MeV/u for different laboratory angles (Fig. 4.2). Our calculations (solid lines) are in reasonable agreement with the experimental data [64] (squares) in absolute magnitude, spectral slope and angular dependence for γ–energies from 30 to 80 MeV. For the same reaction, we present in Fig. 4.3 the energy integrated cross section $d\sigma_\gamma/d\Omega_\gamma$ for photons above 30 MeV (full squares); the experimental points correspond to ^{86}Kr + ^{12}C (squares), ^{86}Kr + ^{109}Ag (circles) and ^{86}Kr + ^{197}Au (dots), where the last two are scaled to the system ^{86}Kr + ^{12}C by the factors 0.134 and 0.131 as in [64]. The good agreement indicates that the proton–neutron bremsstrahlung mechanism dominates the production of energetic photons in heavy–ion collisions.

A detailed phase–space analysis [4] in fact shows that hard photons are primarily produced by nucleons whose momenta originate from the opposite pole caps of the shifted Fermi distributions. Their final state phase space for low momentum components is only partly blocked in contrast to the infinite nuclear matter limit (cf. Fig. 3.4) in the initial phase of the heavy–ion reaction. Furthermore, the

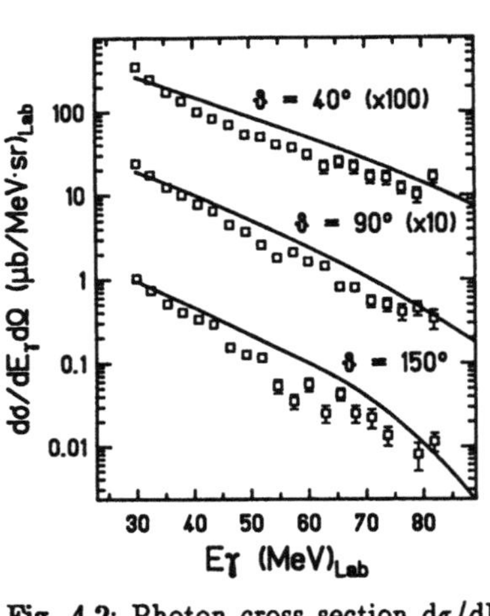

Fig. 4.2: Photon cross section $d\sigma/dEd\Omega$ for the reaction ^{86}Kr + ^{12}C at 44 MeV/u for three laboratory angles. The solid lines are the results of our calculation with the T–matrix approximation from Schäfer et al. [58] (4.4) for the elementary γ–production cross section. The data are taken from Bertholet et al. [64].

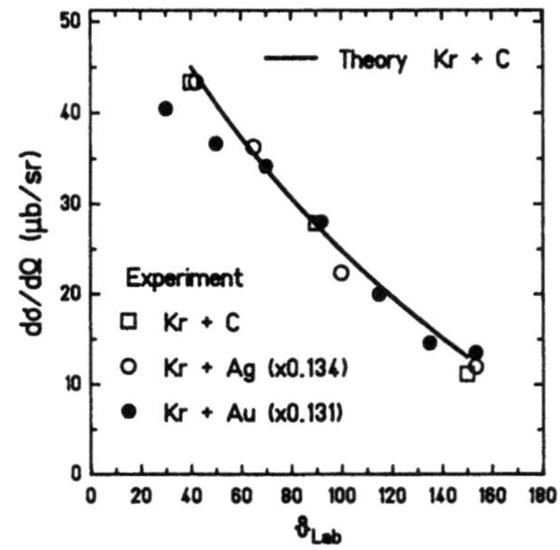

Fig. 4.3: Energy integrated photon cross section $d\sigma/d\omega$ ($\omega \geq 30$ MeV) for reactions Kr + nucleus at 44 MeV/u. Theoretical results (full line) are shown only for ^{86}Kr + ^{12}C. Experimental data are taken from Bertholet et al. [64] and scaled to ^{86}Kr + ^{12}C as indicated in the figure.

source–velocity distribution of the hard photons, which corresponds to the individual nucleon–nucleon c.m.s., is peaked around midrapidity also for asymmetric systems in accordance with experimental observations [4].

Apart from the detailed comparison with double differential photon data it is of further interest to see how the pn bremsstrahlung approach compares with the photon systematics in the literature [65,66]. For this purpose we consider first the scaling of hard photon yields (E_γ>30 MeV) with the mass of projectile and target.

We have performed calculations for various heavy–ion collisions at 40 MeV/u [21] and find that the photon cross section approximately follows a power law of the form

$$\sigma_\gamma(E_\gamma > 30 \text{ MeV}) \sim (A_1 \cdot A_2)^X \qquad (4.5)$$

with x = 0.92 for $A_1 = A_2$ while asymmetric systems scale with a lower value of x close to 0.74. This scaling behavior has to be compared with the semi–empirical expression [65]

$$\sigma_\gamma(E_\gamma > 30 \text{ MeV}) \approx \sigma_R <N_{pn}> P_\gamma \qquad (4.6)$$

where σ_R is the total reaction cross section given by

$$\sigma_R = \pi \cdot (r_0 \cdot (A_P^{1/3} + A_T^{1/3} + 2.0))^2 \cdot (1 - V_c/E) \qquad (4.7)$$

with r_0 = 1.16 fm while V_c denotes the mutual Coulomb barrier. $<N_{pn}>$ is the average number of first pn collisions approximated analytically as

$$<N_{pn}> \approx A_p \frac{5\, A_T^{2/3} - A_P^{2/3}}{5(A_T^{1/3} + A_p^{1/3})^2 \cdot A_P A_T} (Z_P N_T + Z_T N_P) \qquad (4.8)$$

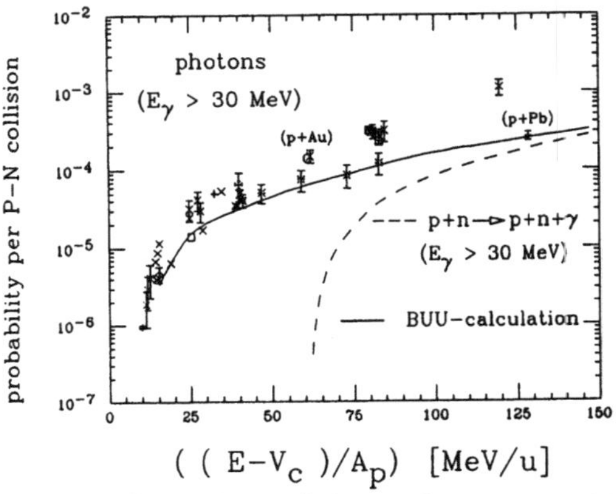

Fig. 4.4: Comparison of the production probability (4.6) for energetic photons (ω > 30 MeV) (solid line) with the experimental data [66] as a function of the bombarding energy per nucleon above the Coulomb barrier. The dashed line shows the result for the elementary process pn→pnγ with no Fermi motion involved.

where A_P, A_T, N_P, N_T, Z_P, and Z_T denote nucleon number, neutron number and proton number of projectile and target, respectively. The scaling law (4.5) compares well with our calculations for C+A. On the other hand, the exponent x = 0.92 from our calculations for A_1= A_2 (as compared to x=0.83 from (4.5)) indicates that there are also volume contributions to the photon yield for heavy symmetric systems or contributions from secondary collisions in addition to first–chance collisions.

In Fig. 4.4 we display our results for $\sigma_\gamma(E_\gamma > 30$ MeV) as a function of bombarding energy in comparison with the experimental results [65,66]. In this case we have evaluated σ_γ (E_γ> 30 MeV) for the system ^{40}Ca + ^{40}Ca and divided the resulting cross section by $\sigma_R \cdot \langle N_{pn} \rangle$ in the same way as the experimental data. We have used the OBE T–matrix approximation (4.4) above 25 MeV/u and the G–matrix approximation (2.38) below 25 MeV/u for the evaluation of the elementary process pn→pnγ in this case, which reproduces the experimental yield reasonably well. A disagreement is seen to exist at the highest energies where we underestimate the cross sections by a factor of 2–3. At present it is not clear whether this discrepancy is due to a failure of the OBE T–matrix model at higher energies for the elementary process pn→pnγ or due to photons originating from π^0 decay which have not properly been subtracted in the experimental analysis. Future experiments will have to clarify the situation.

A further global comparison can be obtained from the slope of the photon spectra measured at 90^0 with respect to the beam axis as a function of bombarding energy which is shown in Fig. 4.5 in comparison with our calculations. Here we again find that the experimental data are well described within our approach. We note, that for the extraction of the experimental slope parameter at 125 MeV/u the photons originating from π^0 decay have been taken into account and subtracted; this is not the case for the data shown in Fig. 4.4.

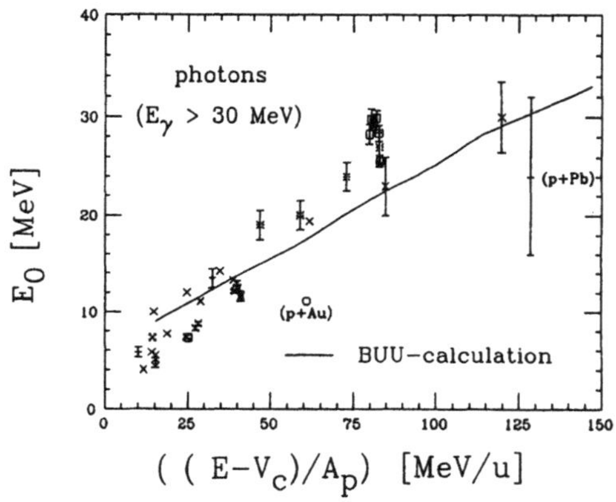

Fig. 4.5: Comparison of the calculated slope parameters E_0 for energetic photons with the experimental values [66] as a function of the bombarding energy per nucleon above the Coulomb barrier.

Summarizing this section, hard photon production up to bombarding energies of about 100 MeV/u can be understood within the picture of incoherent proton–neutron bremsstrahlung within a factor of two which may be the inherent uncertainty arising from different approximations in the transport theory itself (cf. section 3) as well as from ambiguities in the elementary differential production cross section (4.2).

5. Microscopic Analysis of Deeply–Inelastic Collisions

The experimental study of deeply inelastic collisions (DIC) between heavy ions has been a major subject of nuclear physics for about two decades and a variety of models have been developed (for reviews we refer the reader to refs. [67–70]). The only microscopic theory in this respect has been provided by TDHF [6], however, lacking residual nucleon–nucleon collisions. Thus the nuclear stopping power in HIC as well as two–body observables like the variance in mass or charge for the binary reaction products have been severely underestimated in this approximation. As shown in section 2, time–dependent G–matrix theory (TDGMT) is a natural and consistent extension of TDHF since the effective TDHF mean–field theory can be interpreted as a parametrization of Re{G} (2.38) for the effective interaction while assuming Im{G} = 0. One thus would like to know if TDGMT is able to overcome those difficulties encountered with TDHF.

Due to the tremendous numerical complexity of (2.32), however, we will restrict ourselves to the semiclassical limit of TDGMT, i.e. the selfconsistent transport equation (2.35) with $U_m(1)$ and $d\sigma/d\Omega$ given by (2.36) and (2.37), respectively. As shown in section 3, the semiclassical limit breaks down for bombarding energies close to the Coulomb barrier, such that we will consider heavy systems above 14.7

MeV/u where the approximations performed (cf. section 3) are expected to hold reasonably well. We note, that we use the parametrization (2.38) for Re{G}, which is fitted to nuclear groundstate properties, and that the theory is completely determined by the four parameters in (2.39) for Re{G}.

5.1 Time Evolution of the Heavy–Ion Collision

In Fig. 5.1 we show the time evolution of the nucleon density $\rho(x,y=0,z;t)$ in contour lines as a function of time for ^{100}Mo + ^{100}Mo at 14.7 MeV/u and impact parameter b=8 fm. The densities start to overlap at t=40 fm/c and the system reseparates again at t ≈ 440 fm/c into two heavy fragments with strongly distorted surfaces. In the meantime, projectile and target have exchanged particles and transferred relative kinetic energy to surface modes.

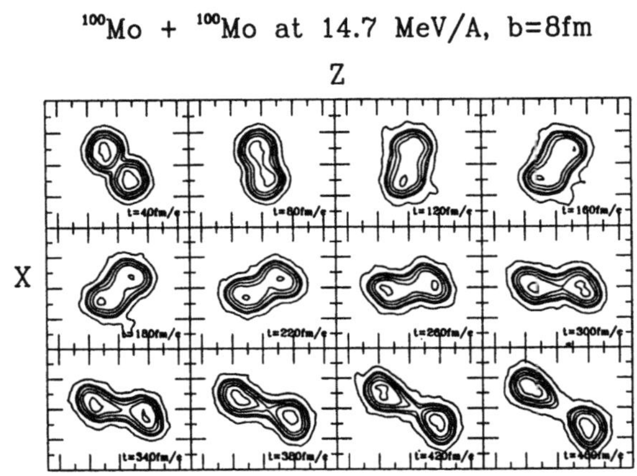

Fig. 5.1: Density distribution $\rho(x,y=0,z;t)$ for ^{100}Mo + ^{100}Mo at 14.7 MeV/u at impact parameter b = 8 fm. The times are given in units of fm/c while the contour lines increase by a factor of two from line to line.

5.2 Damping of Relative Motion

Due to the semiclassical nature of the testparticle simulation we are able to follow the particle trajectories in time and to count how often they have crossed a plane perpendicular (at half distance) to the longitudinal symmetry axis of the dinuclear complex. This determines a distribution $N(b,N_{ex})$ where N_{ex} is the number of exchanges the testparticle has performed. This quantity is shown in Fig. 5.2 for ^{100}Mo + ^{100}Mo at 23.7 MeV/u and b=7 fm and indicates that there is quite a substantial number of particles being exchanged several times. The number of single exchange particles $N^1_{ex}(b)$ is indicated by the hatched area in Fig. 5.2 and given by $N^1_{ex}(b) = N(b,1)$. The total number of exchanges is defined by

$$N^{tot}_{ex}(b) = \sum_{N_{ex}} N_{ex}\, N(b,N_{ex}) \tag{5.1}$$

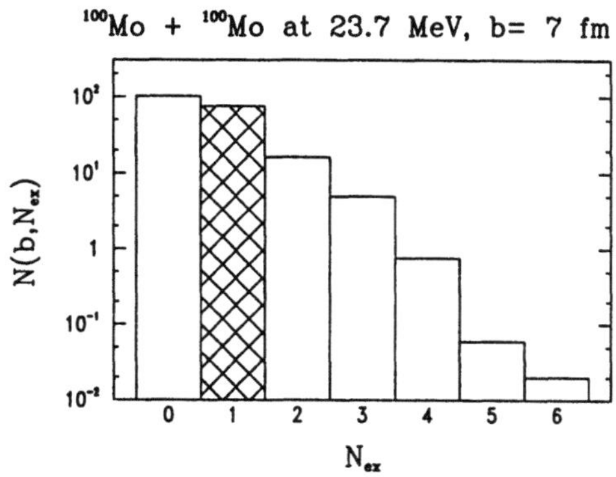

Fig. 5.2: Distribution of particle exchanges versus the number of exchanges for ^{100}Mo + ^{100}Mo at 23.7 MeV/u and b = 7 fm. The hatched area denotes the number of single exchanges N^1_{ex}.

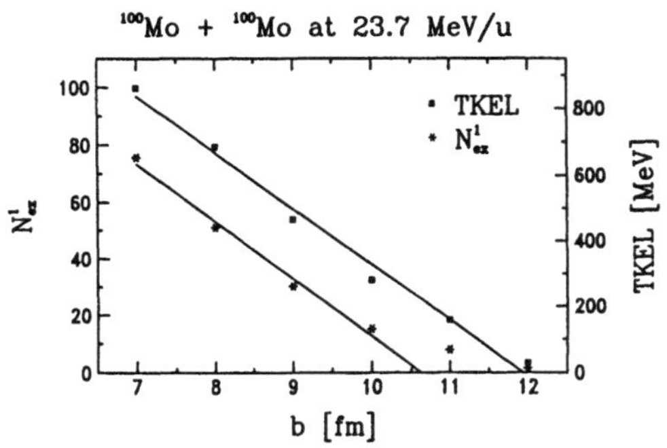

Fig. 5.3: The number of single exchanges and the total kinetic energy loss TKEL as a function of impact parameter b for ^{100}Mo + ^{100}Mo at 23.7 MeV/u.

and related to the fluctuation in mass via

$$\sigma_A(b) = \{N^{tot}_{ex}(b)\}^{1/2} \tag{5.2}$$

assuming σ_A to be entirely determined via the statistical exchange of particles. On the other hand, for $t \to \infty$ – when two separated fragments emerge in the density distribution – we can evaluate their average mass A_f, charge Z_f, momentum p_f, intrinsic energy E^* and intrinsic angular momentum L_f [71]. This allows us to evaluate the final (Coulomb corrected) scattering angle $\vartheta(b)$ as well as the total kinetic energy loss TKEL(b). In Fig. 5.3 we show TKEL(b) and $N^1_{ex}(b)$ as a function of impact parameter b for ^{100}Mo + ^{100}Mo at 23.7 MeV/u. Both quantities are approximately linear functions of b; their scaling indicates that each single exchanged particle contributes with roughly 10 MeV to TKEL. The relative shift of the two curves in b implies that about 200 MeV of TKEL cannot be attributed to particle exchange, especially for 10.5 fm < b < 12 fm. This energy loss is basically due to the excitation of surface modes as put forward by Broglia et al. [72] a long time ago. We note that we find a similar scaling for all systems investigated above 14.7 MeV/u bombardment energy [71].

5.3 Variances in Mass and Energy

Due to a one–to–one relation between impact parameter b and TKEL we can correlate the variance in mass σ_A with TKEL or the total kinetic energy TKE = E_{cm} – TKEL on the basis of pure statistical fluctuations. This correlation is shown in Fig. 5.4 by the full line for ^{100}Mo + ^{100}Mo at 23.7 MeV/u in comparison with the preliminary data from ref. [73]. The agreement is reasonably good up to TKEL ≈ 500 MeV while at higher energy losses we underestimate the data. This leaves room for other mechanisms like random neck decay [74] occuring in more central collision which is expected to enhance the variance in mass.

Since we can extract the average final mass of the remnants $A_f(b)$ for given impact parameter b, $A_f(b)$, and obtain $\sigma_A(b)$ from (5.2), it is straightforward to calculate double differential cross sections $d^2\sigma/dTKE\ dA$ for the systems investigated assuming a gaussian distribution in mass A. In this respect we show the calculated differential cross section in TKE and mass for ^{92}Mo + ^{92}Mo at 19.5 MeV/u in Fig. 5.5 within contour lines increasing by a factor of 2 from line to line. The hatched area indicates the relative shift of the distribution to smaller masses which is due to nonequilibrium particle emission. We note that this shift is very small at 14.7 MeV/u but increases sizably with bombardment energy.

The double differential cross section $d^2\sigma/dTKE\ d\vartheta$ is evaluated according to

$$d^2\sigma/dTKE\ d\vartheta = \pi/(\mu\cdot E_{cm}) \int l\ dl\ \{4\pi^2\sigma^2_{TKE}\ \sigma^2_\vartheta\}^{-1/2}$$
$$\cdot \exp\{-1/2[(TKE-\overline{TKE}(l))^2/\sigma^2_{TKE}(l) + (\vartheta - \overline{\vartheta}(l))^2/\sigma^2_\vartheta(l)]\} \qquad (5.3)$$

where μ denotes the reduced mass of the system, $\overline{TKE}(l)$ and $\overline{\vartheta}(l)$ are the average

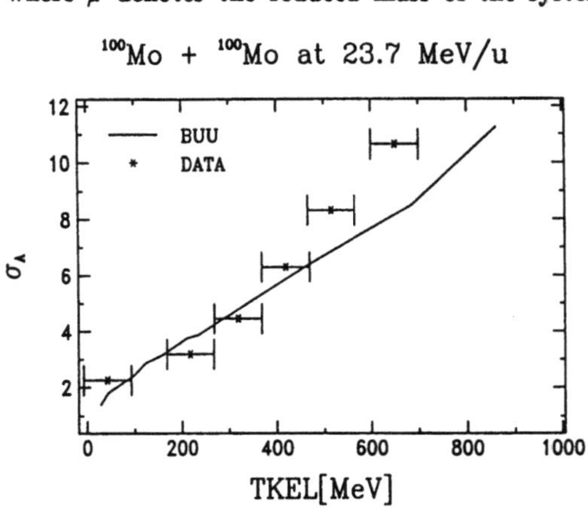

Fig. 5.4: The fluctuation in mass (5.2) versus TKEL (solid line) for ^{100}Mo + ^{100}Mo at 23.7 MeV/u in comparison to the experimental data [73].

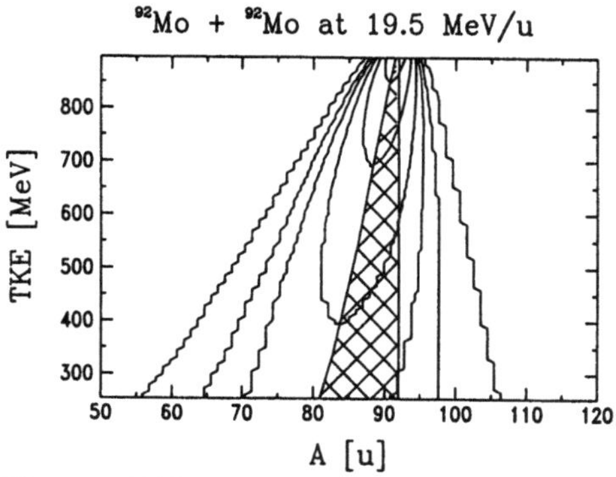

Fig. 5.5: Wilczynski–plot with respect to total kinetic energy TKE and mass A for ^{92}Mo + ^{92}Mo at 19.5 MeV/u. The contour lines increase by a factor of 2 from line to line.

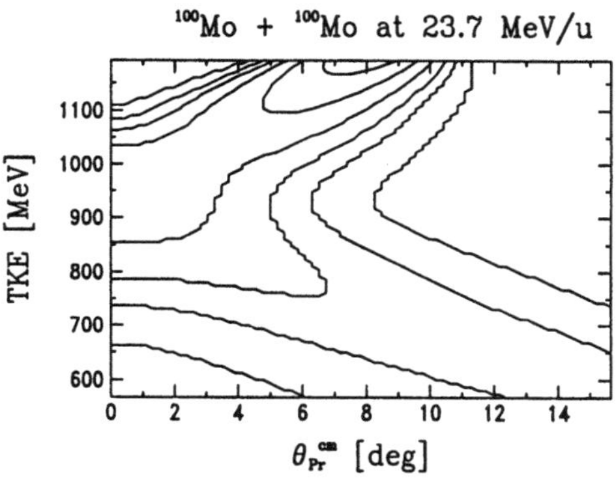

Fig. 5.6: Wilczynski–plot with respect to total kinetic energy (TKE) and scattering angle ϑ for ^{100}Mo + ^{100}Mo at 23.7 MeV/u. The contour lines increase by a factor of 2 from line to line.

total kinetic energy and scattering angle as a funtion of the angular momentum l, respectively, and E_{cm} is the center–of–mass energy. The fluctuations σ^2_{TKE} and σ^2_{ϑ} are determined by

$$\sigma_{TKE}(l) = a_{TKE}\ \sigma_A(l); \quad \sigma_{\vartheta}(l) = a_{\vartheta}\ \sigma_A(l) \tag{5.4}$$

where the coefficients a_i are determined from a linear fit to the functions TKE(A) and ϑ(A) [71]. A resulting Wilczynski–plot for ^{100}Mo + ^{100}Mo at 23.7 MeV/u is shown in Fig. 5.6 in contour lines increasing by a factor of two from line to line, which shows an almost isotropic angular distribution for high energy losses, i.e. TKE < 750 MeV.

We note in passing that the 'sticking limit' for angular–momentum transfer, i.e. $\Delta l/l = 2/5$, is rather well supported by our microscopic studies [71].

Furthermore, we are able to evaluate the nucleon emission as a function of time and to determine the amount of energy carried away by pre–equilibrium particle emission [71]. For the system ^{92}Mo + ^{92}Mo at 19.5 MeV/u this leads to an energy loss of the dinuclear complex which can be correlated to TKEL as

$$E^{em}_{pre}(TKEL) = 0.27\cdot(TKEL - 200\ MeV) \qquad (5.5)$$

for TKEL > 200 MeV while for TKEL < 200 MeV we have $E^{em}_{pre} \approx 0$. This correlation is well in line with experimental observation for this system [75].

Concluding this section we apparently achieve also a quite reasonable description of deeply inelastic heavy–ion collisions within the transport approach described in section 2 that is entirely determined by the four parameters in (2.39) for Re{G}.

6. Summary and Outlook

In this contribution we have presented a nonperturbative approach to the nuclear many–body problem in the nonrelativistic case and derived transport equations of the Vlasov–Uehling–Uhlenbeck type where the real part of the nucleons self–energy and the in–medium nucleon–nucleon cross section are based on the same G–matrix without double counting of interactions. This approach generally provides the framework for the description of intermediate–energy heavy–ion collisions. In fact, the theory requires one to evaluate nonperturbatively an integral equation for each configuration in phase space in order to determine the respective G–matrix while following the nucleon testparticle trajectories in space and time which are subject to random on–shell collision processes. From a fundamental point of view, it is this integral equation that provides the information on the nuclear system at high density and even far from equilibrium. However, it is not clear, if the truncation scheme considered – reaching far beyond actual computations – is sufficient to provide ultimate answers.

We thus have adopted a more pragmatic point of view and parametrized the real part of the G–matrix in terms of Skyrme–type and Yukawa forces for nuclear configurations not too far from the groundstate. The four parameters (2.39) are determined by the saturation condition for nuclear matter and by binding energies of finite nuclei. These parameters fix the nonequilibrium kinetic theory completely as shown in section 2.

The approximations performed in the derivation of the transport theory, i.e. the quasi–particle energy–momentum relation $E(\mathbf{k})$, the local–density approximation and the on–shell scattering assumption, have been found to hold reasonably well for heavy–ion collisions above about 15 MeV/u provided that projectile and target masses are not too low.

We have applied the transport theory to the production of hard photons in heavy–ion collisions at bombarding energies ranging from roughly 15 MeV/u up to 125 MeV/u and shown that presently available data appear to be compatible with a production mechanism via individual on–shell nucleon–nucleon collisions thus giving no evidence for cooperative effects so far. This is a quite important conclusion since it now becomes possible to use these secondary particles as sensitive probes for specific regions of single–particle phase space.

Furthermore, we have investigated deeply–inelastic collisions of Mo+Mo at bombarding energies from 14.7 to 23.7 MeV/u and evaluated double differential cross sections with respect to total kinetic energy (TKE) and mass or scattering angle ϑ. We find that the mechanism for energy transfer is dominated by the nucleon exchange mechanism except for peripheral reactions, where collective surface excitations contribute substantially up to 200 MeV energy losses. On the basis of the particle exchange mechanism we were able to describe the fluctuations in mass as a pure statistical process except for the largest energy losses. In addition, pre–equilibrium particle–decay phenomena appear to be described equally well.

We conclude that transport theories on the basis of the Vlasov–Uehling–Uhlenbeck (VUU) equation should be understood as the semiclassical on–shell limit of time–dependent G–matrix theory which is well founded in nonperturbative nuclear many–body theory. Within simple parametrizations of the G–matrix this transport theory is numerically solvable by the testparticle method and has a high predictive power for many observables in deeply inelastic and intermediate energy heavy–ion reactions [4]. The approach outlined in section 2 is limited in energy due to the nonrelativistic dynamics and the neglect of nucleon inelastic scatterings. For more recent extensions in these directions we refer the reader to refs. [5,43,76].

Acknowledgements: The authors are grateful for many illuminating discussions with V. Metag, A. Gobbi and J. Randrup. The communication on many–body aspects with A. Pfitzner and S.J. Wang has been very fruitful too. Finally, we wish to thank our theoretical colleagues at Giessen, in particular G. Batko, T.S. Biro and M. Schäfer.

References

1) G.F. Bertsch and S. Das Gupta, Phys. Rep. 160, 189 (1988)
2) S. Nagamiya and M. Gyulassy, Adv. Nucl. Phys. 13, 201 (1984)
3) H. Stöcker and W. Greiner, Phys. Rep. 137, 277 (1986)
4) W. Cassing et al., Phys. Rep. 188, 363 (1990)
5) W. Cassing and U. Mosel, Prog. Part. Nucl Phys. 25, 235 (1990)
6) J.W. Negele, Rev. Mod. Phys. 54, 913 (1982)
7) J. Aichelin and G.F. Bertsch, Phys. Rev. C31, 1730 (1985)
8) A. Ainsworth et al., Nucl. Phys.A464, 740 (1987)
9) G.F. Bertsch et al., Phys. Rev. C29, 673 (1984)
10) G.F. Bertsch et al., Phys. Lett. B189, 384 (1987)
11) B. Blättel et al., Phys. Rev. C38, 1767 (1988)
12) D.H. Boal, Phys. Rev. C30, 749 (1984)
13) A. Bohnet et al., Nucl. Phys. A494, 349 (1989)
14) C. Gale et al., Phys. Rev. C35, 1666 (1987)
15) C. Grégoire et al., Nucl. Phys. A465, 317 (1987)
16) R. Heuer et al., Z. Phys. A330, 315 (1988)
17) C.M. Ko et al., Phys. Rev. C31, 2324 (1985)
18) H. Kruse et al., Phys. Rev. Lett. 54, 289 (1985)
19) K. Niita, W. Cassing and U. Mosel, Nucl. Phys. A504, 391 (1989)
20) W. Bauer et al., Phys. Rev. C34, 2127 (1986)

21) T.S. Biro et al., Nucl. Phys. A471, 604 (1987)
22) L.W. Nordheim, Proc. Roy. Soc. (London) A119, 689 (1928)
23) E.A. Uehling and G.E. Uhlenbeck, Phys. Rev. 43, 552 (1933)
24) J.H. Irving and R.W. Zwanzig, J. Chem. Phys. 19, 1173 (1951)
25) L.P. Kadanoff and G. Baym, (1962) London–Amsterdam, Benjamin/Cummings, Quantum Statistical Mechanics
26) W. Botermans and R. Malfliet, Phys. Lett. B171, 22 (1986)
27) W. Botermans and R. Malfliet, Phys. Lett. B215, 617 (1988)
28) W. Cassing, Z. Phys. A327, 447 (1987)
29) W. Cassing, K. Niita and S.J. Wang, Z. Phys. A331, 439 (1988)
30) W. Botermans and R. Malfliet, Phys. Rep. 198 (1990) 115
31) S.J. Wang and W. Cassing, Ann. Phys. 159, 328 (1985)
32) W. Cassing and S.J. Wang, Z. Phys. A328, 423 (1987)
33) W. Cassing and A. Pfitzner, Z. Phys. A337, 175 (1990)
34) K.A. Brueckner, Phys. Rev. 97, 1353 (1955)
35) J. Cugnon, A. Léjeune and P. Grangé, Phys. Rev. C35, 861 (1987)
36) T. Izumoto, S. Krewald and A. Faessler, Nucl. Phys. A341, 319 (1980)
37) S. Krewald, K. Nakayama and J. Speth, Phys. Rep. 161, 103 (1988)
38) A. Léjeune et al., Nucl. Phys. A453, 189 (1986)
39) R. Malfliet, Prog. Part. Nucl. Phys. 21, 207 (1988)
40) K. Chen et al., Phys. Rev. 166, 949 (1968)
41) J. Cugnon et al., Nucl. Phys. A352, 505 (1981)
42) G. Welke et al., Phys.Rev. C38, 2101 (1988)
43) Gy. Wolf et al., Nucl. Phys. A517 (1990) 615
44) C.Y. Wong, Phys. Rev. C25, 1460 (1982)
45) D. Vautherin and D.M. Brink, Phys. Rev. C5, 626 (1972)
46) W. Cassing et al., Phys. Lett. B181, 217 (1986)
47) W. Cassing, Z. Phys. A329, 487 (1988)
48) R.Y. Cusson et al., Phys. Rev. C18, 2589 (1978)
49) W. Cassing, Z. Phys. A326, 21 (1987)
50) W. Cassing, Z. Phys. A327, 87 (1987)
51) W. Rösch et al., Nucl. Phys. A496, 141 (1989)
52) P. Danielewicz, Ann. Phys. 152, 305 (1984)
53) W. Cassing, J. Phys. C2, 185 (1987)
54) D. Vasak et al., Phys. Lett. B93, 243 (1980)
55) D. Vasak et al., J. Phys. G11, 1309 (1985)
56) W. Bauer et al., Nucl.Phys.A456, 159 (1986)
57) R. Shyam and J. Knoll, Nucl. Phys. A426, 606 (1984)
58) M. Schäfer et al., Z. Phys. A (1991) in print
59) V.R. Brown and J. Franklin, Phys. Rev. C8, 1706 (1973)
60) R. Baier et al., Nucl. Phys. B11, 675 (1969)
61) F.P. Brady et al., Phys. Rev. Lett. 20, 750 (1968)
62) D.P. Murdock and C.F. Horowitz, Phys. Rev. C35, 1442 (1987)
63) C.F. Horowitz, Phys. Rev. C31, 1340 (1985)
64) R. Bertholet et al., Nucl. Phys. A474, 541 (1987)
65) H. Nifenecker and J.A. Pinston, Prog. Part. Nucl. Phys. 23, 271 (1989)
66) V. Metag, Nucl. Phys. A488, 483c (1988)
67) W.U. Schröder and J. Huizenga, Treatise on Heavy–Ion Science, Vol. 2, ed. by D.A. Bromley, Plenum (1984), p. 115
68) H. Feldmeier, Rep. Prog. Phys. 50, 1 (1985)
69) K.T.R. Davies et al., in 'Treatise on Heavy–Ion Science', Vol. 3, Plenum Press, New York (1984), p. 3
70) J. Randrup, Nucl. Phys. A307, 319 (1978); A327, 490 (1979); A383, 468 (1982)
71) S. Hannuschke, diploma thesis, university of Giessen 1990, unpublished
S. Hannuschke, W. Cassing, U. Mosel and K. Niita, Nucl. Phys. A (1991)
72) R.A. Broglia and A. Winther, Heavy–Ion Reactions, Vol. 1, Benjamin/Cummings (1981)
73) R.J. Charity and A. Gobbi, private communication
74) U. Brosa and S. Grossmann, J. Phys. G10, 933 (1984)
75) V. Metag, Ann. d. Physik 48 (1991) 121
76) W. Cassing and S.J. Wang, Z. Phys. A337, 1 (1990)

Quantum Chaology: Our Knowledge and Ignorance

M.V. Berry

H.H. Wills Physics Laboratory, Tyndall Avenue, Bristol BS8 1TL, UK

Abstract. A brief review is given of what has been achieved in understanding the discrete spectra of classically chaotic bound quantum systems, emphasising the central role of semiclassical techniques and the unsolved problems associated with the clash between the long-time limit and the semiclassical limit.

We are fairly sure that, apart from a curious special class of cases [1,2], there is no chaos in quantum mechanics. It is obvious, however, that in the semiclassical limit $\hbar \to 0$ (e.g. highly excited states) the quantum behaviour must somehow reflect the nature of the classical trajectories, and in particular must exhibit characteristic effects if the orbits are chaotic. For the study of these quantum signatures of classical chaos I proposed the term 'quantum chaology' [3,4]. Here I provide a brief 'readers' guide' to the main advances there have been in the subject, and emphasize the major unsolved problems. I confine myself to the quantum chaology of the discrete spectra (energy levels and wavefunctions) of time-independent bound Hamiltonians with N (>1) freedoms. Therefore I will not discuss the important areas of the time-development of quantum states [5,6] or chaotic scattering [7,8]. It is hardly necessary to say that even for spectra my account will be partial and deliberately selective.

One thing that is very well known about the density $d(E)$ of energy levels is its average value $\langle d(E) \rangle$ (over energy ranges that are classically small but include many quantum levels). This is given by the Weyl rule [9,10]: one quantum state per classical phase space volume $\hbar^N$. Thus d is proportional to $\hbar^{-N}$ and the level spacing proportional to $\hbar^N$. The first few $\hbar$-corrections to $\langle d \rangle$ are known for quantum billiards (vibrating membranes) [10], including the effects of magnetic flux [11] and relativity [12] (Dirac spinorization). These corrections are routinely used in checking computations of quantum billiard eigenvalues, and form the basis of an extremely accurate error-detection scheme [13]. I have not seen explicit general formulae for the $\hbar$-corrections to $\langle d \rangle$ for general smooth potentials, apart for the one-dimensional (WKB) case [14].

Although the first $\hbar$-corrections for the smoothed level density are known for billiards, nobody has studied the late terms of the series. But we know that this must diverge, because the exact $d(E)$ is a series of Dirac δs, which cannot be represented by a power series in $\hbar$. The approach to quantization, starting from $\langle d \rangle$, is given not by $\hbar$-corrections but by a series of contributions, of order $\exp\{iS_n/\hbar\}$ (i.e. nonanalytic as $\hbar \to 0$), from each classical closed orbit, with action S_n. [15,16,17]. Each closed orbit describes oscillatory clustering of the levels, on an energy scale $\hbar/T_n$, where T_n is the period of the orbit. It is important to appreciate that the clustering associated with a given orbit is classically small (of order $\hbar$) but large in comparison with the quantum level spacing (each 'cluster' includes of order $1/\hbar^{N-1}$ levels). Therefore in order to understand spectra on fine scales (that is on the scale of individual levels)

Eds: Y. Abe, H. Horiuchi, K. Matsuyanagi

within the framework of semiclassical asymptotics, enormously long classical orbits must be involved, with periods of order $1/\hbar^{N-1}$.

There is, however, another framework for studying the fine scales of spectra, namely energy-level statistics [18-22]. Here we abandon the attempt to locate each level precisely, and instead study average properties (fluctuations, correlations, spacing distributions). Now universality emerges: the statistics depend only on the classical chaology, in the following way. For classically integrable systems, the levels are Poisson-distributed [18]; for classically chaotic systems, levels are distributed according to random-matrix ensembles [23] that depend on symmetries - Gaussian unitary (GUE) when there is no symmetry [24-26], Gaussian orthogonal (GOE) when there is time-reversal and the system is Bosonic (e.g. scalar-wave) [21], and Gaussian symplectic when there is time-reversal and the system is Fermionic [27]. Of course there are intermediate cases, where some orbits are regular and some chaotic, and, correspondingly, statistics which interpolate between the four universality classes.

To understand these largely numerical observations about level distributions, the appropriate theoretical framework must be semiclassical, because statistics are not even defined for an individual Hamiltonian unless there are infinitely many levels, and then almost all are highly excited - that is, semiclassical. The semiclassical connection is necessary in order to explain the observed dependence on the dynamics. Without this connection, other theories, such as those based on the 'motion' of levels as parameters vary [28-30], must fail, although they certainly contain interesting mathematics.

The only semiclassical theory available at present is based on the expansion of the spectral fluctuations $d-\langle d\rangle$ as a sum of closed orbits, and we have seen that any explanation of fine-scale spectral structure must involve the very long orbits. At this point we come up against the central theoretical problem in the whole subject [17], namely the fact that our semiclassical formulae break down at long times. This reflects the fundamental quantum-mechanical fact that the semiclassical limit $\hbar\to 0$ and the long-time limit $t\to\infty$ do not commute. We do not know how, or even whether, the closed-orbit sum generates the individual δs in the level density for chaotic systems. This is a serious - perhaps shocking - situation, because it means that we are ignorant of the mechanism of quantization. (Sometimes, singularities in $d(E)$ can be generated by summing selected infinite series of orbits, for example all repetitions of a given one, but these are false singularities [31], not energy levels, even for integrable systems [32].)

Nevertheless, there has been some progress in incorporating long times into the semiclassical framework. The intensities (squares of amplitudes) of the closed orbits are purely classical quantities which for long times obey a sum rule [33], depending only on the classical chaology and reflecting the uniformity of the distribution of long closed orbits in phase space. The requirement that the long orbits must combine (albeit by some unknown mechanism) to give a sequence of δs with the correct density $\langle d\rangle$ gives another sum rule [34,17], involving the interference between pairs of orbits. With these classical and semiclassical sum rules, it was possible to understand a large class of energy-level statistics [34] in a fundamental way, that is in terms of classical mechanics, by means of semiclassical asymptotics. These are the statistics quadratic in $d(E)$, such as the spectral rigidity, the number variance, the two-level correlation function, and the form factor (all defined in [22]).

From this semiclassical theory of some of the statistics, there emerged as an unexpected bonus the explanation [34] of the breakdown of universality at large energy ranges, first observed in computations [35] on an integrable system. Since the universality is a consequence of classical uniformity, and this only holds for long orbits, it will fail at and beyond energy scales where the spectral fluctuations depend on short orbits. This scale is $\hbar/T_{min}$, where T_{min} is the period of the shortest orbit with energy E.

The difficulties associated with long times disappear if the spectrum is smoothed, for example by giving the energy a sufficiently large imaginary part $i\varepsilon$, because then the sum over closed orbits converges - and indeed provides a useful description of the large-scale spectral fluctuations [41,42]). For chaotic systems, however, ε must exceed a value determined [36] by the entropy of the dynamics. Overcoming this 'entropy barrier' and thereby understanding quantization, is thus a problem of analytic continuation. One possible way to solve this is suggested by the analogy [34, 37,17] between the spectra of classically chaotic systems and the imaginary parts of the zeros of Riemann's zeta function. This analogy has already proved productive in the quantum→Riemann direction, by inspiring an analytical and quantitatively-accurate formula [38] for the number variance of the Riemann zeros. Recently [39] it was employed in the Riemann→quantum direction, to obtain a quantum analogue of the Riemann-Siegel formula [40] which enables efficient computation of the zeros; the idea (already underlying the semiclassical sum rule [34]) is the 'resurgence' [17] of the long orbits to give, by resummation, information related to $<d>$ and the short orbits. This new quantum condition is the vanishing of a sum over a large but finite number of closed orbits. Perhaps it will lead to a semiclassical explanation for the higher-than quadratic level statistics, which the earlier theory [34] did not achieve.

Even if the closed orbits can be resummed to give a quantum condition, either explicitly in the form of the 'Riemann-Siegel look-alike' or implicitly as a compactified determinant [43], there remains the awkward fact that the semiclassical theory on which it is based is only a lowest-order one, in the sense that higher terms in $\hbar$, mutiplying the contributions $\exp\{iS_n/\hbar\}$, are ignored (and unknown). This raises the unexplored question of the asymptotic accuracy and meaning of any 'eigenvalues' generated by these methods: the true levels have spacing $\hbar^N$, but the methods have errors $\hbar^2$. It might be possible to get some insight into this difficult problem by studying systems with special symmetries, such as quantum billiards on tilings of the hyperbolic plane [44,42], or the quantum cat maps [45-7], for which the sum over closed orbits is known to be exact.

For wavefunctions, the mixture of knowledge and ignorance is very similar - although for a long time we did not appreciate how close the analogy is. There is a lowest-order description (analogous to the Weyl rule for the smoothed level density) which states that the probability density for an eigenstate of an ergodic system is the projection of the energy surface from phase space onto coordinate space [48] (for integrable systems, tori must be projected). Then some spectacular computations [49,50] showed wavefunctions looking very different from uniform projections of the energy surface, revealing imprints ('scars') of individual closed orbits. These led to the development of a theory [51] in which, just as for the level density, the lowest-order wavefunctions are corrected by characteristic interference fringes near the closed orbits. The theory can also be formulated in phase space [52,17], and the characteristic fringes there, which take the form of hyperbolae with each orbit's invariant manifolds as asymptotes, could explain certain features of computations [53] of the Husimi (smoothed Wigner) functions for eigenstates. We do not know how to understand the (divergent) superposition of very long and overlapping scar patterns. There is reason to think that their net effect might (after some kind of resummation) be to provide a Gaussian random background decorating the smooth background of the lowest-order description, as conjectured in the old theory [48,9].

References

1. B.V. Chirikov, F.M. Izraelev, and D.L. Shepelyansky, PhysicaD **33**, 77 (1988).
2. M.V. Berry, this volume.
3. M.V. Berry, Proc. Roy. Soc. Lond. **A413**, 183 (1987).
4. M.V. Berry, Physics Scripta **40**, 335 (1989).
5. B.V. Chirikov, F.M. Izraelev, and D.L. Shepelyansky, Sov. Sci.Rev. **C2**, 209 (1981).
6. T. Dittrich and U. Smilansky, Nonlinearity (2 papers in press, 1991).
7. B. Eckhardt, PhysicaD **33**, 89 (1988).
8. U. Smilansky, in Les Houches Lecture Series 52 (eds: M. Giannoni and A. Voros) (North-Holland, Amsterdam, in press 1990).
9. M.V Berry, in Les Houches Lecture Series 36 (eds: G. Iooss, R.H.G Helleman and R. Stora) (North- Holland, Amsterdam 1983) 171.
10. H.P.Baltes and E.R. Hilf, *Spectra of Finite Systems* (Mannheim, B-I Wissenschaftsverlag 1976).
11. M.V. Berry, J.Phys.A. **19**, 2281 (1986).
12. M.V Berry and R. Mondragon, Proc. Roy. Soc. Lond. **A412**, 53 (1987).
13. M.V. Berry, J.Phys.A. **20**, 2389 (1987).
14. A.Voros, Ann. Inst. H. Poincaré **39A,** 211 (1983).
15. M.C.Gutzwiller, J.Math.Phys. **12**, 343 (1971).
16. R. Balian and C. Bloch, Ann. Phys. (N.Y.) **69**, 76 (1972).
17. M.V. Berry, in Les Houches Lecture Series 52 (eds: M-J. Giannoni and A. Voros) (North-Holland, Amsterdam, in press 1990).
18. M.V. Berry and M. Tabor, Proc. Roy. Soc. Lond. **A356**, 375 (1977).
19. S.W. McDonald and A.N. Kaufman, Phys. Rev. Lett. **42**, 1189 (1979).
20. M.V Berry, Ann. Phys. (N.Y.) **131**, 163 (1981).
21. O. Bohigas, M-J. Giannoni and C. Schmit, Phys. Rev. Lett. 52, **1** (1984).
22. O. Bohigas, M-J. Giannoni, in *Mathematical and Computational Methods in Nuclear Physics* (eds: J.H. Dehesa, J.M.G. Gomez and A. Polls), Springer Lecture Notes in Physics **209**, 1 (1984).
23. C.E. Porter (ed), *Statistical Theories of Spectra: Fluctuations* (New York, Academic Press 1965)
24. T.H. Seligman, J.J.M. Verbaarschot and M.R. Zirnbauer, J.Phys.A. **18**, 2751 (1985)
25. M.V. Berry and M. Robnik, J.Phys.A. **19**, 649 (1986).
26. M. Robnik and M.V. Berry, J.Phys.A. **19**, 669 (1986).
27. R. Scharf, B. Dietz, M. Kus, F. Haake and M.V. Berry, Europhys. Lett. **5**, 383 (1988).
28. P. Pechukas, Phys. Rev. Lett. **51**, 943 (1983).
29. M.V. Berry in *Chaotic Behavior in Quantum Systems* (ed: G. Casati) (New York, Plenum 1985) 123.
30. K. Nakamura and M. Lakshmanan, Phys.Rev.Lett. **57**,1661 (1986)
31. A. Voros, J. Phys. A. **21**, 685 (1988).
32. J.P. Keating and M.V. Berry, J.Phys.A. **20**, L1139 (1987).
33. J.H. Hannay and A.M. Ozorio de Almeida, J.Phys.A. **17**, 3429 (1984).
34. M.V. Berry, Proc. Roy. Soc. Lond. **A400**, 229 (1985).
35. G. Casati, B.V. Chirikov and I. Guarneri, Phys. Rev. Lett. **54**, 1350 (1985).
36. B. Eckhardt and E. Aurell, Europhys. Lett. **9**, 509 (1989).

37. M.V. Berry, in *Quantum Chaos and Statistical Nuclear Physics*, (eds:`T.H. Seligman and H.Nishioka), Springer Lecture Notes in Physics **263**, 1 (1986)
38. M.V. Berry, Nonlinearity **1**, 399 (1988).
39. M.V. Berry and J.P. Keating, J.Phys.A. (in press 1990)
40. H.M. Edwards *Riemann's Zeta Function* (New York, Academic Press 1974)
41. D. Wintgen, Phys.Rev.Lett. **61**, 1803 (1988).
42. R. Aurich and F. Steiner, Physica D **39**, 169 (1989).
43. E.B. Bogomolny, Comments At. Mol. Phys. (in Press 1990).
44. N.L. Balazs and A. Voros, Physics Reports **143**, 109 (1986).
45. J.H. Hannay and M.V. Berry, Physica **1D**, 267 (1980).
46. J.P. Keating, Nonlinearity (in press 1990/1)
47. J.P. Keating, Nonlinearity (in press 1990/1)
48. M.V. Berry, J.Phys.A. **10**, 2083 (1977).
49. E.J. Heller, Phys.Rev.Lett. **53**, 1515 (1984).
50. E.J. Heller, in Les Houches Lecture Series 52 (eds: M. Giannoni and A. Voros) (North-Holland, Amsterdam, in press 1990).
51. E.B. Bogomolny, Physica **D31**, 169 (1988)
52. M.V. Berry, Proc. Roy. Soc. Lond. **A423**, 219 (1989).
53. P. Leboeuf and M. Saraceno, J.Phys.A. **23**, 1745 (1990).

True Quantum Chaos? An Instructive Example

M.V. Berry

H.H. Wills Physics Laboratory, Tyndall Avenue, Bristol BS8 1TL, UK

Abstract. Any chaotic classical system can be transformed into a quantum system that preserves the chaos, because the classical Liouville equation involving $2N$ phase-space variables $\boldsymbol{q},\boldsymbol{p}$ has the form of a 'Schrödinger equation' with 'coordinates' $\boldsymbol{Q}=\{\boldsymbol{q},\boldsymbol{p}\}$. The feature of this quantum system that allows chaos to persist is linearity of the 'Hamiltonian' in the $2N$ 'momentum' operators conjugate to $\boldsymbol{Q}$.

It is commonly stated that there is no quantum chaos. Various reasons are given: driven quantum systems absorb energy more slowly than their chaotic classical counterparts [1], bound systems have discrete energy levels, the Schrödinger equation is linear, Planck's constant $\hbar$ smooths away classical phase-space fine structure [2-4] or replaces it, effectively, by a discrete lattice [5]. Here I want to show with a curious example that it is in fact possible to construct quantum systems with genuinely chaotic behaviour, and thereby demonstrate that some of the above statements are not true. However, the systems are so exceptional that the reasons why they allow quantum chaos to exist can help us understand why quantum chaos does not occur in situations that are physically more reasonable. The example is in essence the same as that described by Chirikov et al.[6], but I wish to present it somewhat differently and investigate it in a little more detail.

Consider a classical time-independent Hamiltonian $H_c(\boldsymbol{q},\boldsymbol{p})$ with N freedoms, generating bounded and chaotic trajectories in all or part of the $2N$-dimensional phase space. The Liouville equation governing the evolution of classical phase-space densities $\rho(\boldsymbol{q},\boldsymbol{p},t)$ is

$$\frac{\partial\rho}{\partial t}+\dot{\boldsymbol{q}}\cdot\nabla_{\boldsymbol{q}}\rho+\dot{\boldsymbol{p}}\cdot\nabla_{\boldsymbol{p}}\rho=0 \tag{1}$$

As has often been pointed out, the fact that this equation is linear and yet preserves the chaos of the trajectories disposes of the argument that the linearity of the Schrödinger equation is responsible for the absence of chaos in quantum mechanics. Indeed, the Liouville equation becomes a 'Schrödinger equation' if ρ is regarded as a 'wavefunction' depending on $2N$ 'coordinates' $\boldsymbol{Q}$ which are simply the variables of the original phase space:

$$\boldsymbol{Q}\equiv\left\{Q_1\cdots Q_{2N}\right\}=\{\boldsymbol{q},\boldsymbol{p}\} \tag{2}$$

Now let $\boldsymbol{V}(\boldsymbol{Q})$ denote the original classical phase-space velocity and $\boldsymbol{P}$ the momentum operator conjugate to $\boldsymbol{Q}$, i.e

$$\boldsymbol{V}(\boldsymbol{Q})\equiv\{\dot{\boldsymbol{q}},\dot{\boldsymbol{p}}\}=\left\{\nabla_{\boldsymbol{p}}H_c,-\nabla_{\boldsymbol{q}}H_c\right\},\quad \boldsymbol{P}\equiv-\mathrm{i}\hbar\nabla_{\boldsymbol{Q}} \tag{3}$$

Eds: Y. Abe, H. Horiuchi, K. Matsuyanagi

Then on multiplication by $i\hbar$, (1) becomes

$$i\hbar\frac{\partial\rho(Q,t)}{\partial t}=\mathcal{H}_{\mathrm{qu}}\rho(Q,t) \tag{4}$$

involving the Hermitian 'quantum' Hamiltonian operator, depending on $4N$ $\boldsymbol{Q}$ and $\boldsymbol{P}$ operators,

$$\mathcal{H}_{\mathrm{qu}}=\frac{1}{2}\{V(Q)\cdot P+P\cdot V(Q)\} \tag{5}$$

(the expected commutator term is absent because $\mathrm{div}V=0$). See [14] for another problem where a 'doubled phase space' arises.

The 'Schrödinger lookalike' (4) is of course exactly the Liouville equation (1) and so has the same solutions. In particular, the contours of any evolving density develop infinite complexity (in the form of whorls and tendrils [7]) as the trajectories that convect them separate exponentially. Thus these 'wavefunctions' are chaotic: there is no $\hbar$-smoothing, no quantal suppression of chaos.

The eigenfunctions of $\mathcal{H}_{\mathrm{qu}}$ can be chaotic too. These are of two sorts. First are uniform densities on invariant manifolds of the original classical motion (in $\boldsymbol{Q}$ space), which have 'energy' $\mathcal{E}=0$. The manifolds could be whole energy surfaces, or closed orbits. In the latter case, an eigenfunction supported by a very long closed orbit (or combination of closed orbits) can be arbitrarily complicated, because the closed orbits shadow the non-closed, chaotic orbits. Second are densities restricted to individual closed orbits, with ρ varying periodically round the orbit and proportional to

$$\exp\{in(\theta-\omega t)\} \tag{6}$$

where θ is the angle variable (equal θs in equal times) and $T=2\pi/\omega$ is the period. These are easily seen to satisfy (4) with 'energy'

$$\mathcal{E}(n,T)=n\hbar\omega=nh/T \tag{7}$$

These eigenfunctions are also chaotic for long orbits, with the additional complication when n is large that nearby parts of an orbit will have uncorrelated phases $n\theta$. The spectrum has an infinite degeneracy at $\mathcal{E}=0$, and is continuous elsewhere, because each closed orbit exists for a range of values of the original classical energies (i.e. of H_c); the set of states labelled by a given value of H_c forms a dense point spectrum. All states are localized in $\boldsymbol{Q}$.

These wavefunctions should be regarded as the result of quantizing not the original Hamiltonian $H_c(\boldsymbol{Q})$ but the artificial classical Hamiltonian $\mathcal{H}_c(\boldsymbol{Q},\boldsymbol{P})$ corresponding to (5), namely

$$\mathcal{H}_c(Q,P)=V(Q)\cdot P \tag{8}$$

The equations of motion for the $4N$ phase-space variables are

$$\dot{Q}=V(Q),\qquad \dot{P}=-\mathbf{A}(Q(t))P,$$

$$\text{where } \mathbf{A}_{mn} \equiv \frac{\partial V_n}{\partial Q_m}, \text{ and } 1 \le m,n \le 2N \tag{9}$$

In the first of these equations, the original phase-space motion simply reappears, with all its chaos, as motion of the 'coordinates' $\boldsymbol{Q}$, which are independent of the motion of the new 'momenta' $\boldsymbol{P}$. The second equation shows that the $\boldsymbol{P}$ are slaved to the $\boldsymbol{Q}$ by a linear matrix equation which is exactly that governing the linearized small deviations from the orbit $\boldsymbol{Q}(t)$.

Before pointing out the essential peculiarity of $\mathcal{H}_{\mathrm{qu}}$ which enables it to possess chaos, I note some other unusual features of this Hamiltonian and its classical counterpart $\mathcal{H}_{\mathrm{c}}$. First, the quantum states localized in $\boldsymbol{Q}$ on individual invariant manifolds (e.g. an energy surface, or a closed orbit) are not normalizable, because these manifolds have zero measure in $\boldsymbol{Q}$ space. Second, there are two classically conserved quantities: the original energy $H_{\mathrm{c}}(\boldsymbol{Q})$, and the value of the new Hamiltonian $\mathcal{H}_{\mathrm{c}}(\boldsymbol{Q},\boldsymbol{P})$. Third, in the full phase space $\{\boldsymbol{Q},\boldsymbol{P}\}$ the motion is unbounded, because according to (9) the 'small deviations' $\boldsymbol{P}$ grow when the $\boldsymbol{Q}$ motion is chaotic. Fourth it follows from this that the closed orbits in $\boldsymbol{Q}$ are not periodic in $\boldsymbol{Q},\boldsymbol{P}$. It is amusing that the 'amplification factor' for $2N$-1 of the $\boldsymbol{P}$'s can be expressed as a 'nonabelian geometric amplitude factor' [8] depending only on the geometry of the orbit's path through $\boldsymbol{Q}$ space and not on its time-dependence (the exceptional momentum corresponds to a deviation that changes the original energy).

What allows quantum chaos in this class of examples? It is the fact that the Hamiltonian $\mathcal{H}_{\mathrm{qu}}$ is *linear in the momentum operators*. At first this seems an odd assertion, because the classical chaos, whilst not being incompatible with the fact that (8) is linear in $\boldsymbol{P}$, is preserved when $\mathcal{H}_{\mathrm{c}}$ is replaced by any function of itself (e.g. $\mathcal{H}_{\mathrm{c}}^2$) - such transformations leave the orbits unchanged apart from an energy-dependent time rescaling. But in quantum mechanics it is the linearity in $\boldsymbol{P}$ that prevents the spreading of wavepackets which eventually obliterates chaos. This can be seen by comparing the propagators for the trivial one-dimensional quantum Hamiltonians P and P^2:

$$H = P, \quad \langle Q|\exp\{-iHt/\hbar\}|Q'\rangle = \delta(Q - Q' - t) \tag{10}$$

and

$$H = P^2, \quad \langle Q|\exp\{-iHt/\hbar\}|Q'\rangle = \frac{\exp\left\{i(Q - Q')^2/4\hbar t\right\}}{\sqrt{2i\hbar t}} \tag{11}$$

For P and P^2 the orbits are lines of constant P, but the fact, classically trivial, that the velocity of Q is independent of P in the first case, and proportional to P in the second case, has a crucial effect on the Q-propagator. (Because of linear canonical invariance, this argument about what permits quantum chaos is not restricted to Hamiltonians linear in the $2N$ Ps: it also holds for Hamiltonians that depend linearly on any commuting set of $2N$ of the Qs and Ps.)

It is worth remarking that the $\boldsymbol{P}$-linearity of $\mathcal{H}_{\mathrm{qu}}$, and hence the associated quantum chaos, is not destroyed by arbitrary canonical transformations among the original phase-space variables $\boldsymbol{Q}$. In the doubled phase space these are just coordinate transformations, and generate new $\boldsymbol{P}$s depending linearly on the old ones.

Linearity permits quantum chaos but of course does not guarantee it. Obviously there is no chaos in (10), which was intended only to illustrate nonspreading. And if in the classically chaotic and quantally nonchaotic kicked rotator [1,5] the kinetic energy term P^2 is replaced by P, the aim is achieved of making the quantum motion exactly follow the classical motion [9], but at the price of destroying the chaos. Another class of classically chaotic but quantally nonchaotic linear Hamiltonians are those which, as in neutrino billiards [10], act on spinors rather than scalars, - although Pasmanter [11] gives two interesting examples where quantum spinorization does not destroy the chaos.

It is curious that whereas in classical mechanics chaos requires *nonlinearity* in the equations of motion (that is, $H_c(Q)$ must be more than quadratic in some of its variables), the preservation of that chaos into the 'Schrödinger lookalike' via Liouville's equation requires *linearity* of $\mathcal{H}_{qu}$ in $\boldsymbol{P}$.

When the quantum Hamiltonian is not linear in momentum, we have the now-familiar situation: quantum mechanics suppresses the chaos, and the characteristic phenomena of quantum chaology appear [12,13].

Acknowledgment. I thank Dr J. Robbins for some illuminating comments.

References

1. G. Casati, B.V. Chirikov, J. Ford and F.M. Izraelev, in *Stochastic behavior in classical and quantum systems* (eds: G. Casati and J. Ford), Springer Lecture notes in Physics 93, 334 (1979).
2. M.V Berry, in Les Houches Lecture Series 36 (eds: HG. Iooss,R.H.G Helleman and R. Stora) (North- Holland, Amsterdam 1983) 171.
3. M.V. Berry, Phil. Trans. Roy. Soc. Lond. A287, 237 (1977).
4. H.J. Korsch and M.V. Berry, Physica **3D**, 627 (1981).
5. B.V. Chirikov, F.M. Izraelev, and D.L. Shepelyansky, Sov. Sci.Rev. **C2**, 209 (1981).
6. B.V. Chirikov, F.M. Izraelev, and D.L. Shepelyansky, Physica **33D**, 77 (1988).
7. M.V. Berry, N.L. Balazs, M. Tabor and A. Voros, Ann. Phys (N.Y.) **122**, 26 (1979).
8. A. Shapere and F. Wilczek (eds), *Geometric Phases in Physics* (World Scientific,Singapore, 1989).
9. M.V Berry, Physica **10D**, 369 (1984).
10. M.V Berry and R. Mondragon, Proc. Roy. Soc. Lond. **A412**, 53 (1987).
11. R.A. Pasmanter, Phys. Rev. A.**42**, 3622 (1990).
12. M.V. Berry, Proc. Roy. Soc. Lond. **A413**, 183 (1987).
13. M.V. Berry, Physics Scripta **40**, 335 (1989).
14. T.B. Smith, J. Phys. A.**11**, 2179 (1978).

Toward the Fundamental Theory of Nuclear Matter Physics: The Microscopic Theory of Nuclear Collective Dynamics

F. Sakata[1], *T. Marumori*[2], *Y. Hashimoto*[2], *H. Tsukuma*[2], *Y. Yamamoto*[2], *J. Terasaki*[2], *Y. Iwasawa*[2], and *H. Itabashi*[2]

[1]Institute for Nuclear Study, University of Tokyo, Tanashi, Tokyo 188, Japan
[2]Institute of Physics, University of Tsukuba, Ibaraki 305, Japan

Abstract. Since the research field of nuclear physics is expanding rapidly, it is becoming more imperative to develop the microscopic theory of nuclear matter physics which provides us with a unified understanding of diverse phenomena exhibited by nuclei. An establishment of various stable mean-fields in nuclei allows us to develop the microscopic theory of nuclear collective dynamics within the mean-field approximation. The classical-level theory of nuclear collective dynamics is developed by exploiting the symplectic structure of the time-dependent Hartree-Fock (TDHF)-manifold. The importance of exploring the single-particle dynamics, e.g. the level-crossing dynamics in connection with the classical order-to-chaos transition mechanism is pointed out. Since the classical-level theory is directly related to the full quantum mechanical boson expansion theory via the symplectic structure of the TDHF-manifold, the quantum theory of nuclear collective dynamics is developed at the dictation of what is developed in the classical-level theory. The quantum theory thus formulated enables us to introduce the quantum integrability and quantum chaoticity for individual eigenstates. The inter-relationship between the classical-level and quantum theories of nuclear collective dynamics might play a decisive role in developing the quantum theory of many-body problems.

1. Introduction

1.1. Basic Problems in Nuclear Matter Physics

In the light of the recent developments of nuclear physics, new trends of nuclear collective dynamics in the 1990s seem to be focussing on the following three subjects; first: the dynamics of symmetry breaking, i.e. the dynamical relation among many vacua with different symmetries, second: the dynamics of the order-to-chaos transition mechanism in the quantum system, and: the dynamics of the many-body quantum system in terms of geometrical or topological structure of the symplectic manifold. Nowadays, these basic problems are becoming rather common in many other fields of theoretical physics. It is an objective of the present contribution to try to explain how suitable it is to exam-

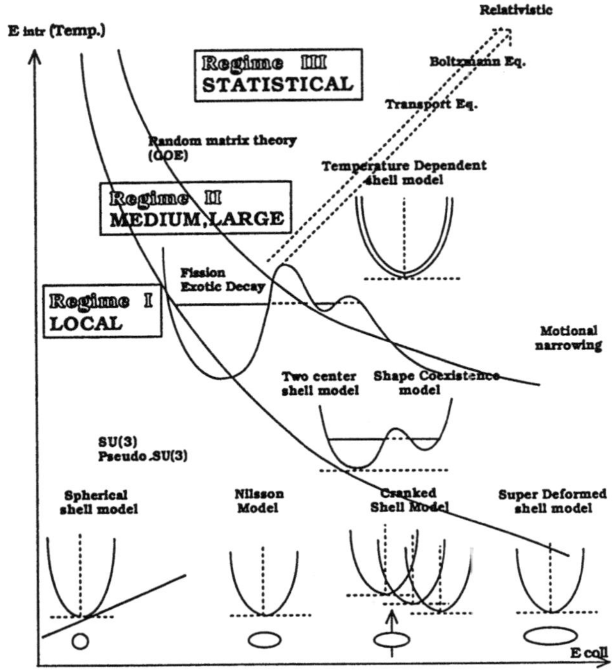

Figure 1. Three characteristic regimes of nuclear matter physics.

ine these fascinating subjects in the field of nuclear physics. With this aim, let us start by summarizing the present status of nuclear physics.

With the aid of recent improved experimental facilities both in accelerators and in complex detector systems, a great variety of phenomena are being extensively studied, ranging from the low-lying collective excited states up to the violent high energy heavy-ion reactions in the so-called BUU or VUU regime [1]. These phenomena might be classified into three characteristic regimes shown in the phase diagram in Fig. 1. The first regime is characterized by ***local phenomena*** which are understood by introducing an appropriate stable mean-field with one local minimum specified by a certain symmetry, special coupling scheme, group theoretical model space or phenomenological collective subspace etc. In the second regime, there are ***medium- or large-amplitude collective phenomena*** which are usually described by a suitable mean-field with many local minima with different symmetries. The third regime consists of ***statistical or probabilistic phenomena*** which are explained by means of thermo-dynamical concepts like the transport equation, stochastic equation, temperature, dissipation, equation of states etc. Since a great store of knowledge on individual phenomena is now available and since our knowledge on each phenomenon is expected to expand rapidly, it becomes quite imperative to develop the fundamental theory of nuclear matter physics which provides us with a unified understanding of diverse phenomena classified into the above-stated three regimes. In other words, we are now at the beginning of a new era of nuclear physics capable of developing a physics of ***evolution of nuclear matter***, i.e. of studying the mutual dynamical relations among various characteristic regimes in addition to the conventional nuclear physics of individual ***states of matter***. Developing such a theory, we have to study the following two basic problems;

I: the dynamics of the evolution from regime 1 to regime 2. In this case, we have to answer how the stable mean-field, i.e. the vacuum with fixed symmetry (or more generally, a certain coupling scheme or group theoretical model space) becomes unstable and how a new mean-field (new vacuum) with a different symmetry becomes stable. This is the first subject stated in the beginning.

II: the dynamics of the evolution from regime 2 to regime 3. In this case, we have to study how the statistical or probabilistic features arise as a consequence of the deterministic dynamics and how a system in a non-equilibrium state reaches a state of equilibration. In other words, we encounter the essential problem of why the irreversible dynamics described by e.g. the transport equation results from the reversible dynamics. Even though there have been plenty of attempts at non-equilibrium statistical physics, in our opinion, the royal road to attack problem II might be to start with exploring the dynamics of the order-to-chaos transition mechanism, i.e. the second subject.

1.2. Quantum Theory of the Many-Body Problem

Now, let us discuss one advantage of nuclear physics in studying the above-stated two basic problems. The first important step in developing the physics of evolution of matter is how to describe the states which are situated far from the "vacuum", like a stable mean-field or thermodynamical state of equilibration, i.e. the states in the large-amplitude region. As is well known, it is not so easy to develop such a quantum theory of many-body problems capable of describing the states in the large-amplitude region. The non-perturbative quantum theory describing the states in the large-amplitude region is one of the main current subjects in a variety of fields of theoretical physics.

In nuclear physics, however, it is becoming more and more clear that some of the excited states are better understood by introducing a suitable mean-field. In the past decade, the most exciting progress in nuclear physics has been an establishment of the "rotating (cranked Nilsson) shell model" as well as the "super-deformed shell model" [2]. It is astonishing that there exists a set of characteristic states which are well isolated from the thousands of neighboring excited states, and are simply characterized by a suitable stable mean-field.

The existence of the various stable mean-fields in the nucleus allows us to explore the nuclear collective dynamics in terms of ***mean-field dynamics***, i.e. the single-particle dynamics within the time-dependent Hartree-Fock (TDHF) theory. Recent theoretical development [3] of nuclear physics has shown that the TDHF equation is formally equivalent to the classical Hamilton's equation of motion, indicating a possibility of developing a "classical-level" dynamics of the quantum theory of many-body problems with the aid of geometrical and topological arguments in the TDHF phase space. This specific feature is an advantage of nuclear physics in studying the third subject stated in the beginning of §1.1. Since the TDHF phase space has symplectic structure, the classical order-to-chaos transition mechanism explored extensively in the classical dynamical system would shed light on the classical-level dynamics in the TDHF theory.

The second advantage of nuclear physics is its clear correspondence between classical-level and quantum theories of microscopic description of many-body problems. Namely, it has been established [4] that the TDHF theory (classical-level theory) is directly related to the full quantum mechanical boson expansion theory through the symplectic structure of the TDHF theory. This theoretical relation suggests to us that the nucleus would provide us with a unique opportunity to construct the full quantum theory of many-body problems at the dictation of what is developed in the classical-level theory. From the above argument, it is expected that nuclear physics could give an answer to an interesting question on an interrelationship between ***classical chaos*** (within the symplectic manifold obtained under the time-dependent mean-field theory) and ***quantum chaos*** (within the full quantum theory).

1.3. Single-Particle Dynamics

Let us discuss what we could learn within the classical-level treatment. Since classical-level dynamics in the TDHF theory is formally equivalent to the classical dynamics, it is possible to explore the classical order-to-chaos transition mechanism in connection with single-particle dynamics. A possibility of studying the classical chaos by using the various experimental facts on individual quantized single-particle orbits gives us a sound basis for discussion of classical chaos in nuclear physics.

An important effect of single-particle dynamics, e.g. the level-crossing mechanism on the collective motion has been the main issue in nuclear physics from its early stages on. Actually, there has been much work on this problem using the ***adiabatic*** or ***diabatic*** single-particle potential models, i.e. the mean-field characterized by a set of shape parameters (with fixed symmetry) put in by hand. Nowadays, it is widely recognized that

i) a potential barrier in between two local minima for the collective motion,
ii) the dynamics of barrier penetration in the fission process or exotic particle decay,
iii) the dissipation process of collective motion,
iv) an appearance and decaying process of super-deformation,
v) a particle alignment mechanism to the rotation axis,
and so on ought to be understood by the microscopic dynamics of single-particle modes of motion, i.e. by the level-crossing dynamics of single-particle states. For example, the importance of level slippage in the nuclear fission process was first pointed out by Hill-Wheeler [5] and recently by Swiatecki [6]. The diabatic level crossing followed by two-body collisions was discussed by Nørenberg [7] in describing the dissipating collective motion in heavy-ion deep inelastic collisions and a concept of successive level-crossing was introduced by Barranco-Broglia-Bertsch [8] in understanding the exotic particle decay process. The importance of diabatic levels defined by eliminating the spurious interaction has been discussed by Tanaka-Suekane [9] and Shimizu-Matsuyanagi [10] and further developed by the Lund group [11] in connection with shape coexisting phenomena. In spite of this important work, the essential role of the level-crossing dynamics in nuclear collective dynamics has still been left unresolved because of the difficulty of going beyond the adiabatic approximation.

Recently, it has been clearly stated by Swiatecki [6] that the level-crossing in single-particle dynamics affects the property of the collective motion and is strongly related to the classical "order-to-chaos" transition mechanism: If the nucleonic motion in the nuclear potential is ***integrable***, we get a diabatic collective potential. If it is ***non-integrable***, we get an adiabatic collective potential. It was also stated by Swiatecki [6] that the order-to-chaos transition in nucleonic motion should be studied by paying careful attention to both the dynamics of symmetry breaking in the single-particle potential and the inclusion of residual interaction among single-particle states. In conventional approaches, the symmetry of the nuclear potential (specified by shape parameters) is imposed artificially from the outset, and the residual interactions (pairing interaction is usually employed) among adiabatic or diabatic single-particle states are put in by hand. If one intends to explore the order-to-chaos transition dynamics of nucleonic motion in the single-particle potential whose symmetry might change during a large-amplitude collective motion like the fission process, one has to take account of the self-consistent dynamical relation between nuclear potential and residual interaction, by going beyond the adiabatic or diabatic approximations.

As is well known, it is considered one of the most difficult problems in the microscopic theory of nuclear collective dynamics to go beyond the adiabatic approximation. If one intends to get over the adiabatic approximation, one has to be eventually involved in the difficult problem of properly treating the ***non-linear dynamics*** between the collective and single-particle modes of motion, which comes from the self-consistent property inherent to such a self-sustained finite many-body quantum system as the nucleus. Let us briefly discuss the above stated non-linearity arising from the self-consistency. The collective coordinate (e.g. deformation δ) is expressed as a coherent sum of particle($a^{\dagger}_{\mu}$)-hole($b^{\dagger}_{i}$) operators. On the other hand, the single-particle orbits in turn should adjust their features self-consistently in accordance with the time-dependent variation of the mean-field expressing the collective motion. Namely, the single-particle operators $a^{\dagger}_{\mu}$ and $b^{\dagger}_{i}$ change their structure depending on the amplitude of δ. Consequently, there appears a strong non-linear dependence between the dynamical change in the symmetry of single-particle potential and the nucleonic motion inside the nuclear potential, which is schematically expressed by

$$\hat{\delta} = \Sigma_{\mu i}\ \chi_{\mu i}\ a^{\dagger}_{\mu}(\delta)\, b^{\dagger}_{i}(\delta) + \dots \,. \tag{1.1}$$

In the 1970s, there were many theoretical attempts [12] to describe the states in the large amplitude region by properly taking account of the above-stated non-linearity and by going beyond the small-amplitude approximation, i.e. the random phase approximation (RPA). The main purpose of these attempts was to dynamically extract the collective path on which the large scale collective motion is bound. Namely, the description of large-amplitude collective states far from the stationary mean-field had been the main issue in these attempts. What was developed is called the adiabatic TDHF (ATDHF) theory and had been applied to the actual phenomena. By using a simple model calculation, it was shown that the ATDHF theory gives a good description, provided the collective motion is relatively slow. It turned out, however,that the extraction of a collective path form the huge dimensional ***coordinate space*** was very difficult and sometime, incompatible with the microscopic equations of motion, if the collective motion under consideration is not sufficiently slow compared to the other non-collective motions.

In 1980, the above difficulty of the ATDHF theory was subjugated by the self-consistent collective coordinate (SCC) method [13], which was proposed on the basis of the "invariance principle of the time-dependent Schödinger equation" introduced by Marumori [14]. The superiority of the SCC method rests on the fact that it intends to extract the collective submanifold out of the TDHF-manifold, i.e. ***phase space*** rather than the ***coordinate space***. What we will discuss in this contribution is a further development of the SCC method, aimed at getting a unified understanding of various phenomena exhibited by the nucleus.

In §2, we will discuss how to define the concept of classical integrability in the context of nuclear collective dynamics within the TDHF theory. We will also discuss the dynamical condition at the onset of classical chaos by paying special attention to its relation to the level crossing dynamics. In §3, we will discuss the quantum theory of nuclear collective dynamics which has been developed by extending the basic idea discussed in §2. The quantum theory allows us to properly introduce the concepts of quantum integrability and quantum chaoticity for individual eigenfunctions. Special attention will be paid to show how the quantum theory of nuclear collective dynamics has been developed under instruction of the single-particle dynamics formulated within the TDHF theory. By using numerical calculation, we will illustrate what the quantum chaos for each eigenfunction means and relate it to the usual discussion of quantum chaos for the nearest neighbor level spacing statistics based on the random matrix theory. Future perspectives of developing the fundamental theory of nuclear matter physics will be discussed in §4.

2. Nuclear Collective Dynamics in the TDHF Theory

2.1. Symplectic Structure of the TDHF-Manifold

Let us discuss the classical-level theory of nuclear collective dynamics within the TDHF theory. Aiming at treating the non-linear dynamics specifically for the nuclear system stated in the previous section and making clear the dynamics of collective motion in connection with the single-particle dynamics, we start with the TDHF equation

$$\delta < \phi | \hat{H} - i\frac{\partial}{\partial t} | \phi > = 0 \ . \tag{2.1}$$

Here and hereafter, we use a convention $\hbar = 1$. In Eq. (2.1), $|\phi>$ denotes a time-dependent single Slater determinant state given by

$$|\phi\rangle = \exp\left\{i\hat{F}\right\}|\phi_0\rangle, \quad \hat{F} = \sum_{\mu,i}\left\{f_{\mu i}\, a_\mu^\dagger b_i^\dagger + f_{\mu i}^{\,*}\, b_i a_\mu\right\} , \tag{2.2}$$

where $|\phi_0>$ denotes a certain Hartree-Fock stationary state satisfying

$$\delta\langle\phi_0|\,\hat{H}\,|\phi_0\rangle = 0 \ . \tag{2.3}$$

In Eq.(2.2), $a_\mu{}^\dagger$ and $b_i{}^\dagger$ represent particle- and hole-operators with respect to $|\phi_0>$

$$a_\mu|\phi_0\rangle = 0 \ , \ \mu = 1,\dots, M ,$$

$$b_i\,|\phi_0\rangle = 0 \ , \ i = 1,\dots, N . \tag{2.4}$$

In Eq. (2.4), M (N) denotes a number of single-particle (-hole) states. Instead of 2MN expressing a number of particle-hole amplitudes $\{f_{\mu i}^*, f_{\mu i}\}$, we hereafter use a number 2K (=2MN). Since $\{f_{\mu i}^*, f_{\mu i}\}$ are not canonical variables, we introduce a set of 2K canonical variables $\{C_{\mu i}^*, C_{\mu i}\}$ defined by

$$C_{\mu i} = \left\{ \frac{\sin\sqrt{f f^\dagger}}{\sqrt{f f^\dagger}} f \right\}_{\mu i}, \qquad C_{\mu i}^* = \left\{ f^\dagger \frac{\sin\sqrt{f f^\dagger}}{\sqrt{f f^\dagger}} \right\}_{i\mu} . \tag{2.5}$$

By means of $\{C_{\mu i}^*, C_{\mu i}\}$ thus defined, the TDHF equation (2.1) is expressed as

$$\delta \langle \phi(C,C^*)| \sum_{\mu,i} \left\{ i\dot{C}_{\mu i} \frac{\partial}{\partial C_{\mu i}} + i\dot{C}_{\mu i}^* \frac{\partial}{\partial C_{\mu i}^*} \right\} - \hat{H} |\phi(C,C^*)\rangle = 0 \ ,$$
$$|\phi(C,C^*)\rangle \equiv |\phi\rangle \ . \tag{2.6}$$

Let us introduce the following one-body operators,

$$\hat{O}_{\mu i}^\dagger \equiv \left\{ \frac{\partial}{\partial C_{\mu i}} e^{i\hat{F}} \right\} \cdot e^{-i\hat{F}}, \qquad \hat{O}_{\mu i} \equiv - \left\{ \frac{\partial}{\partial C_{\mu i}^*} e^{i\hat{F}} \right\} \cdot e^{-i\hat{F}} \ . \tag{2.7}$$

By using Eq. (2.5), it is easily shown that there hold the following "weak" boson-like commutation relations,

$$\langle \phi(C,C^*)| \left[\hat{O}_{\mu i}, \hat{O}_{\nu j}^\dagger \right] | \phi(C,C^*) \rangle = \delta_{\mu\nu} \cdot \delta_{ij} \ ,$$
$$\langle \phi(C,C^*)| \left[\hat{O}_{\mu i}, \hat{O}_{\nu j} \right] | \phi(C,C^*) \rangle = 0 \, . \tag{2.8}$$

With the aid of one-body operators in Eq. (2.7), Eq. (2.6) is rewritten as

$$\delta \langle \phi(C,C^*)| \sum_{\mu,i} \left\{ i\dot{C}_{\mu i} \hat{O}_{\mu i}^\dagger - i\dot{C}_{\mu i}^* \hat{O}_{\mu i} \right\} - \hat{H} |\phi(C,C^*)\rangle = 0 \ . \tag{2.9}$$

Taking the variations in Eq. (2.9) with respect to $\{C_{\nu j}^*, C_{\nu j}\}$ for $|\phi(C,C^*)\rangle$, we get

$$\langle \phi(C,C^*)| [\ \hat{O}_{\nu j}^\dagger \, , \ (\sum_{\mu,i} \left\{ i\dot{C}_{\mu i} \hat{O}_{\mu i}^\dagger - i\dot{C}_{\mu i}^* \hat{O}_{\mu i} \right\} - \hat{H} \ ) \ ] |\phi(C,C^*)\rangle = 0 \ ,$$

$$\langle \phi(C,C^*)| [\ \hat{O}_{\nu j} \, , \ (\sum_{\mu,i} \left\{ i\dot{C}_{\mu i} \hat{O}_{\mu i}^\dagger - i\dot{C}_{\mu i}^* \hat{O}_{\mu i} \right\} - \hat{H} \ ) \ ] |\phi(C,C^*)\rangle = 0 \ . \tag{2.10}$$

By using the weak boson-like commutation relations in Eq. (2.8), Eq. (2.10) reduces into a set of classical Hamilton's equations of motion

$$i\dot{C}_{\mu i} = \frac{\partial H}{\partial C^*_{\mu i}} = \{C_{\mu i}, H\}_{P.B.} ,$$

$$i\dot{C}^*_{\mu i} = -\frac{\partial H}{\partial C_{\mu i}} = \{C^*_{\mu i}, H\}_{P.B.} , \tag{2.11}$$

where $H = H(C,C^*) \equiv <\phi(C,C^*)|\hat{H}|\phi(C,C^*)>$ and

$$\{A , B\}_{P.B.} = \sum_{\mu i}\left(\frac{\partial A}{\partial C_{\mu i}}\frac{\partial B}{\partial C^*_{\mu i}} - \frac{\partial B}{\partial C_{\mu i}}\frac{\partial A}{\partial C^*_{\mu i}}\right) . \tag{2.12}$$

Equation (2.11) clearly shows a symplectic structure of the **TDHF *manifold*** M^{2K} : $\{C_{\mu i}{}^*, C_{\mu i}\}$ [3].

2.2. Extraction of Collective Submanifold

Since the TDHF equation (2.1) is formally equivalent to the classical canonical equations of motion with K-degrees of freedom, Eq. (2.1) defines a trajectory in M^{2K} under a specific initial condition. For the classical dynamical system organized by Eq.(2.11), one may imagine two extreme cases. If an ensemble of trajectories with slightly different initial conditions develops in almost the same way, forming a bundle, the system is classified to be in the ***collective regime*** as shown in Fig.2a. In this case, one could introduce a representative trajectory as well as a ***collective submanifold*** Σ^{2L} to which many trajectories are accumulating. As shown in Fig. 2a, the representative trajectory is travelling on Σ^{2L}. On the other hand, the system is classified to be in the ***chaotic regime***, if many trajectories with almost the same initial conditions develop quite differently, having lost their initial focussing (Fig.2b). In the latter case, one could not introduce any submanifold within M^{2K}. From the above discussion, one may expect that the transition mechanism between "regular" and "chaotic" motions may be studied by extracting the collective submanifold Σ^{2L}

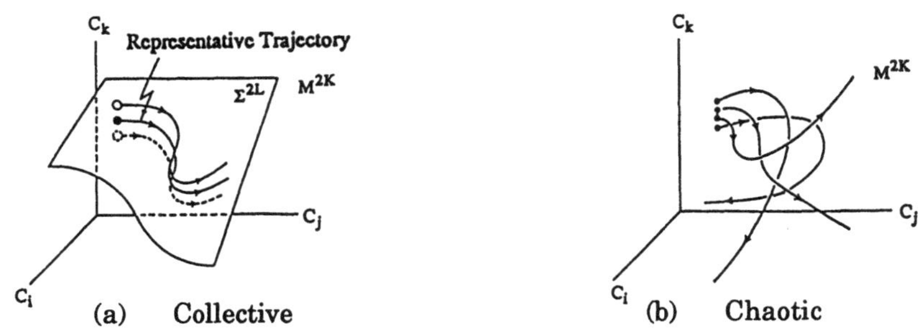

Figure 2. Schematic figures representing the collective and chaotic behaviors of ensemble of TDHF trajectories in the TDHF-manifold M^{2K}. The sheet in (a) represents the collective submanifold Σ^{2L}.

out of M^{2K}. In the collective regime where many trajectories are accumulating on a certain submanifold Σ^{2L}, there must be an optimal coordinate system called the ***dynamical canonical coordinate*** (DCC) system [15] where the minimum number of relevant (collective) variables $\{\eta_a,\eta_a^*:a=1,...,L<<K\}$ is required in describing Σ^{2L}. The remaining canonical variables $\{\xi_\alpha,\xi_\alpha^* : \alpha= L+1,...,K\}$ in the DCC system are called the irrelevant (intrinsic) variables. The coordinate system $\{C_j,C_j^*\}$ used in describing Eq. (2.11) is called the ***initial canonical coordinate*** (ICC) system, because this coordinate system has nothing to do with a specific correlation in the Hamiltonian $H(C,C^*)$.

The extraction of Σ^{2L} is achieved by determining a canonical transformation between the ICC and DCC systems. Since the transformation should be canonical, there must hold the following canonical variable conditions;

$$\sum_j\left\{C_j\frac{\partial C_j^*}{\partial\eta_a^*}-C_j^*\frac{\partial C_j}{\partial\eta_a^*}\right\}=\eta_a+2i\frac{\partial \boldsymbol{S}}{\partial\eta_a^*},\quad a=1,...,L,$$
$$\sum_j\left\{C_j\frac{\partial C_j^*}{\partial\xi_\alpha^*}-C_j^*\frac{\partial C_j}{\partial\xi_\alpha^*}\right\}=\xi_\alpha+2i\frac{\partial \boldsymbol{S}}{\partial\xi_\alpha^*},\quad \alpha=L+1,...,K. \tag{2.13}$$

Here and hereafter, the double indexes (μi) specifying the particle-hole states are simply expressed by the single indexes (j). In Eq. (2.13), $\boldsymbol{S}$ denotes a generating function of the canonical transformation between the ICC and DCC systems. In order to fix the transformation, it is decisive to introduce the following expansion scheme: Since $\{\eta_a,\eta^*_a\}$ are supposed to describe large-amplitude collective motions inside Σ^{2L}, whereas $\{\xi_\alpha,\xi^*_\alpha\}$ describe small fluctuations around Σ^{2L}, it is reasonable to introduce a power series expansion with respect to irrelevant variables. For example,

$$C_j=\sum_n C_j^{(n)}=[C_j]+\sum_\alpha\left\{\xi_\alpha\left[\frac{\partial C_j}{\partial\xi_\alpha}\right]+\xi_\alpha^*\left[\frac{\partial C_j}{\partial\xi_\alpha^*}\right]\right\}+\cdots, \tag{2.14}$$

$$H=\sum_n H^{(n)}=[\,H\,]+\sum_\alpha\left\{\xi_\alpha\left[\frac{\partial H}{\partial\xi_\alpha}\right]+\xi_\alpha^*\left[\frac{\partial H}{\partial\xi_\alpha^*}\right]\right\}+\cdots, \tag{2.15}$$

$$\boldsymbol{S}=\sum_n \boldsymbol{S}^{(n)}=[\,\boldsymbol{S}\,]+\sum_\alpha\left\{\xi_\alpha\left[\frac{\partial \boldsymbol{S}}{\partial\xi_\alpha}\right]+\xi_\alpha^*\left[\frac{\partial \boldsymbol{S}}{\partial\xi_\alpha^*}\right]\right\}+\cdots, \tag{2.16}$$

where the symbol [g] for any function $g(\eta,\eta^*;\xi,\xi^*)$ denotes an expansion function on the expansion surface Σ^{2L}, i.e. $[g] = g(\eta,\eta^*;\xi=0,\xi^*=0)$. In Eqs. (2.14, 15,

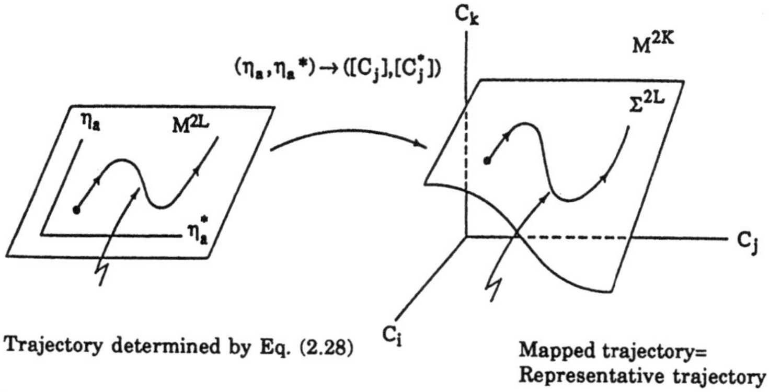

Figure 3. Diffeomorphic mapping defined by the functions $\{[C_j],[C_j^*]\}$.

16), n denotes a power of irrelevant variables contained in each term. Since the lowest order Hamiltonian $H^{(0)} = [H]$ contains only the relevant variables, it is identified to be the collective Hamiltonian $H_{coll}= H^{(0)}$. In the following discussion, we will consider up to the second order Hamiltonian $H^{(2)}$. However, the generalization to any order is straightforward.

Now, let us define Σ^{2L} within M^{2K}. This task is achieved by determining the functional forms of $\{[C_j],[C^*_j]\}$ with respect to $\{\eta_a,\eta^*_a\}$, because we have the following diffeomorphic mapping after getting $\{[C_j],[C^*_j]\}$;

Diffeomorphic Mapping ;

$$M^{2L}:\{\eta_a,\eta_a^*\} \to \Sigma^{2L}:\{[C_j],[C_j^*]\} \text{ embedded in } M^{2K}, \qquad (2.17)$$

which defines Σ^{2L} in M^{2K} as is shown in Fig.3. According to the self-consistent collective coordinate (SCC) method [13], the basic equations for defining Σ^{2L} within M^{2K} are given by the following two requirements

Requirement 1: There should hold the lowest order ***canonical variable condition;***

$$\text{(I)} \quad \sum_j \left\{[C_j]\frac{\partial[C_j^*]}{\partial\eta_a^*} - [C_j^*]\frac{\partial[C_j]}{\partial\eta_a^*}\right\} = \eta_a + 2i\frac{\partial[S]}{\partial\eta_a^*}, \qquad (2.18)$$

which is derived from Eq.(2.13).

Requirement 2: There should hold the ***equation of collective submanifold;***

$$\text{(II)} \quad \left[\frac{\partial H}{\partial C_j^*}\right] = \sum_a \left\{\frac{\partial[H]}{\partial\eta_a^*}\frac{\partial[C_j]}{\partial\eta_a} - \frac{\partial[H]}{\partial\eta_a}\frac{\partial[C_j]}{\partial\eta_a^*}\right\}. \qquad (2.19)$$

With the aid of these two requirements together with the "boundary condition of Σ^{2L}" derived from the initial condition of the representative trajectory under discussion, we uniquely determine Σ^{2L}. An explicit method of getting the diffeomorphic mapping is found elsewhere [13].

2.3 Self-Consistent Constrained Hartree-Fock Theory

In the previous subsection, we have discussed how to extract the integral surface of the representative trajectory out of the classical phase space M^{2K}. Making clear what is done by the SCC method within the single-particle dynamics, we start by reformulating the SCC method within the original TDHF theory. This reformulation clearly shows that the SCC method just corresponds to the self-consistent constrained Hartree-Fock theory, where the constraining operators are dynamically defined following the microscopic dynamics of the system. In marked contrast with the conventional constrained Hartree-Fock theory where the constraining operators are put in by hand, the SCC method intends to extract the constraining operators whose instantaneous directions just coincide with that of the representative trajectory. Consequently, the SCC method reformulated in the single-particle dynamics gives a foundation of the single-particle states in the nuclear potential which changes its shape depending on the amplitude of collective variables $\{\eta_a,\eta^*_a\}$, e.g. the single-particle states in the Nilsson model and the Routhian in the rotating shell model.

By using the canonical variables $\{C_j,C^*_j\}$, the TDHF equation (2.1) is expressed as

$$\delta\left\langle\phi(C,C^*)\right|\left(i\frac{\partial}{\partial t}-\hat{H}\right)\left|\phi(C,C^*)\right\rangle=0\ . \tag{2.20}$$

In accordance with the discussion in § 2.2, the equation of motion of the representative trajectory is described by

$$\begin{aligned}&\delta\left\langle\phi([C],[C^*])\right|\left(i\frac{\partial}{\partial t}-\hat{H}\right)\left|\phi([C],[C^*])\right\rangle\\&=\delta\left\langle\phi([C],[C^*])\right|\left(i\sum_a\left\{\dot{\eta}_a\frac{\partial}{\partial\eta_a}+\dot{\eta}^*_a\frac{\partial}{\partial\eta^*_a}\right\}-\hat{H}\right)\left|\phi([C],[C^*])\right\rangle=0\ .\end{aligned} \tag{2.21}$$

In Eq. (2.21), the canonical variables $\{C_j,C^*_j\}$ are sandwiched by the symbol [ ], because the representative trajectory is bounded on the collective submanifold Σ^{2L} and the time dependence of $\{[C_j],[C^*_j]\}$ is supposed to be described by the collective variables $\{\eta_a,\eta^*_a\}$ alone. In Eq. (2.21),

$$\left|\phi([C],[C^*])\right\rangle=\exp\left\{i[\hat{F}]\right\}\left|\phi_0\right\rangle\ ,\ \ [\hat{F}]=\sum_{\mu,i}\left\{[f_{\mu i}]a^\dagger_\mu b^\dagger_i+[f^*_{\mu i}]b_i a_\mu\right\}. \tag{2.22}$$

Let us introduce the following one-body collective operators defined by

$$\hat{O}_a^\dagger = \left(\frac{\partial}{\partial \eta_a} e^{i[\hat{F}]}\right) \cdot e^{-i[\hat{F}]} \ , \quad \hat{O}_a = -\left(\frac{\partial}{\partial \eta_a^*} e^{i[\hat{F}]}\right) \cdot e^{-i[\hat{F}]} \ . \tag{2.23}$$

Substituting Eq. (2.23) for Eq. (2.21), we get

$$\delta\left\langle \phi([C],[C^*])\right| \left(i\sum_a \left\{ \dot{\eta}_a \hat{O}_a^\dagger - \dot{\eta}_a^* \hat{O}_a \right\} - \hat{H} \right) \left| \phi([C],[C^*])\right\rangle = 0. \tag{2.24}$$

Now, let us discuss how to express the canonical variable condition (2.18) in the present case. The collective variables $\{\eta_a, \eta_a^*\}$ are certified to be canonical, provided there hold the following relations;

$$\left\langle \phi([C],\ [C^*])\right| \hat{O}_a^\dagger \left| \phi([C],\ [C^*])\right\rangle = \frac{1}{2}\eta_a^* \ - i\frac{\partial [S]}{\partial \eta_a}. \tag{2.25}$$

Equation (2.25) represents the ***canonical variable condition*** and is equivalent to Eq. (2.18). Using Eq. (2.25), we derive the following "weak" boson-like commutation relation

$$\left\langle \phi([C],[C^*])\right| [\hat{O}_b \ , \ \hat{O}_a^\dagger] \left| \phi([C],[C^*])\right\rangle = \delta_{ab} \ ,$$
$$\left\langle \phi([C],[C^*])\right| [\hat{O}_b \ , \ \hat{O}_a] \left| \phi([C],[C^*])\right\rangle = 0. \tag{2.26}$$

It should be noticed that these relations hold at any value of (η_a, η_a^*), whereas the conventional RPA modes satisfy the boson-like commutation relation at one point of the stationary Hartree-Fock state $|\phi_0>$. In marked contrast with the conventional RPA modes, the collective operators defined in Eq. (2.23) contain the particle-particle and hole-hole operators with respect to the state $|\phi([C],[C^*])>$ in addition to the particle-hole operators. This desirable property of the collective operators assures that $\{\hat{O}_a, \hat{O}_a^\dagger\}$ are ***global*** operators, whereas the RPA modes are ***local*** operators. Namely, the existence of particle-particle and hole-hole terms makes it possible to describe the structural change of collective operators $\{\hat{O}_a, \hat{O}_a^\dagger\}$, depending on the amplitudes of $\{\eta_a, \eta_a^*\}$.

Taking the variation of $|\phi([C],[C^*])>$ in Eq. (2.24) with respect to η_b and η_b^*, we get

$$\left\langle \phi([C],[C^*])\right| \ [\hat{O}_b, \left(i\sum_a \left\{ \dot{\eta}_a \hat{O}_a^\dagger - \dot{\eta}_a^* \hat{O}_a \right\} - \hat{H} \right)] \ \left| \phi([C],[C^*])\right\rangle = 0 \ ,$$

$$\left\langle \phi([C],[C^*])\right| \ [\hat{O}_b^\dagger, \left(i\sum_a \left\{ \dot{\eta}_a \hat{O}_a^\dagger - \dot{\eta}_a^* \hat{O}_a \right\} - \hat{H} \right)] \ \left| \phi([C],[C^*])\right\rangle = 0 \ . \tag{2.27}$$

By using Eq. (2.26), Eq. (2.27) reduces into the "classical" equations of motion

$$i\dot{\eta}_a = \left\langle \phi([C],[C^*]) \right| [\hat{O}_a,\hat{H}] \left| \phi([C],[C^*]) \right\rangle = \frac{\partial[H]}{\partial\eta_a^*} = \frac{\partial H_{coll}}{\partial\eta_a^*},$$
$$i\dot{\eta}_a^* = \left\langle \phi([C],[C^*]) \right| [\hat{O}_a^\dagger,\hat{H}] \left| \phi([C],[C^*]) \right\rangle = -\frac{\partial[H]}{\partial\eta_a} = -\frac{\partial H_{coll}}{\partial\eta_a} , \qquad (2.28)$$

which describe the representative trajectory of the bundle on the collective submanifold Σ^{2L}. Substituting Eq. (2.28) for Eq. (2.27), we get

$$\delta\left\langle \phi([C],[C^*]) \right| \left(\sum_a \left\{ \frac{\partial[H]}{\partial\eta_a}\hat{O}_a^\dagger + \frac{\partial[H]}{\partial\eta_a^*}\hat{O}_a \right\} - \hat{H} \right) \left| \phi([C],[C^*]) \right\rangle = 0 , \qquad (2.29)$$

which represents the ***equation of collective submanifold*** and is equivalent to Eq. (2.19). If one takes a variation in Eq. (2.29) toward the relevant direction, one gets the equation of collective motion in Eq. (2.28). If one takes a variation toward the irrelevant direction perpendicular to collective directions, one has

$$\delta_\perp\left\langle \phi([C],[C^*]) \right| \left(\sum_a \left\{ \frac{\partial[H]}{\partial\eta_a}\hat{O}_a^\dagger + \frac{\partial[H]}{\partial\eta_a^*}\hat{O}_a \right\} - \hat{H} \right) \left| \phi([C],[C^*]) \right\rangle$$
$$= -\ \delta_\perp\left\langle \phi([C],[C^*]) \right| \hat{H} \left| \phi([C],[C^*]) \right\rangle = 0 . \qquad (2.30)$$

In Eq. (2.30), $\delta_\perp$ means variations perpendicular to the collective directions. In the language of classical-level theory in §2.2, Eq. (2.30) just corresponds to the maximal-decoupling condition given by

$$H^{(1)} = 0 \text{ , i.e. } \left[\frac{\partial H}{\partial \xi_\alpha}\right] = \left[\frac{\partial H}{\partial \xi_\alpha^*}\right] = 0 . \qquad (2.31)$$

From Eq.(2.31), it is clear that the collective submanifold Σ^{2L} is defined so as to satisfy the ***stationary condition*** with respect to variations towards the irrelevant directions.

At this point, it should be remembered that the Hartree-Fock state just corresponds to the stationary ***point*** of the total Hamiltonian (potential energy) in the coordinate space: { $Re(C_j)$, $Re(C_j^*)$}, satisfying the condition

$$\frac{\partial H(Re(C), Re(C^*))}{\partial Re(C)} = 0 , \qquad (2.32)$$

whereas the collective submanifold corresponds to the stationary ***surface*** of the total Hamiltonian in the phase space M^{2K} :{C_j, C^*_j}. In this sense, the SCC method is a natural extension of the Hartree-Fock theory from the stationary ***point*** to the stationary 2L-dimensional ***surface*** Σ^{2L}.

Here, it is worthwhile mentioning that the conventional constrained Hartree-Fock theory satisfies the stationary condition toward the directions perpendicular to the constraining operators. It is also the case in Eq. (2.30). Namely, Eq. (2.30) is just a generalization of the constrained Hartree-Fock equation given by

$$\delta\langle\omega|\ \omega\hat{J}-\hat{H}\ |\omega\rangle=0\ \ ,\ \text{i.e.}\ \ \delta_{\perp}\langle\omega|\,\hat{H}\ |\omega\rangle=0\ ,$$
$$\delta\langle q|\ q\hat{Q}-\hat{H}\ |q\rangle=0\ \ ,\ \text{i.e.}\ \ \delta_{\perp}\langle q|\,\hat{H}\ |q\rangle=0\ . \qquad (2.33)$$

In Eq. (2.33), $\delta_{\perp}$ means variations perpendicular to the collective directions, i.e. $\hat{J}$ and $\hat{Q}$. In the conventional approach, the constraining operators $\hat{J}$ and $\hat{Q}$ have fixed microscopic structure and are put in by hand. On the contrary, the microscopic structure of collective operators $\{\hat{O}_a,\hat{O}_a{}^{\dagger};a=1,...,L\}$ in Eq. (2.30) change depending on the position of the representative trajectory in Σ^{2L} and they are determined dynamically by means of the SCC method which properly takes account of the non-linear, microscopic single-particle dynamics in the TDHF theory. Namely, the SCC method can extract the dynamical constraining operators, on which the system under discussion develops by itself.

It should be also noticed that the state $|\phi([C],[C^*])\rangle$ in Eq. (2.29) satisfies the Hartree-Fock stationary condition. Namely, the auxiliary Hamiltonian in the generalized moving frame defined by

$$\hat{H}_{\text{moving}} = \hat{H} - \left\{\ \omega\hat{J}\ \right\}_{\text{generalize}}\ , \qquad (2.34)$$

with the generalized cranking term

$$\left\{\ \omega\hat{J}\ \right\}_{\text{generalize}} = \sum_{a}\left\{\frac{\partial[H]}{\partial\eta_a}\hat{O}_a^{\dagger}+\frac{\partial[H]}{\partial\eta_a^{*}}\hat{O}_a\right\}\ , \qquad (2.35)$$

has the following property; the Hartree-Fock Hamiltonian (, i.e. the single-particle part of the auxiliary Hamiltonian in Eq. (2.34) with respect to the single Slater determinant state $|\phi([C],[C^*])\rangle$) has no particle-hole components with respect to $|\phi([C],[C^*])\rangle$. This is due to the fact that there holds the stationary relation (2.29). In this case, one may introduce a set of single-particle states at any point on $|\phi([C],[C^*])\rangle$, i.e. at any value of $\{\eta_a,\eta_a{}^*\}$. This $\{\eta_a,\eta_a{}^*\}$-dependence of the single-particle states is due to the existence of particle-particle and hole-hole components in the dynamical constraining operators $\{\hat{O}_a,\hat{O}_a{}^{\dagger}\}$ and may give a microscopic foundation of the single-particle states (Routhian) in the nuclear potential which depends on collective coordinates $\{\eta_a,\eta_a{}^*\}$, e.g. the Nilsson model, rotating shell model etc.

2.4. Level-Crossing Dynamics and Onset of Classical Chaos

In §2.2, we have discussed how to extract the collective submanifold Σ^{2L} on which the representative trajectory of the bundle is travelling. In §2.3, we have discussed that this extraction is equivalent to dynamically defining the constraining collective operators of the self-consistent constrained Hartree-Fock theory. Namely the collective operators thus defined are global operators and are well defined at any point of the collective submanifold, changing their microscopic structure. Moreover, they coincide at any instant with the direction on which the system develops by itself.

At this point, let us discuss that the DCC system defined by the SCC method is not unique, because the separation between relevant and irrelevant variables is invariant under any canonical transformation among relevant variables and among irrelevant variables [16]. Even though this ambiguity does not change any discussion in §§2.2 and 2.3., this freedom of choice allows us to ask ***whether or not the system is integrable*** [17].

In order to get a unique DCC system, let us first determine the canonical transformation among collective variables, i.e.

$$\text{Canonical Transformation;} \quad \{\eta_a^*, \eta_a\} \leftrightarrow \{\eta_a'^*, \eta_a'\} \ . \tag{2.36}$$

This transformation is determined by

Requirement 3: The resultant collective Hamiltonian after the transformation in Eq. (2.36) should depend on collective action variables only.

With the aid of the canonical transformation in Eq.(2.36) thus determined, the Hamiltonian is transformed as

$$H_{coll}(\eta,\eta^*) + H(2) \leftrightarrow H'_{coll}(N_1,\ldots,N_L) + H'(2) \quad ; N_a = \eta_a^* \eta_a = \eta_a'^* \eta_a' \ , \tag{2.37}$$

where the collective Hamiltonian H'_{coll} is expressed by collective action variables $\{N_a\}$ alone. This canonical transformation always exists as far as the Birkhoff-Gustavson normal-form expansion method [18] converges. Secondly, we determine a canonical transformation among irrelevant variables, i.e.

$$\text{Canonical Transformation} \quad (N_a, \xi_\alpha, \xi_\alpha^*) \leftrightarrow (\tilde{N}_a, \tilde{\xi}_\alpha, \tilde{\xi}_\alpha^*) \ . \tag{2.38}$$

This transformation is fixed by imposing

Requirement 4: There should hold the ***separability condition*** given by

$$\{H''_{coll}, H''(2)\}_{P.B.} = 0 \ . \tag{2.39}$$

Here $H''_{coll.}$ and $H''(2)$ denote the resultant collective and intrinsic Hamiltonians, respectively. Corresponding to the canonical transformation in Eq. (2.38), the Hamiltonian is transformed as

$$H'_{coll.}(N)+H'(2) \leftrightarrow H''_{coll.}(\tilde{N})+H''(2) \; ; \; H''(2) = \Sigma_\alpha \omega_\alpha(\tilde{N})\,\tilde{n}_\alpha \, ,$$
$$\tilde{N}_a = N_a \;\; , \;\; \tilde{n}_a = \tilde{\xi}^*_\alpha \xi_\alpha \; . \tag{2.40}$$

It is obvious from the above discussion that the last coordinate system is defined such that both the collective and intrinsic Hamiltonians are expressed by only the action variables for collective and intrinsic degrees of freedom. Since the resultant Hamiltonian clearly shows an integrability of the system up to the second order Hamiltonian, the canonical coordinate system $(\tilde{N},\tilde{\xi},\tilde{\xi}^*)$ is called the DCC system with ***maximum integrable-form representation*** [17].

Here, it should be mentioned that a canonical transformation can't meet Requirement 4 if there holds a ***resonance condition.*** In the language of single-particle dynamics, the resonance condition is satisfied at the level-crossing point. Consequently, an essentially non-integrable interaction manifests itself at the level-crossing point.

In our classical-level theory, validity of the collective submanifold Σ^{2L} determined by Requirements 1 and 2 is studied by both the ***stability*** and ***separability*** conditions. As stated above, violation of the separability condition (2.39) means an appearance of non-integrable motion, i.e. an onset of classical chaos. As is easily understood, the resonance point just corresponds to the "weak stability point". According to Prigogine [19], when there appear many weak stability points, the **physics of being** , i.e. the dynamics described by a single trajectory organized by the Hamilton equation of motion should be replaced by the **physics of becoming**, i.e. the dynamics described by a distribution function following the Liouville equation. Consequently, the level-crossing dynamics may play a decisive role in studying the important question of why statistical or probabilistic features result from deterministic dynamics. This point will be discussed in §4.

The stability condition of Σ^{2L} is evaluated by calculating the small-amplitude oscillations toward irrelevant directions at each point on Σ^{2L}. If there holds the stability condition,

$$\omega^2_\alpha(N) > 0 \;\; , \tag{2.41}$$

the representative trajectory is running along the valley of the contour map of the Hamiltonian. In this case, many trajectories are expected to accumulate around the representative trajectory, indicating the collectivity of the extracted submanifold. If the condition (2.41) is not satisfied, the representative trajectory is travelling on the ridge-line of the Hamiltonian, showing the stochasticity of the system.

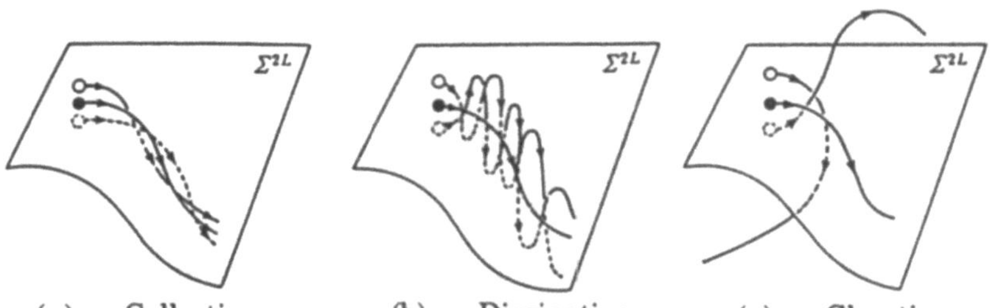

Figure 4. Characteristic behavior of TDHF trajectories starting from the vicinity of collective submanifold Σ^{2L}.

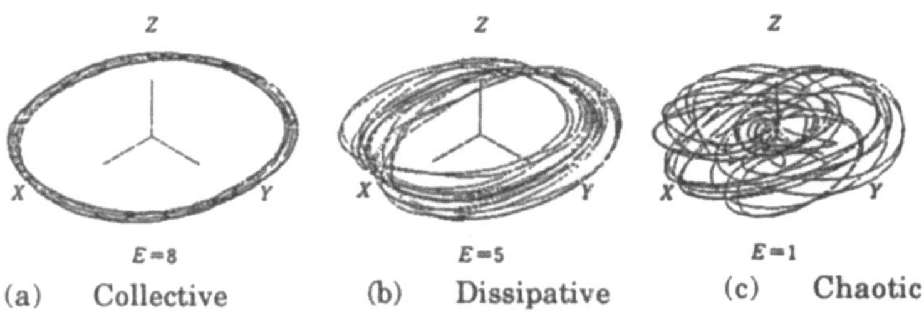

Figure 5. TDHF trajectories of the SU(3) model Hamiltonian starting from the vicinity of Σ^{2L} for three different cases. Detailed explanation is found in ref. [20].

With the aid of both the stability and separability conditions, we may divide the collective submanifold Σ^{2L} into three characteristic regions.

	Collective Region	**Dissipative Region**	**Chaotic Region**
Stability	$\omega^2 >> 0$	$\omega^2 > 0$	$\omega^2 \leq 0$
Separability	well separated	weakly separated	strongly mixed

In the collective region, a TDHF trajectory starting from the vicinity of Σ^{2L} is expected to stay always in a tiny region near Σ^{2L}. In the dissipative region, a TDHF trajectory found near Σ^{2L} at certain instant may oscillate toward the irrelevant direction (perpendicular to the submanifold) with a fairly large amount of deviation from Σ^{2L}. In the chaotic region, the trajectory occasionally travelling near Σ^{2L} does not come back in its neighborhood. These typical situations are illustrated in Fig. 4. With the aid of a simple soluble model Hamiltonian with SU(3) symmetry, we have evaluated [20] the stability and separability conditions and divided the collective submanifold into three characteristic regions. By calculating the TDHF trajectory starting from the vicinity of the collective submanifold, we get three characteristic trajectories shown in Fig. 5. As is clearly seen from Fig. 5, the stability and separability conditions give important information on the extracted submanifold Σ^{2L}.

This section is summarized by stating that the introduction of the collective submanifold Σ^{2L} associated with the DCC system for a given trajectory and that of the stability and separability conditions for Σ^{2L} are decisive for studying whether the system under discussion is regular or chaotic and how the regular system transfigures into the chaotic system. If the collective subspace Σ^{2L} satisfies both the stability and separability conditions, the auxiliary Hamiltonian defined in Eq. (2.34) has physical significance. In this case, an introduction of the $\{\eta_a, \eta_a^*\}$-dependent Hartree-Fock single-particle states defined in Eq. (2.29) is also justified. If Σ^{2L} does not satisfy the separability condition, the $\{\eta_a, \eta_a^*\}$-dependent single-particle states have no physical meaning, because the adiabatic treatment (or more generally the separate treatment) of the collective (relevant) degrees of freedom from the irrelevant degrees of freedom is not justified. In this way, one may evaluate the validity of various stable-mean fields. It is also clear from the above discussion that the order-to-chaos transition of collective motion is directly related to the properties of single-particle states.

3. Quantum Theory of Nuclear Collective Dynamics

3.1. How to Define Quantum Chaoticity for Individual Eigenfunctions

In this section, we will discuss the quantum theory of many-body problems which has been developed under instruction of what was developed in the classical-level theory summarized in §2. Namely, we will try to extend the concepts of the DCC system, separability condition and stability condition in §2 to the quantum many-body system. This extension allows us to introduce the concepts of quantum integrability and quantum chaoticity for individual eigen-functions, which is one of the current issues in a variety of fields of physics [21].

Since the even-number many-fermion system is transformed into the boson system by means of the boson expansion theory [22], it is general enough to formulate a quantum theory of nuclear collective dynamics [23] in the boson system which is described by K-kinds of boson operators

$$\{B_j, B_j^\dagger;\ j = 1,\ldots,K\}\ . \tag{3.1}$$

The representation using $(B, B^\dagger)$ is called an ***initial representation*** (IR). Let us consider a special situation where a set of collective excited states constitutes an approximately invariant subspace (i.e. collective subspace) $\mathcal{D}^{2L}$ of the Hamiltonian. In this case, there must be an optimal representation called ***dynamical representation*** (DR) where the minimum number of relevant operators is required in describing $\mathcal{D}^{2L}$. Let

$$\{\alpha_j,\alpha_j^\dagger;\ j=1,\dots,L<<K\} \tag{3.2a}$$

be a set of relevant (collective) operators necessary for spanning $\mathcal{D}^{2L}$ and let

$$\{\beta_j,\beta_j^\dagger;\ j=L+1,\dots,K\} \tag{3.2b}$$

be the remaining operators called the irrelevant (intrinsic) operators.

The extraction of an approximately invariant subspace is achieved by finding a non-linear unitary transformation between IR and DR given by

$$\alpha_j^\dagger = e^{F(B^\dagger,B)}B_j^\dagger e^{-F(B^\dagger,B)}\ ;\quad j=1,\dots,L\ ,$$
$$\beta_j^\dagger = e^{F(B^\dagger,B)}B_j^\dagger e^{-F(B^\dagger,B)}\ ;\quad j=L+1,\dots,K\ . \tag{3.3}$$

In order to determine a generating function $F(B,B^\dagger) = -F^\dagger(B,B^\dagger)$ of the unitary transformation, we apply the following expansion scheme. Since the relevant operators are supposed to describe large-amplitude collective motions within $\mathcal{D}^{2L}$ whereas the irrelevant operators represent small-amplitude fluctuations around $\mathcal{D}^{2L}$, it is reasonable to introduce an expansion in power series of the irrelevant boson operators. For example, the Hamiltonian in the DR is expanded as

$$H(B,B^\dagger) = \mathcal{H}_{coll.}(\alpha,\alpha^\dagger) + \Delta\mathcal{H}(\alpha,\alpha^\dagger;\beta,\beta^\dagger)\ , \tag{3.4}$$

The right hand side of Eq. (3.4) is expressed in the normal ordered form with respect to the operators $\{\alpha,\alpha^\dagger;\beta,\beta^\dagger\}$ of the DR. In Eq. (3.4), $\mathcal{H}_{coll.}$ corresponds to the lowest order term consisting of only the relevant operators and is identified to be the collective Hamiltonian. The remaining part is expanded as

$$\Delta\mathcal{H} = \Sigma_{n=1}\mathcal{H}(n)\ , \tag{3.5}$$

where n denotes a power of the irrelevant bosons contained in $\mathcal{H}(n)$. It was shown [23] that the unitary transformation in Eq.(3.3) is determined by requiring the maximal-decoupling condition given by

$$\mathcal{H}(1) = 0\ , \tag{3.6}$$

together with a boundary condition of $\mathcal{D}^{2L}$, i.e. the structure of $\alpha_j^\dagger$ in its small-amplitude limit. The explicit procedure of getting the non-linear unitary transformation is found elsewhere [23].

After getting the DR, one may introduce (m,n)-basis states defined as

$$|m,n) = \prod_{j=1}^{L}\frac{1}{\sqrt{m_j!}}(\alpha_j^\dagger)^{m_j}\cdot\prod_{j=L+1}^{K}\frac{1}{\sqrt{n_j!}}(\beta_j^\dagger)^{n_j}\,|0)\ , \tag{3.7}$$

where m (n) collectively expresses the relevant (irrelevant) quantum numbers. By means of (m,n)-basis states, the collective subspace $\mathcal{D}^{2L}$ is expressed as

$$\mathcal{D}^{2L} : \{ |m, n=0) \} , \tag{3.8}$$

with no intrinsic excitations. As is discussed elsewhere [23], the validity of the (m,n)-basis states, i.e. that of the collective subspace $\mathcal{D}^{2L}$ thus defined, is studied by both the separability and stability conditions. The separability condition is given by

$$[\mathcal{H}_{coll.}(\alpha,\alpha^{\dagger}) , \mathcal{H}(2)] = 0 , \tag{3.9}$$

which is just a quantum analogue of Eq. (2.39). The stability condition is given in such a way that the small-amplitude irrelevant motions described by the RPA-type equation with $\mathcal{H}(2)$ should remain in a tiny region around $\mathcal{D}^{2L}$.

Here, it should be noticed that the number of commuting observables, i.e. the number of good quantum numbers specifying individual eigenstates, has a special importance in the quantum mechanics. Especially, the (m,n)-basis states defined in Eq.(3.7) are specified by K-kinds of quantum numbers associated with the DR. Without any loss of generality, it was shown [24] that the dynamical correlations coming from a part of $\Delta\mathcal{H}$ violating the stability condition can be incorporated into the basis states under the restriction of conserving the number of quantum numbers. Namely, it is always possible to introduce the (μ,ν)-basis states

$$|\mu,\nu) = |\mu_1,\cdots,\mu_L;\nu_{L+1},\cdots,\nu_K) , \tag{3.10}$$

which are still specified by K-kinds of quantum numbers. Since the (μ,ν)-basis states thus introduced are the best, one may choose in incorporating the correlations into the basis states as much as possible without decreasing the number of quantum numbers, the (μ,ν)-basis states are nothing but a quantum analogue of the DCC system with ***maximum integrable-form representation***. The validity of the (μ,ν)-basis states is estimated by the separability condition alone, because they are defined so as to automatically satisfy the stability condition.

Let {| i)} be a set of exact eigenstates of the total Hamiltonian H(B,B†). From the above discussion, it is clear that the exact eigenstate | i) is no longer specified by one of the (μ,ν)-basis states when the separability condition is violated. Namely the exact eigenstate is not specified by K-kinds of quantum numbers any more but by a smaller number of quantum numbers in accordance with the violation of the separability condition. In other words, there occurs a ***dissolution of quantum numbers*** when the separability condition does not hold. In the classical theory, a dissolution of integrability simply means an onset of chaotic motion. In the quantum system, therefore, it is reasonable to state that a dissolution of quantum numbers may indicate an onset of "quantum chaos". In accordance with the above definition of quantum chaos, we may classify the exact eigenstates $\{|i\rangle\}$ into three characteristic cases with the aid of the (μ,ν)-basis states;

Quantum integrable motion ; If one finds one of the (μ,ν)-basis states for a given eigenstate $|i\rangle$ satisfying $|\langle\mu,\nu|i\rangle|^2 \cong 1$, then $|i\rangle$ is classified to be in "quantum integrable motion" because it is specified by K-kinds of quantum numbers.

Quantum KAM motion; If $|i\rangle$ is described perturbatively starting from the most appropriate (μ,ν)-basis state, then it is specified to be in "quantum KAM motion" because it is still characterized by K-kinds of asymptotic quantum numbers.

Quantum chaotic motion ; If $|i\rangle$ is not described perturbatively starting from any (μ,ν)-basis states, then it is regarded to be in "quantum chaotic motion" because $|i\rangle$ is not characterized by K-kinds of quantum numbers any more.

3.2. Application of Quantum Theory

In the previous subsection, we have introduced the DR as well as the separability and stability conditions. We have further defined (μ,ν)-basis states which allow us to characterize the quantum integrability and quantum chaoticity for individual eigenfunctions. In order to illustrate the quantum theory of nuclear collective dynamics developed in §3.1, let us apply it to a simple soluble three-level SU(3) model Hamiltonian by Li, Klein and Dreizler [25] with slight modification given as

$$\hat{H} = \sum_{i=0}^{2} \varepsilon_i \hat{K}_{ii} + \frac{V_1}{2}(\hat{K}_{10}\hat{K}_{10} + \text{h.c.}) + \frac{V_2}{2}(\hat{K}_{20}\hat{K}_{20} + \text{h.c.}) ,$$

$$\hat{K}_{ij} \equiv \sum_{m=1}^{N} C^{\dagger}_{im} C_{jm} \ . \tag{3.11}$$

There are three single-particle levels with energies $\varepsilon_0<\varepsilon_1<\varepsilon_2$ and each level has N-fold degeneracy. Hereafter, we treat the even number N-particle system where the lowest level with i=0 is completely occupied. Recently, the spectral statistics of the SU(3) model Hamiltonian have been extensively studied by Meredith, Koonin and Zirnbauer [26]. More recently, the structure of eigenfunctions of the same Hamiltonian has been explored [27] in terms of classical trajectories by using Husimi distribution [28].

By employing the generalized Holstein-Primakoff boson expansion theory [22], the fermion Hamiltonian in Eq.(3.11) is transformed into the boson Hamiltonian

$$H(B,B^\dagger) = (\varepsilon_1 - \varepsilon_0)B_1^\dagger B_1 + (\varepsilon_2 - \varepsilon_0)B_2^\dagger B_2$$
$$+\frac{V_1}{2}(B_1^\dagger\sqrt{N - B_1^\dagger B_1 - B_2^\dagger B_2}B_1^\dagger\sqrt{N - B_1^\dagger B_1 - B_2^\dagger B_2} + \text{h.c.})$$
$$+\frac{V_2}{2}(B_2^\dagger\sqrt{N - B_1^\dagger B_1 - B_2^\dagger B_2}B_2^\dagger\sqrt{N - B_1^\dagger B_1 - B_2^\dagger B_2} + \text{h.c.})\,. \tag{3.12}$$

The "physical" boson space corresponding to the fermion space is given by

$$|m,n) = \frac{1}{\sqrt{m!}}\frac{1}{\sqrt{n!}}(B_1^\dagger)^m(B_2^\dagger)^n|0) \quad , \text{ with } 0 \le m+n \le N\,. \tag{3.13}$$

In this paper, we will consider a set of "collective" excited states whose small-amplitude motion near the stationary state $|\,0\,)$ is described by a single kind of boson operators $\{B_1{}^\dagger, B_1\}$. Since the maximal-decoupling condition (3.6) is satisfied by the Hamiltonian in Eq.(3.12), the representation using the boson operators $\{B_i{}^\dagger, B_i; i= 1,2\}$ just corresponds to the DR for the present case. Namely, a set of basis states in Eq.(3.13) provides us with the (m,n)-basis states defined in Eq.(3.7).

According to the theory discussed in §3.1, the chaoticity of each eigenstate is visualized by expanding it in terms of the (μ,ν)-basis states;

$$|i) = \sum_{\mu\nu} a^i_{\mu\nu}|\mu,\nu)\ . \tag{3.14}$$

The $\mu\nu$-representation of $\{|\,i\,)\}$ is shown in Fig.6, where the size of a solid circle is proportional to the square of expansion coefficients $|a_{\mu\nu}{}^i|^2$. The parameters used in calculating Fig.6 are N=10, ε_1-ε_0=1 and ε_2-ε_0=$\sqrt{2}$. According to the special symmetry of the interaction in Eq. (3.11), the Hamiltonian conserves the evenness or oddness of the population in the relevant and irrelevant boson numbers. In our numerical calculation, we consider the subspace with even number operators for both the relevant and irrelevant bosons. If the strength of

(a)V_1=V_2=-0.022 (b)V_1=-0.11,V_2=-0,33 (c)V_1=V_2=-0.33

Figure 6. Fractional parentage plots for the case with N=10. (a) quantum integrable motion, (b) quantum KAM motion and (c) quantum chaotic motion.

the interaction is relatively weak, each eigenstate $|i\rangle$ is characterized by one of the (μ,ν)-basis states. In this case (Fig.6a), the system may be classified to be in "quantum integrable motion", because each eigenstates are specified by K=2 quantum numbers. If the force strength becomes large (Fig.6c), the expansion coefficients are distributed over a wide range of the (μ,ν)-basis states, indicating the dissolution of quantum numbers associated with the DR. In this case, the system may be classified to be in "quantum chaotic motion", because it is hard to characterize individual eigenstates by K=2 quantum numbers. In the intermediate case (Fig.6b), the expansion coefficients spread out over many (μ,ν)-basis states. Since the diagonal components still remain the largest, the system in Fig. 6b may be classified to be in "quantum KAM motion". These figures are called the ***fractional parentage plots*** (FPP). Since the FPP clearly shows the chaoticity of individual eigenfunctions, it is regarded as a quantum analogue of the Poincaré section map which clearly shows the chaoticity of individual trajectories.

Let us discuss the relation between the present definition of quantum chaos for each eigenstate and the usual definition of quantum chaos for an ensemble of eigenstates. As is well known, there is a well established method for describing the statistical aspects of an ensemble of nuclear excited states near the neutron threshold. By using the random matrix theory [29] which is based on some probabilistic assumptions, the level statistics of neutron resonance states has been discussed in detail [30]. As a relevant discussion to the quantum chaos, the following empirical rule is frequently stated: If a classical system is integrable, its quantum correspondent system shows the Poisson distribution in the nearest neighbor level spacing statistics. If it is non-integrable or chaotic,the corresponding quantum system shows the Wigner distribution. Basing on this argument, it is widely recognized that the quantum chaos is characterized by the Wigner distribution and the order-to-chaos transition mechanism in the quantum system may be studied by employing the level statistics. Here it should be noticed that the discussion based on the Gaussian Orthogonal Ensemble (GOE) treats statistical aspects of an ensemble of eigenstates, whereas the classical order-to-chaos transition is discussed by means of dynamical properties exhibited by individual trajectories. Since the quantum chaos defined in §3.1 is given for individual eigenfunctions, it is quite interesting to discuss how the nearest neighbor level spacing statistics relates to the quantum chaos defined above for each eigenfunction.

In order to get the nearest neighbor level spacing statistics, we calculated the Hamiltonian in Eq. (3.12) with N=60, ε_1-ε_0=1 and ε_2-ε_0= 2. The final dimension of the Hamiltonian matrix to be diagonalized is N_d=496 which might be sufficiently large for discussing a variety of statistical aspects of an ensemble of eigenstates. In Fig.7, the FPP is shown for three different cases. In this figure,

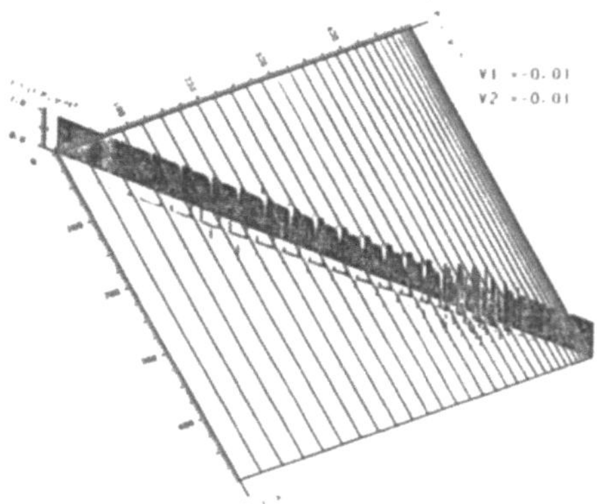

(a) Case A $V_1=V_2=-0.01$

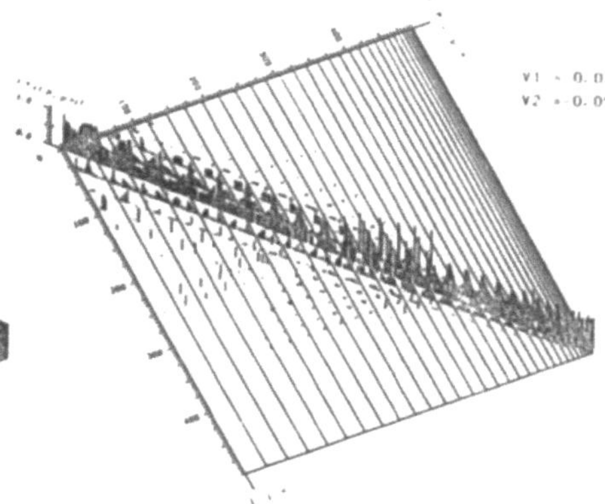

(b) Case B. $V_1=-0.01$, $V_2=-0.05$

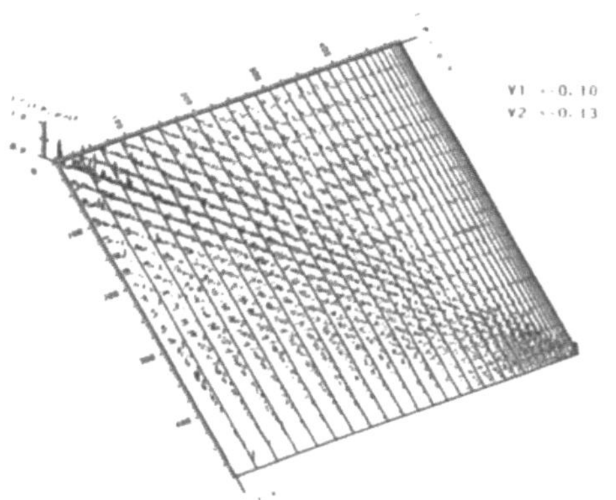

(c) Case C $V_1=-0.1$, $V_2=-0.13$

Figure 7. Fractional parentage plots for the case with N=60. Three different parameter sets correspond to (a) quantum regular, (b) intermediate and (c) quantum chaotic situation, respectively.

the squares of expansion coefficients are shown by the lengths of vertical segments. As is easily seen from Fig.7, three different parameter sets used in our calculation just correspond to the quantum integrable, intermediate and quantum stochastic cases, respectively. Corresponding to these three parameter sets, we have calculated the nearest neighbor level spacing statistics which are shown in Fig. 8. In Case A (Fig.8a), the level statistics gives a strong peak at the origin, indicating an "integrability" of the system. With increased effects of V_1- and V_2-interactions, the peak at the origin becomes less pronounced. Namely, the Poisson type distribution in Case A is disturbed toward the Wigner type distribution when the system shows "quantum KAM" character as in Case B. In Case C where almost all eigenstates are classified to be in quantum chaotic motion, the peak at the origin disappears. From these numerical calculations, we may conclude that our definition of quantum chaos seems to be compatible with the usual discussion based on the GOE.

Here it should be remembered that the random matrix theory relies on some probabilistic assumptions; the matrix elements should be random and

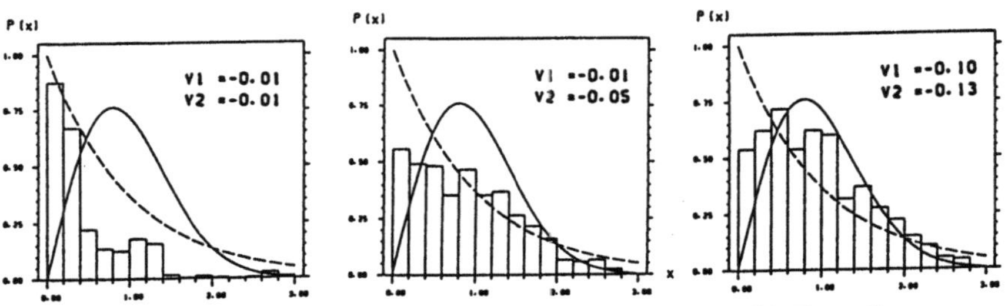

(a) Case A (b) Case B (c) Case C

Figure 8. Nearest neighbor level spacing statistics for all eigenstates N_d=496. Solid and dashed curves show the Wigner and Poisson distributions.

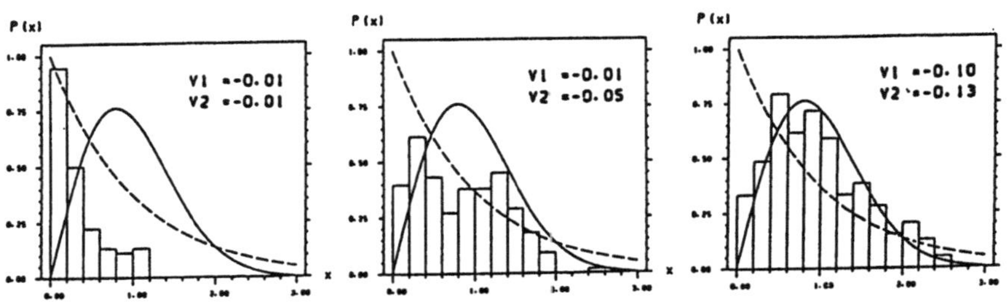

(a) Case A (b) Case B (c) Case C

Figure 9. Nearest neighbor level spacing statistics for intermediate 200 eigenstates.

the resultant distribution should be representation independent. In this treatment, the boundary effects specific for such a finite system as the nucleus are not properly considered. In order to eliminate the boundary effects, we divided the total eigenstates N_d=496 into three groups specified by their energy eigenvalues: the lowest 200, the intermediate 200 and the highest 96. By extracting 200 intermediate eigenstates, we have calculated the nearest neighbor level spacing statistics shown in Fig.9 where the transition from Poisson to Wigner distributions is much more pronounced than the case shown in Fig.8. From the above numerical calculation, it is concluded that an appearance of the Wigner type distribution function does not directly correspond to (even though it is compatible with) the dissolution of quantum numbers associated with the DR. In other words, the dissolution of quantum numbers indicating the individual chaoticity does not necessarily mean an appearance of probabilistic feature which is the basic assumption of the random matrix theory.

Let us discuss how to get an experimental indication of an onset of quantum chaos besides the nearest neighbor level spacing. Detailed information on the wave functions is obtained by evaluating many kinds of transition probabilities between many eigenstates. According to the special symmetry of the interaction in Eq. (4.1), we consider "collective" transition probabilities given by

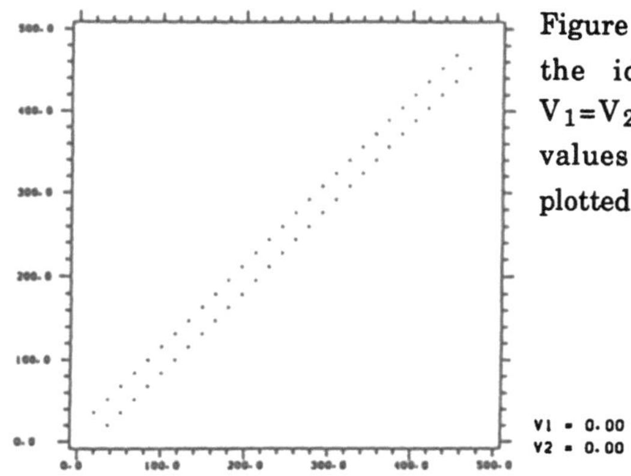

Figure 10. Simulated Eγ-Eγ plots for the ideal integrable case with $V_1 = V_2 = 0$. A couple of sequential values (I_{ijk}, I_{jkl}) and (I_{jkl}, I_{ijk}) are plotted in xy-plane.

$$T_{i,j} \equiv | \, (i|B_1B_1|j) \, |^2 \; . \tag{3.15}$$

However, it is not easy to deduce some general conclusion out of numerous data $\{T_{i,j}\}$, whose number is $O(N_d^2)$. Aiming at extracting some physical information from the numerous experimental data, there has been proposed the E_γ- E_γ plot [31] which has aroused many theoretical discussions on the rotational damping mechanism [32]. Namely, we get some statistical aspect of the rotational motion out of two successive transitions. By means of two successive matrix elements and energy eigenvalues, we may introduce the following quantity:

$$I_{ijk} = (E_i - E_j)T_{i,j} - (E_j - E_k)T_{j,k} \;, \quad \text{with } E_i > E_j > E_k \; . \tag{3.16}$$

Plotting a couple of sequential values (I_{ijk}, I_{jkl}) as well as (I_{jkl}, I_{ijk}) in the xy-plane, we may evaluate statistical features of the system in terms of local information obtained from two successive transition probabilities. From the above discussion, it is quite interesting to study whether or not the stochasticity occurring in the individual eigenfunctions induces some irregularity in the (I_{ijk}, I_{jkl})-plot. We start with a specific situation where all the eigenstates are classified to be in "ideal integrable" motion, i.e. a case with no interaction $V_1 = V_2 = 0$. In this case, we get two straight dotted lines in the xy-plane shown in Fig.10. By switching on the effects coming from the V_1- and V_2-interactions, we get various (I_{ijk},I_{jkl})-plots illustrated in Fig.11. As is seen from Fig.11a which corresponds to Case A in Fig.7, there exists a pronounced regularity when all the eigenstates are classified to be in quantum regular motion. On the other hand, it is hard to see any regularity in Fig.11c which represents Case C in Fig.7. In between these two extreme cases, we find a remnant of regularity in Fig.11b. It is striking that there still remains information on the stochasticity of individual eigenfunctions after accumulating numerous local information

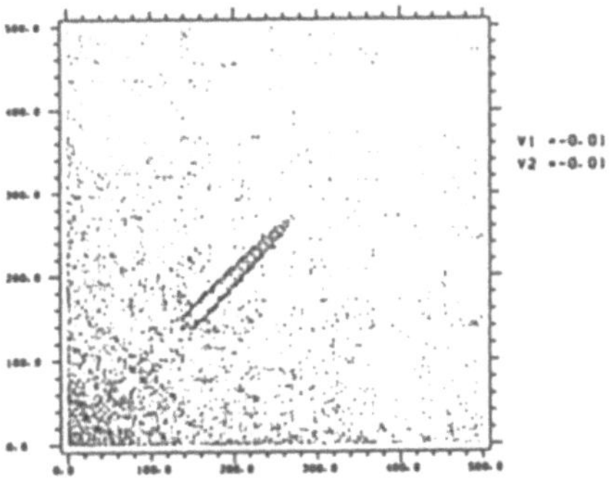

(a) Case A

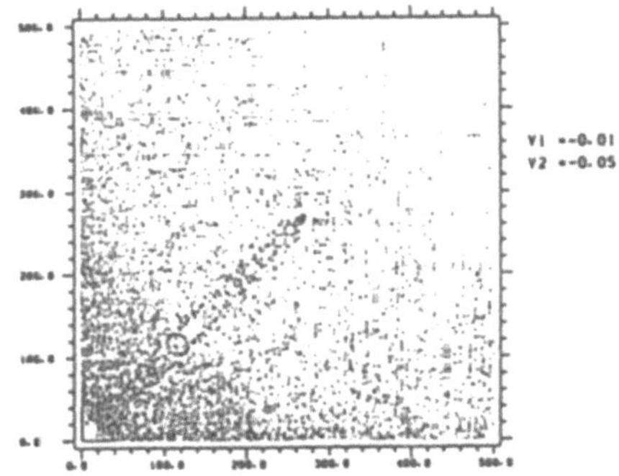

(b) Case B

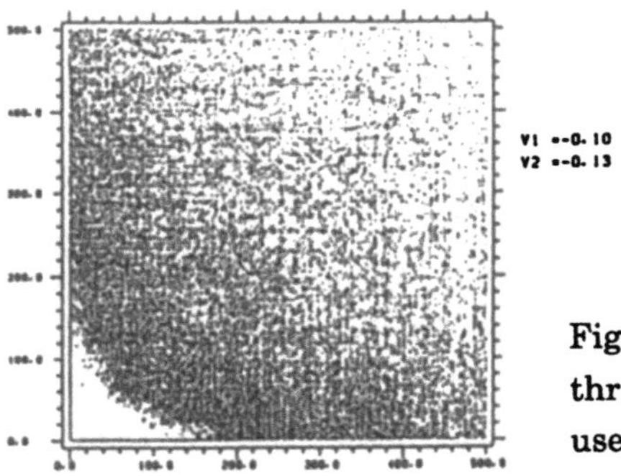

(c) Case C

Figure 11. Simulated Eγ-Eγ plots for three characteristic parameter sets used in Fig. 7.

between two eigenstates connected by the collective operator. As far as the present model Hamiltonian is concerned, it is clearly seen that the (I_{ijk},I_{jkl})-plot gives important information on the dissolution of quantum numbers. Here it should be noticed that the motional narrowing mechanism [32] was studied numerically by calculating a strength function for the system with two degrees of freedom very similar to the present model Hamiltonian. It is a very interesting problem to explore whether the rotational damping is induced by the dissolution of quantum numbers specifying the collective rotational motion or by the motional narrowing mechanism.

4. Future Perspective

As stated in §1, the aim of our project is to develop the fundamental theory of nuclear matter physics which provides us with a unified understanding of diverse phenomena exhibited by nuclei. We are still at a very early stage of the development and there remain many things to be explored. In this section, we will mention some topics which are now under investigation.

a) Relation between classical- and quantum-chaos

In §3, we have discussed the quantum theory of nuclear collective dynamics which was developed at the dictation of what had been constructed in the classical-level dynamics within the TDHF theory discussed in §2. In these theories, the DCC system with maximum integrable-form representation and (μ,ν)-basis states associated in DR play a central role in characterizing the classical-chaos and quantum-chaos, respectively. Furthermore, the violation of the separability condition indicates an onset of chaos in both classical and quantum systems.

Here, it should be remembered that the DCC system with maximum integrable-form representation is defined perturbatively by means of the Birkoff-Gustavson normal-form expansion method, whereas the (μ,ν)-basis states are obtained by diagonalization. From the practical point of view, we get the latter more precisely than the former. Namely, one may evaluate the separability condition more precisely in the quantum system than in the classical-level system. As is well known, the information on the wave function in the boson Hilbert space is transformed into the phase space by using the so-called Husimi function [28]. With the aid of this mapping technique, one may get the various structures of the phase space from that of the Hilbert space.

In Fig. 12, the projected Husimi distributions of the eigenfunctions onto the coordinate space are shown for both the quantum integrable and quantum chaotic motions. Since the quantum integrable motion has a well localized distribution like a torus, whereas the quantum chaotic motion has an overall spreading distribution like a stochastic sea, these figures give another justification of our definition of quantum chaos based on the DR. However, further theoretical and experimental studies are necessary whether or not the quantum chaos defined in §3 is really the physical object .

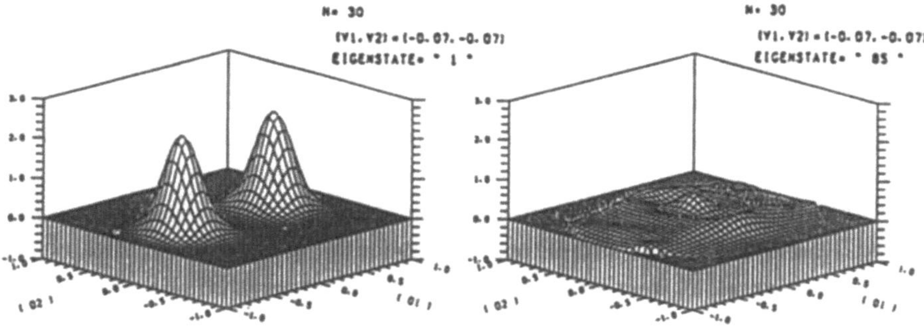

(a) Quantum integrable motion (b) Quantum chaotic motion

Figure 12. Projected Husimi distribution of the eigenfunction onto the coordinate space Q_1-Q_2.

Recently, properties of eigenfunctions in the SU(3) model have been discussed in terms of classical trajectories by Leboeuf and Saraceno [27] by using the Husimi distribution. Even though our primary interest in the context of nuclear physics is to study the inter-relationship between the classical order-to-chaos transition mechanism in the classical-level dynamics (within the TDHF-theory which does not correspond to the classical-limit $\hbar \rightarrow 0$) and quantum order-to-chaos in the quantum system, the interrelationship between quantum and classical-limit dynamics is also an interesting question [33]. The latter problem is one of the main subjects in many fields of physics [34].

b) More Studies on the Single-Particle Dynamics

We have so far applied our classical-level theory to the simple model Hamiltonian with only two degrees of freedom and with one local minimum. In order to explore various interesting questions in connection with the level-crossing dynamics stated in §1, we are now applying our classical-level theory to a more complex and realistic Hamiltonian with at least two local minima. Especially, the interrelationship between the single-particle dynamics (level-crossing dynamics in the self-consistent constrained Hartree-Fock theory discussed in §2.3) and the occurrence of the chaos in the collective motion (described by the dynamical collective operators defined by the SCC method, i.e. the self-consistent constraining operators) induced by violation of the separability condition is of special importance for the further development of the single-particle dynamics.

c) Microscopic Dynamics of the Dissipation Mechanism

As discussed in §2.4, the appearance of many resonance points prevents us from using the dynamics of a single trajectory and suggests use of the dynamics of an ensemble of trajectories. With the aid of the DCC system where the total system is dynamically divided into the relevant and irrelevant degrees of freedom, we have proposed the theory of a nuclear pre-master equation [15]. Namely, we have derived the coupled master equation for the relevant and irrelevant partial distribution functions from the Liouville equation in the TDHF-manifold. In this formulation, there appear the correlation and response functions for both the relevant and irrelevant degrees of freedom. What we know about these functions are their behavior near the state of equilibration within the linear response theory and the dissipation-fluctuation theory. Namely, we get a simple physical meaning of these quantities, if the irrelevant degrees of freedom could be regarded as probabilistic or thermo-dynamical quantities. The primary subject one has to study for the states far from equilibration is the specific role of response and correlation functions in a case where the irrelevant degrees of freedom cannot be treated as probabilistic or thermo-dynamical quantities.

This work is supported by the Ministry of Education, Science and Culture, Japan, International Scientific Research Program (Joint Research) 01044037.

References

1) See for example, G.F. Bertsch and S.Das Gupta, Phys. Rep. 160, No.4 (1988). W. Cassing, Lecture in this volume.
2) See for example, S. Åberg, H. Flocard and W. Nazarewicz, Lund-Mph-90/05 and to appear in Ann. Rev. Nucl. Part. Sci. 40 (1990), and references therein.
3) P. Kramer and M. Saraceno, Lecture notes in Phys. No. 140 (Springer, Berlin, Heidelberg, 1981).
F. Sakata, T. Marumori, Y. Hashimoto and T. Une, Prog. Theor. Phys. 70 (1983) 424.
4) E.R. Marshalek and G. Holzwarth, Nucl. Phys. A191 (1972) 438.
J.P. Blaizot and E.R. Marshalek, Nucl. Phys. A309 (1978) 422,453.
5) D.L. Hill and J.A. Wheeler, Phys. Rev. 89 (1953) 1102.
6) W.J. Swiatecki, Nucl. Phys. A488 (1988) 375c.
7) W. Nørenberg, Phys. Lett. 104B (1981) 107.
8) F. Barranco and R.A. Broglia,Proc.Int.Conf. on Nuclear Shapes (Greece, 1987) 253.
F. Barranco, R.A. Broglia and G.F. Bertsch, Phys.Rev. Lett. 60 (1988) 507.
9) Y. Tanaka and S. Suekane, Prog. Theor. Phys. 66 (1983) 1693.
10) Y.R. Shimizu and K. Matsuyanagi, Prog. Theor. Phys. 67 (1982) 1637; 70 (1983) 144.
11) T. Bengtsson, Nucl. Phys. A496 (1989) 56.
R. Bengtsson and W. Nazarewicz, Z. Phys. A334 (1989) 269.
12) M. Baranger and M. Vénéroni, Ann. Phys. 114 (1978) 123.
F. Villars, Nucl. Phys. A285 (1977) 269.
K. Göke and P.G. Reinhard, Ann. Phys. 112 (1978) 328.
D.J. Rowe and R. Bassermann, Can. J. Phys. 54 (1976) 1941.
G. Holtzwarth and T. Yukawa, Nucl. Phys. A219 (1974) 125.
13) T. Marumori, T. Maskawa, F. Sakata and A. Kuriyama, Prog. Theor. Phys. 64 (1980) 1294.
14) T. Marumori, Prog. Theor. Phys. 57 (1977) 112.
T. Marumori, A. Hayashi, T. Tomoda, A. Kuriyama and T. Maskawa, Prog. Theor. Phys. 63 (1980) 1576.
15) F. Sakata, M. Matsuo, T. Marumori and Y. Zhuo, Ann. of Phys. 194 (1989) 30.
16) M. Yamamura and A. Kuriyama, Suppl. Prog. Theor. Phys. 93 (1987).

17) T. Marumori, Y. Hashimoto, K. Iwasawa and F. Sakata, Univ. of Tsukuba Preprint NT-Rep. (1990) No.2.
18) G.D. Berkhoff, Dynamical Systems (Am. Math. Soc. 1927, New York) Vol.4.
F.G. Gustavson, Astron.J. 71 (1966) 670.
19) I. Progogine, From Being to Becoming (W.H. Freedom and Co., 1980, New York).
20) K. Muramatu, F. Sakata, Y. Yamamoto and T. Marumori, Prog. Theor. Phys. 77 (1987) 347.
21) M.L. Ge, Proc. First Lanzhou Summer School (May-June, 1990,Lanzhou, P.R. China). To be published by Springer-Verlage.
D.H. Feng, Proc. First Lanzhou Summer School (May-June, 1990,Lanzhou, P.R. China). To be published by Springer-Verlage.
22) See, for example, P. Ring and P. Schuck, The Nuclear Many-Body Problem (Springer-Verlag, 1980) and references therein.
23) F. Sakata, T. Marumori and M. Ogura, Prog. Theor. Phys. 76 (1986) 400.
24) F. Sakata, Y. Yamamoto, T. Marumori, S. Iida and H. Tsukuma, Prog. Theor. Phys. 82 (1989) 965.
25) S.Y. Li, A. Klein and R.M. Drizler, J. Math. Phys. 11 (1970) 975.
26) D.C. Meredith, S.E. Koonin and M.R. Zirnbauer, Phys. Rev. A37 (1988) 3499.
27) P. Leboeuf and M. Saraceno, Phys. Rev. 41 (1990) 4614.
P. Leboeuf and M Saraceno, J. Phys. A23 (1990) 1745.
28) K. Husimi, Proc. Phys. Math. Soc. Jpn 22 (1940) 264.
29) C E. Porter, Statistical Theories of Spectra: Fluctuations (Acad. New York, 1965).
30) See, for example, O. Bohigas and M.G. Giannoni, Lecture Notes in Phys. Vol. 209 (Springer,Berlin, Heidelberg, 1984) 1.
31) B. Herskind, Proc. Int. Conf. on Nucl. Phys. (Tipografia, Compositori, Bologna, 1983) 117.
J.D. Garrett, G.B. Hagemann and B. Herskind, Ann. Rev. Nucl. Part. Sci. 36 (1986) 454 and references therein.
32) B. Lauritzen, R.A. Broglia, W. E. Ormand and T. Døssing, Phys. Lett. B207 (1988) 238.
R.A. Broglia, T. Dossing, B. Lauritzen and B.R. Mottelson, Phys. Rev. Lett. 58 (1987) 326.
33) M.V. Berry, Lecture in this Volume.
34) E.J. Heller, Lecture Notes in Phys. 263 (Springer, Berlin, Heidelberg, 1986) 162.

Mean Field Dynamics and Low-Lying Collective Excitations

M. Matsuo
Yukawa Institute for Theoretical Physics, Kyoto University,
Kyoto 606, Japan

1. Introduction

Mean field is one of the most basic concepts of nuclei. The concept pervades various collective properties of the nucleus and the associated single-particle motion. It applies not only to the spherical shell model but also to the various types of deformations as illustrated by the Nilsson model for quadrupole deformation and the pair gap field in the Hartree-Bogoliubov theory. Its current development extends to the rapidly-rotating as well as the superdeformed mean-field models which have played a central role in detailed spectroscopy of high-spin states[1]. On the basis of the mean fields model, we can extract information about the mean field and the single-particle configurations from the experimental data. Thus our knowledge on the nuclear mean field as well as the associated single-particle states is now quite rich and detailed, as far as they are static or uniformly rotating.

Collective excitations in nuclei concern the dynamics of the mean field and the associated nucleon motion. We should note however that the relation between the mean field and the single-particle motion is no longer simple as far as the mean fields evolves in time. The change in the mean field associated with the collective motion often becomes considerable, as shown by the low-lying quadrupole vibration, especially the fission process and heavy ion collisions. In these large-amplitude collective motions, the nucleons rearrange their configuration in the course of the mean field motion, and at the same time, the nucleon configuration affects in return the mean field motion. The mechanism of this dynamical interplay between the mean field and the nucleon configuration is expected to cause *non-linear* properties in the large-amplitude collective motions.

The time-dependent Hartree-Fock (TDHF) method[2,3] provides us with the most general and consistent description of the mean field dynamics on the microscopic level. Thus the TDHF method can incorporate the non-linear mean field dynamics mentioned above. Furthermore, this method is semiclassical in its mathematical structure and it defines a non-linear classical Hamilton system, and hence it allows us to study the order-to-chaos transition in the nuclear mean field dynamics. In this symposium, Sakata discusses recent development toward the quantum mechanical implications of the chaos associated with the nuclear collective dynamics [4].

In this paper, I also discuss the non-linear properties in the nuclear collective dynamics by rather emphasizing its origin, i.e. the interplay mechanism between the large-amplitude mean field motion and the associated nucleon configuration. In particular, I discuss how this mechanism influences the non-linearities which are actually observed in the low-lying collective excitations. The phenomena dealt with here are the anharmonicity of the two-phonon excited quadrupole vibrational states in deformed and soft spherical nuclei, the transition of the mean field from spherical to deformed shape in $N \simeq 90$ transitional nuclei, and the shape coexistence phenomena, which has been recently observed in semi-magic nuclei.

2. The Selfconsistent Collective Coordinate Method: Microscopic Theory of Mean Field Dynamics

Let me start with briefly reviewing a theoretical framework, the Selfconsistent Collective Coordinate (SCC) method formulated by Marumori, Maskawa, Sakata and Kuriyama [5], which has enabled us to study the non-linear mean field dynamics on a microscopic basis.

2.1 TDHB method in the canonical-variables representation

The most consistent mean field theory for the nucleon many-body system is the TDHF method [3], in which the many-body wave function is described by a time-dependent Slater determinant $\det\varphi_i(x_j;t) = \langle x_1 x_2 \cdots x_A|\phi(t)\rangle$. A complete parametrization of the determinantal TDHF state vector $|\phi(t)\rangle$ is, according to the Thouless theorem, given by[6,7]

$$|\phi(t)\rangle = \exp\sum_{i>j}\left\{\Gamma_{ij}(t)a_i^\dagger a_j^\dagger - \Gamma_{ij}^*(t)a_j a_i\right\}|\phi_0\rangle \tag{2.1}$$
$$\equiv e^{i\hat{G}(t)}\,|\phi_0\rangle \ ,$$

where $|\phi_0\rangle$ is a determinantal state vector of reference, and $a_i^\dagger$ and a_i are creation and annihilation operators of particles and holes with respect to the reference state $|\phi_0\rangle$ (It is convenient to choose the HF ground state for $|\phi_0\rangle$. Label i denotes the single-particle levels in the mean field associated with the HF ground state). The expression (2.1) of the TDHF state vector is also applied to the time-dependent Hartree-Bogoliubov (TDHB) method[3] by taking the pairing correlation into account by simply regarding $a_i^\dagger, a_i$ as the Bogoliubov quasiparticle operators. Here and hereafter we assume the TDHB treatment.

The parametrization (2.1) in terms of the complex c-numbers $\{\Gamma_{ij}, \Gamma^*_{ij}\}$ has a firm mathematical basis because it is a *generalized coherent state* [8] defined for the Lie algebra composed by the single-particle operators $\{a^\dagger_i a^\dagger_j, a_j a_i, a^\dagger_i a_j\}$ [9]. Thanks to this mathematical structure, the expression (2.1) provides us with the following two advantages which are essential for the microscopic theory of the mean field dynamics.

i) *A symplectic structure is embedded in the TDHB state vector* [10,6,7]. It is therefore always possible to introduce a set of the canonical variables which parametrize the TDHB state vector. For example, if we introduce a parametrization $|\phi(\mathbf{C}^*, \mathbf{C})\rangle$ of the TDHB state vector defined by

$$\Gamma_{ij}(\mathbf{C}^*, \mathbf{C}) = \left[\mathbf{C}\frac{1}{\sqrt{\mathbf{C}^\dagger\mathbf{C}}}\sin^{-1}\sqrt{\mathbf{C}^\dagger\mathbf{C}}\right]_{\mathbf{ij}} \quad \text{and } c.c. \;, \tag{2.2}$$

the new parametrization satisfies

$$\langle\phi(\mathbf{C}^*, \mathbf{C})|\,\partial\,|\phi(\mathbf{C}^*, \mathbf{C})\rangle = \frac{1}{2}\sum_{i>j}\left(C^*_{ij}\partial C_{ij} - C_{ij}\partial C^*_{ij}\right) \quad , \tag{2.3}$$

with respect to the new set of parameters $\{C^*_{ij}, C_{ij}\} = \{[\mathbf{C}^*]_{ij}, [\mathbf{C}]_{ij}\}$. Here the bi-linear form in the right hand side explicitly indicates the symplectic structure of the parametrization.

The equation of motion of the TDHB method, which is the variational time-dependent Schrödinger equation

$$\delta\,\langle\phi(t)|\, i\frac{\partial}{\partial t} - \hat{H}\,|\phi(t)\rangle = 0 \quad , \tag{2.4}$$

with $\delta\,|\phi(t)\rangle = (a^\dagger_i a^\dagger_j, a_j a_i, a^\dagger_i a_j)\,|\phi(t)\rangle$ reduces to the equation of motion for the new variables $\{C^*_{ij}, C_{ij}\}$:

$$i\frac{d}{dt}C^*_{ij} = \frac{\partial\mathcal{H}}{\partial C_{ij}} \quad \text{and} \quad i\frac{d}{dt}C_{ij} = -\frac{\partial\mathcal{H}}{\partial C^*_{ij}} \tag{2.5}$$

where $\mathcal{H} = \langle\phi(\mathbf{C}^*, \mathbf{C})|\,\hat{H}\,|\phi(\mathbf{C}^*, \mathbf{C})\rangle$ is the energy carried by the determinantal state vector $|\phi(\mathbf{C}^*, \mathbf{C})\rangle$ with respect to the microscopic Hamiltonian $\hat{H}$. The equation of motion (2.5) is essentially the Hamilton equation where C^*_{ij}, C_{ij} are the canonical variables conjugate to each other. The Hamilton equation of motion (2.5) contains all the information of the nuclear mean field and the accompanying nucleon configuration.

ii) *There is a correspondence between the canonical-variables parametrization of the TDHB state vector and the generalized Holstein-Primakoff boson mapping* [11] of the single-particle operators $\{a_i^\dagger a_j^\dagger, a_j a_i, a_i^\dagger a_j\}$ [12,13,6,7]. The canonical representations of the single-particle operators are calculated as

$$(a_i^\dagger a_j^\dagger)_{\mathrm{C}} \equiv \langle\phi(\mathbf{C}^*,\mathbf{C})|\, a_i^\dagger a_j^\dagger \,|\phi(\mathbf{C}^*,\mathbf{C})\rangle = \left[\mathbf{C}^\dagger\sqrt{1-(\mathbf{C}^\dagger\mathbf{C})^{\mathrm{T}}}\right]_{ji} \quad ,$$
$$(a_j a_i)_{\mathrm{C}} \equiv \langle\phi(\mathbf{C}^*,\mathbf{C})|\, a_j a_i \,|\phi(\mathbf{C}^*,\mathbf{C})\rangle = \left[\sqrt{1-(\mathbf{C}^\dagger\mathbf{C})^{\mathrm{T}}}\,\mathbf{C}\right]_{ij} \quad ,$$
$$(a_i^\dagger a_j)_{\mathrm{C}} \equiv \langle\phi(\mathbf{C}^*,\mathbf{C})|\, a_i^\dagger a_j \,|\phi(\mathbf{C}^*,\mathbf{C})\rangle = \left[\mathbf{C}^\dagger\mathbf{C}\right]_{ij} \quad . \tag{2.6}$$

On the other hand, the boson mapping of these operators gives the following expression

$$(a_i^\dagger a_j^\dagger)_{\mathrm{Boson}} = \left[\mathbf{B}^\dagger\sqrt{1-(\mathbf{B}^\dagger\mathbf{B})^{\mathrm{T}}}\right]_{ji}$$
$$(a_j a_i)_{\mathrm{Boson}} = \left[\sqrt{1-(\mathbf{B}^\dagger\mathbf{B})^{\mathrm{T}}}\,\mathbf{B}\right]_{ij}$$
$$(a_i^\dagger a_j)_{\mathrm{Boson}} = \left[\mathbf{B}^\dagger\mathbf{B}\right]_{ij} \tag{2.7}$$

where $[\mathbf{B}^\dagger]_{ij}, [\mathbf{B}]_{ij} = B_{ji}^\dagger, B_{ij}$ are the boson operators satisfying $[B_{ij}, B_{kl}^\dagger] = \delta_{ik}\delta_{jl}$ $(i > j, k > l)$. The apparent correspondence between Eq. (2.6) and Eq. (2.7) implies that the canonical variables $\{C_{ij}^*, C_{ij}\}$ are classical correspondents of the quantum mechanical boson operators $\{B_{ij}^\dagger, B_{ij}\}$ in the boson mapping theory. Thanks to this quantum-classical correspondence, one can "quantize" the TDHB theory and obtain the quantum eigenenergy of the system although the TDHB method describes the time-dependence of the system.

2.2 The Selfconsistent Collective Coordinate Method

The Hamilton equation of motion (2.5) which is equivalent to the original TDHB equation (2.4) describes time-evolution of the TDHB state vector $|\phi(t)\rangle$ under an arbitrarily chosen initial condition. As a general property of the multi-dimensional classical Hamilton system, the system exhibits both chaotic time-evolution and a regular one which is typically a periodic motion. In this connection, the collective motion which has been described in phenomenological models in terms of a few macroscopic parameters (such as quadrupole deformation parameters) should be related with the regular motion because the chaotic motion essentially covers the whole phase space of the system. The SCC method [5] formulates this idea to extract the collective motion embedded in the huge-dimensional Hamilton equation of motion (2.5).

According to the SCC method, the determinantal state vector representing the collective motion is parametrized by collective variables (η, η^*) through a parametrization of Γ_{ij} and Γ^*_{ij} in the expression (2.1):

$$
\begin{aligned}
|\phi(\eta^*,\eta)\rangle &= e^{i\hat{G}(\eta^*,\eta)}\,|\phi_0\rangle \quad , \\
i\hat{G}(\eta^*,\eta) &= \sum_{i>j}\left\{\Gamma_{ij}(\eta^*,\eta)a_i^\dagger a_j^\dagger - \Gamma^*_{ij}(\eta^*,\eta)a_j a_i\right\} \quad .
\end{aligned}
\tag{2.8}
$$

The number of the collective variables should be very small (e.g. 5 for the quadrupole shape motion) in comparison with the number of the parameters $\{\Gamma^*_{ij}, \Gamma_{ij}\}$ necessary for the complete parametrization of the TDHB state vector. Thus the parametrization (2.8) defines a submanifold (called the collective submanifold) in the phase space $\{\Gamma^*_{ij}, \Gamma_{ij}\}$ associated with the TDHB state vector. The time-evolution of the state vector (2.8) travelling on the collective submanifold should be a solution of the TDHF equation of motion (2.4). This condition, which implies that the collective submanifold is an invariant submanifold of the TDHF equation of motion (2.4) or (2.5), determines the collective submanifold.

The basic equations of the SCC method are summarized as follows.

i) *Equations of collective motion*

$$
i\frac{d\eta}{dt} = \frac{\partial\mathcal{H}_{\rm coll}(\eta^*,\eta)}{\partial\eta^*} \quad \text{and} \quad i\frac{d\eta^*}{dt} = -\frac{\partial\mathcal{H}_{\rm coll}(\eta^*,\eta)}{\partial\eta} \quad , \tag{2.9}
$$

where the collective Hamiltonian $\mathcal{H}_{\rm coll}$ as a function of the collective variables is defined by

$$
\mathcal{H}_{\rm coll} = \langle\phi(\eta^*,\eta)|\,\hat{H}\,|\phi(\eta^*,\eta)\rangle \quad . \tag{2.10}
$$

With the collective coordinate and momentum defined by $q = (\eta+\eta^*)/\sqrt{2}$ and $p = (\eta - \eta^*)/\sqrt{2}i$, Eq. (2.9) reduces to the familiar form of the canonical equations

$$
\frac{dq}{dt} = \frac{\partial\mathcal{H}_{\rm coll}}{\partial p} \quad \text{and} \quad \frac{dp}{dt} = -\frac{\partial\mathcal{H}_{\rm coll}}{\partial q} \quad . \tag{2.11}
$$

ii) *Equation of the collective submanifold*

$$
\delta\,\langle\phi(\eta^*,\eta)|\,\hat{H} - \frac{\partial\mathcal{H}_{\rm coll}}{\partial\eta^*}\hat{O}^\dagger - \frac{\partial\mathcal{H}_{\rm coll}}{\partial\eta}\hat{O}\,|\phi(\eta^*,\eta)\rangle = 0 \tag{2.12}
$$

where $\hat{O}^\dagger = \left(\frac{\partial}{\partial\eta}e^{i\hat{G}}\right)e^{-i\hat{G}}$, $\hat{O} = -\left(\frac{\partial}{\partial\eta^*}e^{i\hat{G}}\right)e^{-i\hat{G}}$.

iii) *Canonical-variables condition*

$$
\begin{aligned}
\langle\phi(\eta^*,\eta)|\,\frac{\partial}{\partial\eta}\,|\phi(\eta^*,\eta)\rangle &= \frac{1}{2}\eta^* \quad , \\
\langle\phi(\eta^*,\eta)|\,\frac{\partial}{\partial\eta^*}\,|\phi(\eta^*,\eta)\rangle &= -\frac{1}{2}\eta \quad .
\end{aligned}
\tag{2.13}
$$

These equations are consistently solved[5,14] by means of the Taylor expansions of the unknown functions $\Gamma_{ij}(\eta^*,\eta)$, $\Gamma^*_{ij}(\eta^*,\eta)$ and $\mathcal{H}_{\text{coll}}(\eta^*,\eta)$, i.e.,

$$\begin{aligned}\Gamma_{ij}(\eta^*,\eta) &= \Gamma^{(1)}_{ij}(\eta^*,\eta) + \Gamma^{(2)}_{ij}(\eta^*,\eta) + \cdots, \\ \Gamma^{(n)}_{ij}(\eta^*,\eta) &= \sum_{r+s=n} \Gamma^{(r,s)}_{ij}(\eta^*)^r(\eta)^s \quad ,\end{aligned} \tag{2.14}$$

$$\begin{aligned}\mathcal{H}_{\text{coll}}(\eta^*,\eta) &= \mathcal{H}^{(2)}_{\text{coll}}(\eta^*,\eta) + \mathcal{H}^{(3)}_{\text{coll}}(\eta^*,\eta) + \cdots, \\ \mathcal{H}^{(n)}_{\text{coll}}(\eta^*,\eta) &= \sum_{r+s=n} h^{(r,s)}(\eta^*)^r(\eta)^s \quad .\end{aligned} \tag{2.15}$$

The lowest order term $\Gamma^{(1)}_{ij}(\eta^*,\eta)$ and $\mathcal{H}^{(2)}_{\text{coll}}(\eta^*,\eta)$ represent a RPA normal mode of the mean field motion around the HF ground state $|\phi_0\rangle$ under the linear approximation. Higher order terms represent the non-linearity of the mean field dynamics associated with the large-amplitude collective motion of nucleus. Equations for the higher order terms are derived by the Taylor expansion of the basic equations by means of the formulae such as

$$\begin{aligned}e^{-i\hat{G}}\hat{H}e^{i\hat{G}} =& \hat{H} + [\hat{H}, i\hat{G}^{(1)}] \\ &+ [\hat{H}, i\hat{G}^{(2)}] + \frac{1}{2}[[\hat{H}, i\hat{G}^{(1)}], i\hat{G}^{(1)}] + \cdots \quad ,\end{aligned} \tag{2.16}$$

where $i\hat{G}^{(n)} = \sum\left(\Gamma^{(n)}_{ij}a^\dagger_i a^\dagger_j - \Gamma^{*(n)}_{ij}a_j a_i\right)$.

Thus the SCC method determines consistently the collective Hamiltonian $\mathcal{H}_{\text{coll}}(\eta^*,\eta)$ and the TDHB state vector $|\phi(\eta^*,\eta)\rangle$ which represents the microscopic nucleon configuration associated with the collective motion.

3. Anharmonic Vibrations

The low-lying quadrupole motions provide us with typical examples of non-linearity of the nuclear collective motions because they often accompany the spectra which strongly deviate from the harmonic vibrational patterns. Here I will mainly discuss the anharmonicity of the two-phonon states of the γ-vibration in well-deformed nuclei. I shall also mention recent investigation of the two-phonon states of quadrupole vibration of typical soft spherical nuclei Se and Ru as well as the transitional Sm nuclei.

3.1 Anharmonicity of the γ-Vibration

Anharmonicity of the doubly-excited γ-vibrational states in deformed nuclei has been discussed by many authors since extensive spectroscopy of ^{168}Er

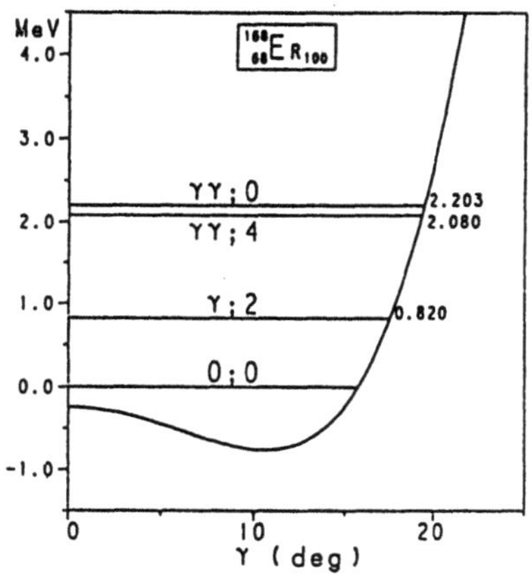

Fig.1 The prediction by the SCC method for two- and one-phonon excited states of the γ-vibration in ^{168}Er as well as the collective potential[21].

was carried out in the early 1980s [15-25]. Many of the microscopic theories now predict large-anharmonicity of the γ-vibration [19-25]. This anharmonic vibration therefore provides us with a good example for which we can study the mechanism of the non-linearity associated with the mean-field dynamics. Here I discuss it in terms of the SCC method [19-21].

Realistic mean fields associated with γ-vibration can be described by the quadrupole deformed modified oscillator potential, the monopole and the quadrupole pairing fields. The effective Hamiltonian generating these fields consists of the Nilsson single particle potential and the separable-type two-body forces that are the doubly-stretched QQ interaction, the monopole and the quadrupole pairing interactions. The TDHB state vector representing the γ-vibrations are parametrized by the collective variables (η^*, η) and $(\tilde{\eta}^*, \tilde{\eta})$, which correspond to the γ-vibrational RPA modes with angular momentum projection $K = 2, -2$, respectively. Thus the collective Hamiltonian constructed up to the fourth order of the SCC method has the following form:

$$\begin{aligned}\mathcal{H}_{\rm coll} =& \mathcal{H}^{(2)}_{\rm coll} + \mathcal{H}^{(4)}_{\rm coll}\\ =&\alpha(\eta^*\eta + \tilde{\eta}^*\tilde{\eta}) + \beta(\eta^*\tilde{\eta}^* + \eta\tilde{\eta})\\ &+ b_1(\eta^*\eta^*\tilde{\eta}^*\tilde{\eta}^* + \tilde{\eta}\tilde{\eta}\eta\eta) + b_2(\eta^*\eta^*\tilde{\eta}^*\eta + \eta^*\tilde{\eta}^*\tilde{\eta}^*\tilde{\eta} + \eta^*\eta\eta\tilde{\eta} + \tilde{\eta}^*\eta\tilde{\eta}\tilde{\eta})\\ &+ b_3(\eta^*\eta^*\eta\eta + \tilde{\eta}^*\tilde{\eta}^*\tilde{\eta}\tilde{\eta}) + b_4\eta^*\eta\tilde{\eta}^*\tilde{\eta} \end{aligned} \tag{3.1}$$

where the coefficients $\alpha, \beta, b_1, \cdots, b_4$ are all obtained as the Taylor-expanded solution of the basic equations of the SCC method. Energy spectra including the multiple vibrational excitations are obtained by replacing the canonical variables $(\eta^*, \eta, \tilde{\eta}^*, \tilde{\eta})$ in the classical collective Hamiltonian by collective boson operators $(B^\dagger, B, \tilde{B}^\dagger, \tilde{B})$ with the normal ordering[26] — "quantization" — and by diagonalizing the quantized Hamiltonian in the collective boson Hilbert space.

Figure 1 is a result for ^{168}Er, for which a strong anharmonicity is expected from the experiments. The calculated two-phonon states $(\gamma\gamma; K = 0, 4)$ show strong anharmonicity of the γ-vibration consistent with the experiments, as is seen in the energy ratio $E(\gamma\gamma;K)/E(\gamma) = 2.54, 2.69$ which deviate considerably from the harmonic vibration limit $E(\gamma\gamma;K)/E(\gamma) = 2$.

The anharmonicity is mainly governed by the fourth-order non-linear terms of the collective Hamiltonian. Analysis of the associated coefficients b_1, b_2, b_3, b_4 clarifies the microscopic origin of the anharmonicity of the collective vibration. For this analysis let us look into the operator $i\hat{G}(\eta^*, \eta, \tilde{\eta}^*, \tilde{\eta})$ which generates the TDHF state vector $|\phi(\eta^*, \eta, \tilde{\eta}^*, \tilde{\eta})\rangle = e^{i\hat{G}}|\phi_0\rangle$ associated with the collective motion. The operator $i\hat{G}$, which is a linear combination of two-quasiparticle creations and annihilations $a_i^\dagger a_j^\dagger$ and $a_i a_j$, is also expanded in terms of the RPA normal modes of excitation, as

$$i\hat{G}(\eta^*, \eta, \tilde{\eta}^*, \tilde{\eta}) = \sum_\lambda \left\{ G_\lambda(\eta^*, \eta, \tilde{\eta}^*, \tilde{\eta})\hat{X}_\lambda^\dagger - G_\lambda^*(\eta^*, \eta, \tilde{\eta}^*, \tilde{\eta})\hat{X}_\lambda \right\} \tag{3.2}$$

in terms of the phonon creation and annihilation operators of the RPA normal modes (labeled by λ)

$$\hat{X}_\lambda^\dagger = \sum_{i<j} \left\{ \psi_\lambda(ij) a_i^\dagger a_j^\dagger + \varphi_\lambda(ij) a_j a_i \right\} \quad , \quad \hat{X}_\lambda = \left(\hat{X}_\lambda^\dagger \right)^\dagger \quad , \tag{3.3}$$

$K^\pi = 0^+$:	β-vib., pairing-vib., pairing-rotation, 2qp exc.
$K^\pi = \pm 2^+$:	γ-vibration, two-quasiparticle(2qp) excitations
$K^\pi = \pm 4^+$:	2qp excitations.

In the small amplitude limit of the TDHB state vector, the collective variables $(\eta^*, \eta, \tilde{\eta}^*, \tilde{\eta})$ generate the γ-vibrational normal modes ($\lambda = \gamma, \tilde{\gamma}$) of the RPA. As the amplitude of $(\eta^*, \eta, \tilde{\eta}^*, \tilde{\eta})$ increases, all the normal modes of RPA can contribute to the sum in Eq. (3.2) due to the non-linear effect which couples the γ-vibrational RPA mode to the other kinds of RPA modes (where not only the collective solution but also the non-collective two-quasiparticle excitations are included). The anharmonicity coefficients b_1, b_2, b_3, b_4 can be divided into two parts *kinematical anharmonicity* and *dynamical anharmonicity*:

$$b_i = b_i^{\text{kin}} + b_i^{\text{dyn}} \quad . \quad (i = 1, 2, 3, 4) \tag{3.4}$$

The kinematical anharmonicity arises from the Pauli exclusion effect interacting between the γ-vibrational RPA phonons and is obtained by truncating the sum in Eq. (3.2) to the γ-vibrational RPA phonons. The dynamical anharmonicity is the contribution from all the other RPA phonon components in $i\hat{G}$. This components arises from the dynamical coupling between the different kinds of the RPA phonons (mode-mode coupling). The dynamical anharmonicity contribution to the fourth-order coefficients b_1, b_2, b_3, b_4 is expressed in terms of the diagrams shown in Fig. 2.

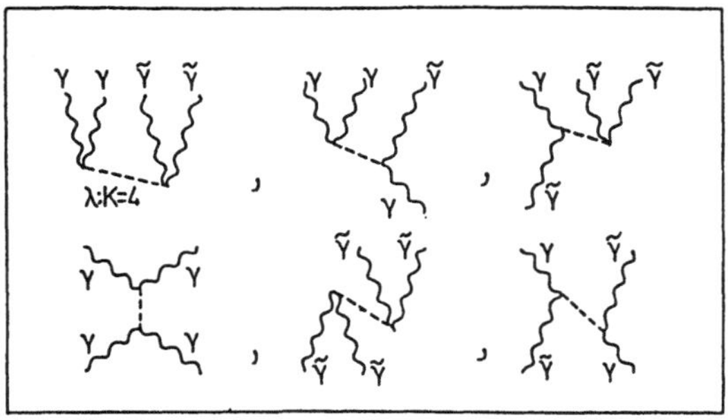

Fig.2 The diagrammatic representation of the dynamical part b_i^{dyn} of the fourth-order anharmonicity coefficients[20]. Only the contribution arising from the mode-mode couplings to the $K^\pi = 4^+$ modes are listed here since the contributions of the $K^\pi = 0^+$ modes have essentially the same structure. The wavy lines represents the γ-vibrational RPA normal modes while the intermediate dashed line denotes the $K^\pi = 4^+$ RPA modes.

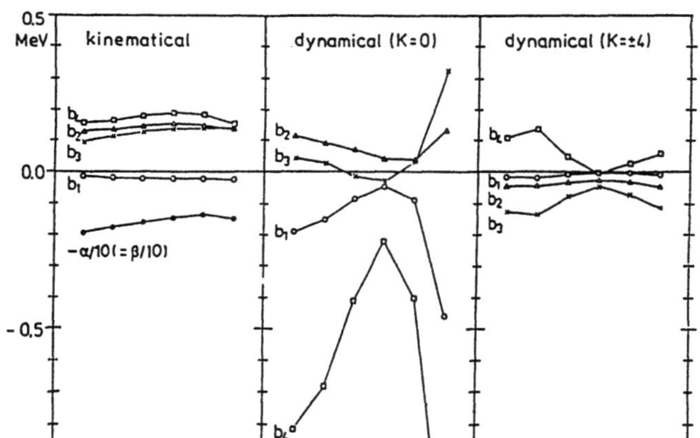

Fig.3 The divided contribution of the fourth-order anharmonicity coefficients b_i for Er isotopes (N=92∼102)[20]. The kinematical part b_i^{kin} (left), the dynamical part $b_i^{\mathrm{dyn}}(K^\pi{=}0^+)$ which arises from the mode-mode coupling to the $K^\pi{=}0^+$ modes (center), and the dynamical part $b_i^{\mathrm{dyn}}(K^\pi{=}\pm4^+)$ arising from $K^\pi{=}\pm4^+$ modes (right).

Figure 3 shows a divided contribution of the anharmonicity coefficients b_1, b_2, b_3, b_4 for several Er isotopes. This figure indicates that the dynamical anharmonicity which arises from the coupling between the RPA normal modes in the TDHB state vector has a sizable effect which is comparable with or larger than the kinematical anharmonicity stemming from the Pauli exclusion. What is apparent and I would emphasize most in Fig. 3 is that the dynamical anharmonicity coefficients change sizably depending on the neutron number of

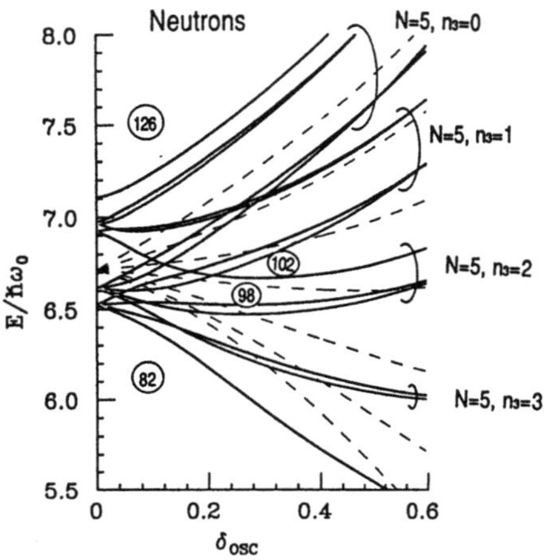

Fig.4 The sub-shell structure of the neutron Nilsson single-particle energy levels in deformed nuclei (deformation parameter $\delta_{osc} > 0$). Only the levels between $N = 82$ and 126 gaps are plotted for simplicity. The mode-mode couplings are weak in ${}^{166}_{68}\mathrm{Er}_{98}$ whose Fermi surface is located within the $N = 5, n_3 = 2$ subshell while they are strong in ${}^{170}_{68}\mathrm{Er}_{102}$ whose Fermi surface is above and below the sub-shells $N = 5, n_3 = 2$ and $N = 5, n_3 = 1$, respectively.

the isotopes. This strong neutron number dependence arises from a subshell effect associated with the deformed mean field. The single-particle levels in the quadratically deformed mean field (Nilsson potential) show a kind of sub-shell structure: the single-particle levels which have the same asymptotic quantum numbers N and n_3 in the same energy region (Fig.4). On the other hand, the quadrupole field $Q_{2\pm2} \propto r^2 Y_{2\pm2}$ associated, in the leading order, with the γ-vibration has matrix elements that are concentrated within the levels having the same quantum numbers for N and n_3. Thus the response of the nucleons to the quadrupole field $Q_{2\pm2}$ strongly depend on the position of the Fermi level relative to the subshells. Precise analysis of the subshell effects make it clear that the mode-mode couplings depend also on the pairing correlation which characterizes the nucleon configuration around the Fermi surface [20]. Thus the dynamical anharmonicity arising from the mode-mode coupling is sensitive to how the nucleons occupies the single-particle levels in the mean field.

3.1 Quadrupole Vibration of Soft Spherical Nuclei

The dynamical anharmonicity effect, i.e., the effect of the mode-mode couplings, on the anharmonic spectra of the quadrupole vibrational excitation in soft spherical nuclei has been analysed from various viewpoints [27]. Recent investigations based on the SCC method by Takada, Yamada and Tsukuma [28-30] and by Aiba [31] demonstrate its essential importance more explic-

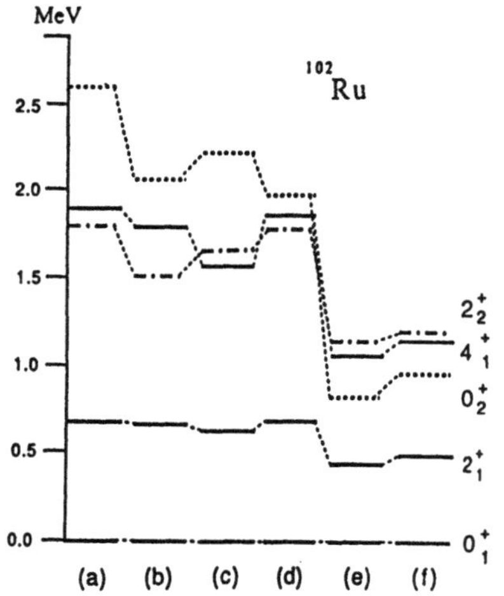

Fig.5 The effect of the mode-mode couplings on the low-lying collective states in the soft spherical nuclei ^{102}Ru [31]. The mode-mode couplings are neglected in the left most column (a) while the all possible couplings to the $J^\pi = 0^+, 2^+, 4^+$ modes are taken into account in the right most column (f). See Ref.[31] for detail.

itly. Aiba analyzed how the two-phonon triplets in Ru and Se isotopes are influenced by the mode-mode coupling between the collective vibrational RPA mode with $J^\pi = 2^+$ and the other RPA modes with $J^\pi = 0^+, 2^+, 4^+$, and he found, as shown in Fig. 5 that the mode-mode coupling plays a crucial role in reproducing the experimentally observed spectra, especially the 0_2^+ state having characteristically the lowest excitation energy among the two-phonon triples. Takada, Yamada and Tsukuma have demonstrated the crucial role of the mode-mode couplings in reproducing the transitional behavior of the low-lying yrast spectra from the vibrational patterns to the rotational one in $N \simeq 90$ nuclei.

4. Single-Particle Energy Levels and Nucleonic Configuration

As emphasized above, the nucleon configuration in the mean field characterizes the mean field dynamics. It is therefore of great importance in understanding the mean field dynamics to look into how the nucleon configuration changes in the course of the collective motion.

The SCC method can be utilized also to define the single-particle states associated with the *time-dependent* mean field, which composes a useful basis for specifying the nucleon configuration. I shall describe the procedure of defining the single-particle states according to Yamada[32].

A necessary element for the definition of the single-particle states is the *single-particle Hamiltonian in the moving frame of reference* [32,33] which is defined by

$$\hat{h}'_{sp} = \hat{h}_{sp} - \frac{\partial \mathcal{H}_{\rm coll}}{\partial \eta^*}\hat{O}^\dagger - \frac{\partial \mathcal{H}_{\rm coll}}{\partial \eta}\hat{O} \,. \tag{4.1}$$

The single-particle Hamiltonian $\hat{h}'_{sp}$ in the moving frame contains not only the mean-field, which is derived from $\hat{H}$ in reference to the TDHB state vector $|\phi(\eta^*,\eta)\rangle$ and represented by $\hat{h}_{sp}$, but also the generalized "cranking" term $-\left(\frac{\partial\mathcal{H}_{\rm coll}}{\partial\eta^*}\hat{O}^\dagger+\frac{\partial\mathcal{H}_{\rm coll}}{\partial\eta}\hat{O}\right)$, which arises from the time-dependence of the mean field. Here $\hat{O}^\dagger=\left(\frac{\partial}{\partial\eta}e^{i\hat{G}}\right)e^{-i\hat{G}}$ and $\hat{O}=-\left(\frac{\partial}{\partial\eta^*}e^{i\hat{G}}\right)e^{-i\hat{G}}$ are the generating operators for the derivative of the TDHB state vector. The generalized cranking terms reduce to the familiar cranking term $\omega_{\rm rot}J_x$ in the case of the uniform rotation.

The single-particle Hamiltonian $\hat{h}'_{sp}$ in the moving frame satisfies the consistency with respect to the equation of the collective submanifold (2.12) or the TDHB equation of motion (2.4) (note that $\hat{h}_{sp}$ without the generalized cranking term does not) because there, the stationary condition holds

$$\delta\left\langle\phi(\eta^*,\eta)\right|\hat{h}'_{sp}\left|\phi(\eta^*,\eta)\right\rangle=0\ , \tag{4.2}$$

for the TDHB state vector $|\phi(\eta^*,\eta)\rangle$ at each point of the collective variables (η^*,η). Single-quasiparticle states in the moving frame are thus defined by diagonalizing the single particle Hamiltonian $\hat{h}'_{sp}$ in the moving frame. They satisfy

$$\hat{h}'_{sp}=\sum_\mu E_\mu(\eta^*,\eta)a^\dagger_\mu(\eta^*,\eta)a_\mu(\eta^*,\eta)\ , \tag{4.3}$$

as well as the vacuum condition $a_\mu(\eta^*,\eta)\left|\phi(\eta^*,\eta)\right\rangle=0$ which represents that the TDHB state vector $|\phi(\eta^*,\eta)\rangle$ is the occupied state of the quasiparticle states. Here $E_\mu(\eta^*,\eta)$ is the energy level for the quasiparticle states $a^\dagger_\mu(\eta^*,\eta)$, $a_\mu(\eta^*,\eta)$.

It is useful to single out the pairing correlation in order to look into the nucleon configuration. For this purpose the mean field Hamiltonian $\hat{h}'^{(\Delta N=0)}_{sp}$ in which the pairing correlation is eliminated can be defined by the number conserving part of $\hat{h}'_{sp}$. The single-nucleon levels defined with respect to $\hat{h}'^{(\Delta N=0)}_{sp}$ are given by

$$\hat{h}'^{(\Delta N=0)}_{sp}=\sum_\mu e_\mu(\eta^*,\eta)c^\dagger_\mu(\eta^*,\eta)c_\mu(\eta^*,\eta)\ , \tag{4.4}$$

where $e_\mu(\eta^*,\eta)$ can be compared with the single-particle energy levels in the constrained Hartree-Fock theory or the phenomenological Nilsson model. The occupation probability of the single-particle level μ is given by

$$\rho_\mu(\eta^*,\eta)=\left\langle\phi(\eta^*,\eta)\right|c^\dagger_\mu(\eta^*,\eta)c_\mu(\eta^*,\eta)\left|\phi(\eta^*,\eta)\right\rangle \tag{4.5}$$

which takes an intermediate value between 0 and 1 because of the pairing

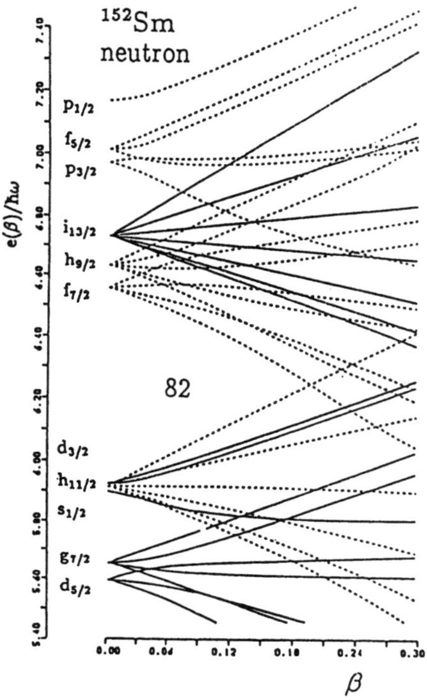

Fig.6 The single-particle energy levels $e_\mu(\beta)$ as a function of the quadrupole deformation β for ^{152}Sm, calculated selfconsistently by means of the SCC method[32].

correlation. The occupation probability tells us important information about the nucleon configuration of the TDHB state $|\phi(\eta^*, \eta)\rangle$.

Yamada calculated the single-quasiparticle and -particle energy levels $E_\mu(\eta^*, \eta)$ and $e_\mu(\eta^*, \eta)$ and the occupation probability $\rho_\mu(\eta^*, \eta)$ associated with the large amplitude motion in $N \simeq 90$ transitional Sm isotopes (Fig.6 and7) [32]. He employs the same microscopic Hamiltonian (spherical Nilsson potential + QQ interaction + monopole and quadrupole pairing interactions) as the one previously used in successful description of the vibration-rotation transition of the low-lying spectra [28-30].

In Fig. 6 we see a sizable deformation splitting similar to that in the phenomenological Nilsson model. This represents that the calculated single-particle Hamiltonian describes well the mean-field associated with the quadrupole deformation. On the other hand, the single-particle occupation probability $\rho_\mu(\eta^*, \eta)$ shown in Fig. 7 directly tells us the microscopic configuration of the TDHB state vector $|\phi(\eta^*, \eta)\rangle$ accompanying the collective motion. It is seen in Fig. 7 that the prolate [oblate] single-particle states with down- [up-] sloping levels get more [less] occupied as the quadrupole deformation ($\sim \beta$) increases. The most remarkable feature, according to Yamada, is the decrease in the occupation probability of the oblate s.p. states which stem from $h_{11/2}$ below the $N = 82$ shell. (The phenomenological Nilsson + BCS model also describes this property through the mechanism that the $h_{11/2}$ oblate orbits come

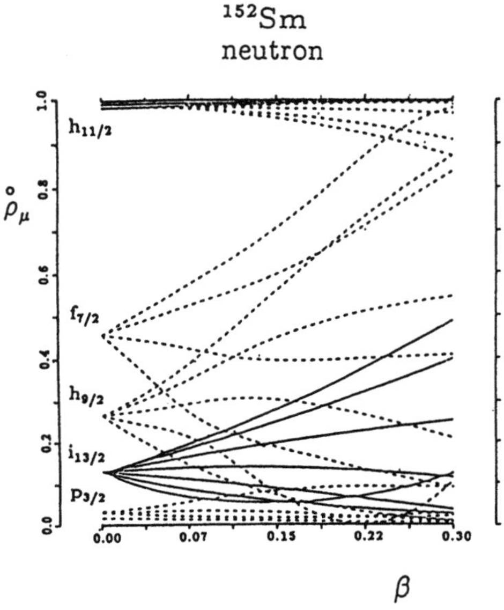

Fig.7 The single-particle occupation probabilities $\rho_\mu(\beta)$ for ^{152}Sm, calculated selfconsistently by means of the SCC method [32].

close to the Fermi level for a large value of β. This is the major shell mixing in the shell model terminology.) He emphasizes here these results would never be obtained if he truncated the operator $i\hat{G}(\eta^*, \eta)$ generating the TDHB state vector within the components of the quadrupole collective RPA phonon. All the RPA phonons $(\hat{X}_\lambda^\dagger, \hat{X}_\lambda)$ with $J^\pi = 0^+, 2^+, 4^+$ begin to take part in describing the single-particle levels and occupations as the deformation evolves. This point explains why the mode-mode couplings between the collective $J^\pi = 2^+$ RPA phonon and all the other $J^\pi = 0^+, 2^+, 4^+$ modes was inevitable in his previous calculation [28-30] which reproduced the vibration-rotation transition of the low-lying spectra in the transitional Sm isotopes.

5. Level Crossing and Diabatic Approach

Nucleon configuration plays more decisive role in the dynamics of nuclear mean field when the pairing correlation (pairing gap) is relatively weak in comparison with the single-energy level spacing. I shall illustrate the dynamics in this situation in connection with the shape-coexistence phenomenon, which is experimentally observed in neutron deficient semi-magic nuclei Sn [34,35] and Pb [36,37].

In the semi-magic nuclei for which the proton shells of the spherical mean field are closed, an oblate occupied level and a prolate unoccupied level cross with each other as the prolate deformation of the mean field evolves. If the nucleons occupying the oblate orbit jump to the prolate orbit at the level crossing point, the nucleon configuration drives the mean field toward prolate deformation. Thus the interchange of configurations bring about another stationary HB state which has prolate deformation coexisting with the spherical HB state. Thus the shape-coexistence phenomenon is related to how the nucleons changes when the single-particle levels cross with each other.

A conventional approach of the nucleon configuration is the adiabatic treatment which assumes the adiabaticity condition that the nucleons occupy the lowest-energy configuration at instant deformation of the mean field (Born-Oppenheimer approximation). The adiabatic nucleon configuration changes its structure at the level-crossing point, and it happens very drastically in the situation of weak pairing. This situation violates the adiabaticity condition. Actually Bengtsson and Nazarewicz have recently discussed the inapplicability of the adiabatic treatment in reference to the shape-coexistence phenomenon [38].

Since the interactions between the collective motions which have different nucleon configurations are a central problem in the shape coexistence phenomenon, an alternative *diabatic* treatment which deals with the nucleon configurations explicitly should be effective. In the diabatic treatment, the nucleon configuration retains their structure and changes only smoothly as the mean field deformation evolves (diabatic basis for the configurations).

The SCC method can be utilized to formulate the *diabatic* treatment of the mean field dynamics associated with the level crossing [39,40]. The TDHB state vectors

$$|\phi_i(\eta^*,\eta)\rangle = e^{i\hat{G}_i(\eta^*,\eta)}|\phi_i\rangle \qquad (i=1,2,\cdots) \tag{5.1}$$

which are constructed for each HB state (labeled by i) having different deformation and nucleon configuration change their structure as smoothly as described by polynomials of the collective variables (η^*,η). Thus the TDHB state vector $|\phi_i(\eta^*,\eta)\rangle$ defines a diabatic basis. The interaction between the different diabatic basis $|\phi_i(\eta^*,\eta)\rangle$ is given by $\langle\phi_i(\eta^*,\eta)|\,\hat{H}\,|\phi_j(\eta^*,\eta)\rangle \equiv \mathcal{H}_{ij}(\eta^*,\eta)$. The collective Schrödinger equation which describes the collective motion as well as the dynamics of the interacting configurations ($i=1,2$) at the level crossing is expressed by

$$\begin{pmatrix} H_1(B^\dagger,B) & H_{12}(B^\dagger,B) \\ H_{21}(B^\dagger,B) & H_2(B^\dagger,B) \end{pmatrix} \begin{pmatrix} |\psi_1\rangle \\ |\psi_2\rangle \end{pmatrix} = E \begin{pmatrix} |\psi_1\rangle \\ |\psi_2\rangle \end{pmatrix} , \tag{5.2)]}$$

where $H_1(B^\dagger,B)$ and $H_2(B^\dagger,B)$ are "quantized" from the collective Hamiltonians $\mathcal{H}_i(\eta^*,\eta) = \langle\phi_i(\eta^*,\eta)|\,\hat{H}\,|\phi_i(\eta^*,\eta)\rangle$ describing the collective motion of each diabatic basis, and $H_{12}(B^\dagger,B)$ and H_{21} $(B^\dagger,B)$ are derived from the interaction Hamiltonian $\mathcal{H}_{ij}(\eta^*,\eta)$ between the diabatic basis.

Let us illustrate the dynamics of the level crossing in terms of a schematic model which simulates the shape coexistence in the semi-magic nuclei [40].

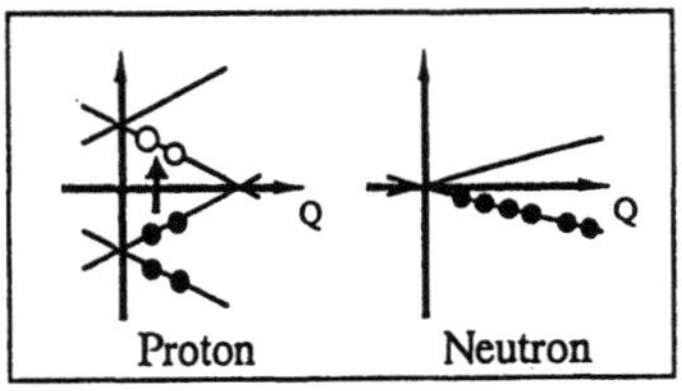

Fig.8 The single-particle level structure of the O(4) model[40].

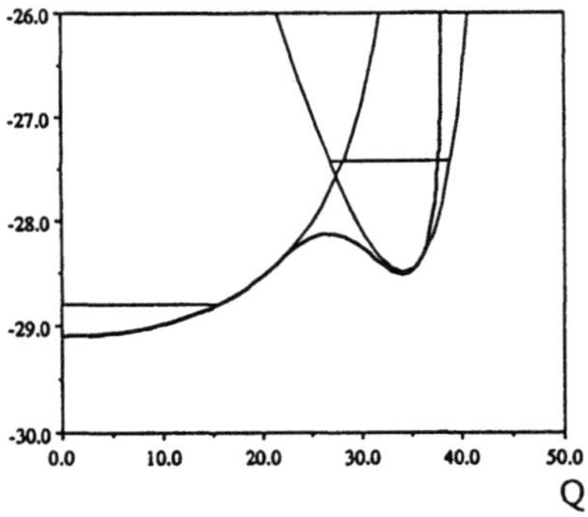

Fig.9 The potential energy surface of the O(4)-model calculated by the adiabatic CHF method (thick curve) and the diabatic collective potentials calculated by the SCC method (thin curve) as functions of the deformation[40]. Estimated zero-point energies associated with the diabatic collective motions are indicated by the horizontal lines.

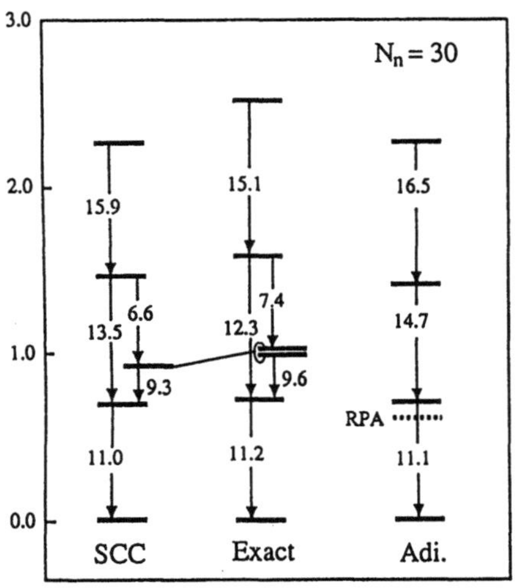

Fig.10 The low-lying states of the O(4) model for three kinds of the calculations[40]. "SCC" denotes the solution of the coupled-configuration collective Schrödinger equation (5.2) based on the diabatic treatment. "Adi" denotes the Born-Oppenheimer approximation using the adiabatic CHF energy surface. The second excited states in "SCC" and "Exact" are the coexisting deformed states.

The model contains a partially occupied single j-shell with large degeneracy for neutrons. The proton single particle levels consists of two j-shells one of which is occupied and the other is unoccupied because of an energy gap between them. The quadrupole-quadrupole force as well as the pairing interaction acts among them while a simplified form of the quadrupole operator is adopted. The model is characterized by $O(4) \times O(4) \times O(4)$ symmetry and it is exactly solvable by numerical diagonalization. Figure 8 shows the single-particle level structure of this model. At a certain value of the quadrupole deformation, a proton occupied oblate level crosses with an unoccupied prolate level, which is the relevant level crossing to the shape coexistence of the model as shown in Fig. 9.

Figure 10 compares the exactly calculated excitation spectra with those obtained by two different theoretical approaches: the spectra denoted by "SCC" are the solution of the coupled-configuration collective Schrödinger equation (5.2), which is based on the *diabatic* treatment of the nucleon configuration, and the other denoted by "Adi" is the solution of the Born-Oppenheimer approximation which assumes the adiabatic configuration calculated by the CHF method. As seen in Fig. 10, the adiabatic approximation does not reproduce the coexisting deformed state while the diabatic approach gives quantitatively nice approximation. This schematic shape existence model indicates that the level crossing plays a key role is the dynamics of the nuclear collective dynamics.

6. Concluding Remarks

In this paper reviewing some recent theoretical investigations of the low-lying collective excited states, I take the viewpoint of the *microscopic* mean field dynamics which emphasizes the interplay mechanism between the mean field and the associated nucleon configuration.

As we see in the two-phonon excited states in deformed and spherical nuclei, the mode-mode couplings between the RPA normal modes strongly influence the anharmonicity of the collective quadrupole vibrations. They are governed by the single-particle level structure and the nucleon configuration. On the contrary, as mentioned in §4, the single-particle levels and configuration along the deformation change is successfully described only by taking into account the mode-mode coupling effects. The interplay between the mean field motion and the nucleon configuration (single-particle levels) emerges more typically in the shape coexistence phenomenon, where we see the dynamics of the interacting configurations associated with the single-particle level crossing.

Last, I would like to emphasize that the viewpoint of the microscopic mean field dynamics is not restricted to the low-lying collective states, but it has a possibility to cover various properties of the nuclear collective dynamics. For example, the dissipation of such large-amplitude collective motions as fissions and deep inelastic heavy ion collisions is related to the single-particle level crossings which are encountered successively during the mean field motion [41-43]. Here I would cite also a recent development along this line which sheds light on the level crossing from the viewpoint of quantum chaos [44].

Acknowledgements

The author expresses sincere thanks to K. Matsuyanagi, K. Yamada, K. Aiba and T. Fukui for valuable help in preparing this paper. He also acknowledges the discussions with those participating in the 1988-90 research project "Non-linear Dynamics of Nuclear Collective Motions" organized by the Yukawa Institute for Theoretical Physics.

References

[1] See, for example, *The variety of nuclear shapes, Proc. Int. Conf. on Nuclear Shapes* (World Scientific 1988), ed. J.D. Garrett *et.al.*

[2] P.A.M. Dirac, Proc. Camb. Phil. Soc. **26**, 376 (1930)

[3] See, for example, P. Ring and P. Schuck, *The Nuclear Many-Body Problem* (Springer, Berlin, Heidelberg 1980).

[4] F. Sakata, talk presented in this symposium.

[5] T. Marumori, T. Maskawa, F. Sakata and A. Kuriyama, Prog. Theor. Phys. **64**, 1294 (1980).

[6] F. Sakata, T. Marumori, Y. Hashimoto and T. Une, Prog. Theor. Phys. **70**, 424 (1983).

[7] M. Yamamura and A. Kuriyama, Prog. Theor. Phys. Suppl. No.93 (1987).

[8] A.M. Perelomov, Sov. Phys. Usp. **20**, 703 (1977).

[9] H. Kuratsuji and T. Suzuki, Phys. Lett. **92B**, 19 (1980); Prog. Theor. Phys. Suppl. No.74&75, 209 (1983).

[10] P. Kramer and M. Saraceno, *Lecture Notes in Physics* No.140 (Springer, Berlin, Heidelberg 1981).

[11] D. Jansen, F. Dönau, S. Frauendorf and R.V. Jolos, Nucl. Phys. **A172**, 145 (1971).

[12] E.R. Marshalek and G. Holtzwarth, Nucl. Phys. **A191**, 438 (1972).

[13] A. Kuriyama and M. Yamamura, Prog. Theor. Phys. **66**, 2130 (1981).

[14] M. Matsuo, Prog. Theor. Phys. **76**, 372 (1986).

[15] A. Bohr and B. R. Mottelson, Physica Scripta **25**, 28 (1982).

[16] T.S. Dumitrescu and I. Hamamoto, Nucl. Phys. **A383**, 205 (1982).

[17] V.G. Soloviev and N. Yu. Shirikova, Z. Phys. **A301**, 263 (1981); Sov. J. Nucl. Phys. **36**, 799 (1982); Z. Phys. **A334**, 149 (1989).

[18] N. Yoshinaga, Y. Akiyama and A. Arima, Phys. Rev. Lett. **17**, 1116 (1986).

[19] M. Matsuo and K. Matsuyanagi, Prog. Theor. Phys. **74**, 1227 (1985).

[20] M. Matsuo and K. Matsuyanagi, Prog. Theor. Phys. **76**, 93 (1986).

[21] M. Matsuo and K. Matsuyanagi, Prog. Theor. Phys. **78**, 591 (1987).

[22] R. Piepenbring and M.K. Jammari, Nucl. Phys. **A481**, 81 (1988).

[23] M.K. Jammari and R. Piepenbring, Nucl. Phys. **A487**, 77 (1988).

[24] S.A. Scotts and T. Tamura, Phys. Rev. **C40**, 2342 (1989).

[25] V.O. Nesterenko, Z. Phys. **A335**, 147 (1990).

[26] M. Matsuo and K. Matsuyanagi, Prog. Theor. Phys. **74**, 288 (1985).

[27] As reviews, T. Marumori, K. Takada and F. Sakata, Prog. Theor. Phys. Suppl. No. 71, 1(1981); K. Matsuyanagi, in *Nuclear Physics* (North-Holland, 1982), p.29, ed. C.H. Dasso.
A recent development in the framework of the boson expansion theory is seen in, H. Sakamoto and T. Kishimoto, Nucl. Phys. **A486**, 1(1988).

[28] T. Takada, K. Yamada and H. Tsukuma, Nucl. Phys. **A496**, 224 (1989).

[29] K. Yamada, T. Takada and H. Tsukuma, Nucl. Phys. **A496**, 239 (1989).

[30] K. Yamada and T. Takada, Nucl. Phys. **A503**, 53 (1989).

[31] H. Aiba, Prog. Theor. Phys. **83**, 358 (1990); RCNP preprint 006, June 1990, to be published in Prog. Theor. Phys.

[32] K. Yamada, Preprint YITP/K-889, October 1990.

[33] F. Sakata, T. Marumori, K. Muramatsu and Y. Hashimoto, Prog. Theor. Phys. **74**, 51 (1985).

[34] J. Bron, W.H.A. Hesselink, A. Van Poelgeest, J.J.A. Zalmstra, M.J. Uitzinger, H. Verheul, K. Heyde, M. Waroguier, H. Vincx and P. Van Isacker, Nucl. Phys. **A318**, 335 (1979).

[35] A. Bäcklin, N.G. Jonsson, R. Julin, J. Kantele, M. Luontama, A. Passoja and T. Poikolainen, Nucl. Phys. **A351**, 490 (1981).

[36] P. Van Duppen, E. Coenen, K. Deneffe, M. Huyse, K. Heyde and P. Van Isacker, Phys. Rev. Lett. **52**, 1974 (1984); Phys. Rev. **C35**, 1861 (1987).

[37] J. Penninga, W.H.A. Hesselink, A. Balanda, A. Stolk, H. Verheul, J. Van Klinken, H.J. Riezebos and M.J.A. de Voigt, Nucl. Phys. **A471**, 535 (1987).

[38] R. Bengtsson and W. Nazarewicz, Z. Phys. **A334**,269(1989).

[39] Y.R. Shimizu and K. Matsuyanagi, Prog. Theor. Phys. **74**, 1346 (1985).

[40] T. Fukui, M. Matsuo and K. Matsuyanagi, Preprint KUNS 1024, July 1990.

[41] D.L. Hill and J.A. Wheeler, Phys. Rev. **89**, 1102 (1953).

[42] L. Wilets, Phys. Rev. **115**, 372 (1959).

[43] G. Schütte, Phys. Rep. **80**, 113 (1981).

[44] W.J. Swiatecki, Nucl. Phys. **A488**, 375c (1988).

Rotational Motion in Warm Atomic Nuclei

T. Døssing and B. Herskind

The Niels Bohr Institute, University of Copenhagen,
Blegdamsvej 17, DK-2100 København Ø, Denmark

Properties of warm nuclear states are discussed. First, the γ decay flow deexciting the states is discussed qualitatively, with emphasis on the competition between E1 and E2 transitions, and the phase transition between undamped and damped rotation. The formalism of rotational damping is briefly reviewed, and subsequently applied to a realistic calculation of the rotational damping width Γ_{rot} for the nucleus ^{168}Yb, using the cranked Nilsson potential. The calculated values are found to be in agreement with recent preliminary data.

1 Introduction

The rotating atomic nucleus has been the subject of intense studies during the last twenty years. Experimentally, nuclear states with angular momentum up to about 60 units have been identified. These specific states of highest known angular momentum are superdeformed states with collective rotation around one of the minor axes. [1]

The standard way to form a rotating nucleus in experiments is by a heavy ion fusion reaction. Thus, the angular momentum is created together with a substantial excitation energy. After its formation, the rotating nucleus rapidly disposes most of its excitation energy by boiling off neutrons, without losing much of the angular momentum. Figure 1 displays the decay by γ rays following the neutron decay in a nucleus which is well deformed, but not superdeformed. With increasing angular momentum, the energy of the lowest lying state, denoted the "yrast state", increases, on the average proportional to $I(I+1)$, as for a rigid rotation. The neutron decay terminates when the last neutron has brought the nucleus to a point in excitation energy which is lower than the yrast energy plus the neutron separation energy. This line is labelled "Neutron Separation Energy" on the figure. A neutron emitted from states with energy below this line will have to carry some angular momentum to go to daughter states above the yrast line, and the associated barrier penetration will severely cut down the neutron decay width. The nucleus then decays by γ rays.

Eds: Y. Abe, H. Horiuchi, K. Matsuyanagi

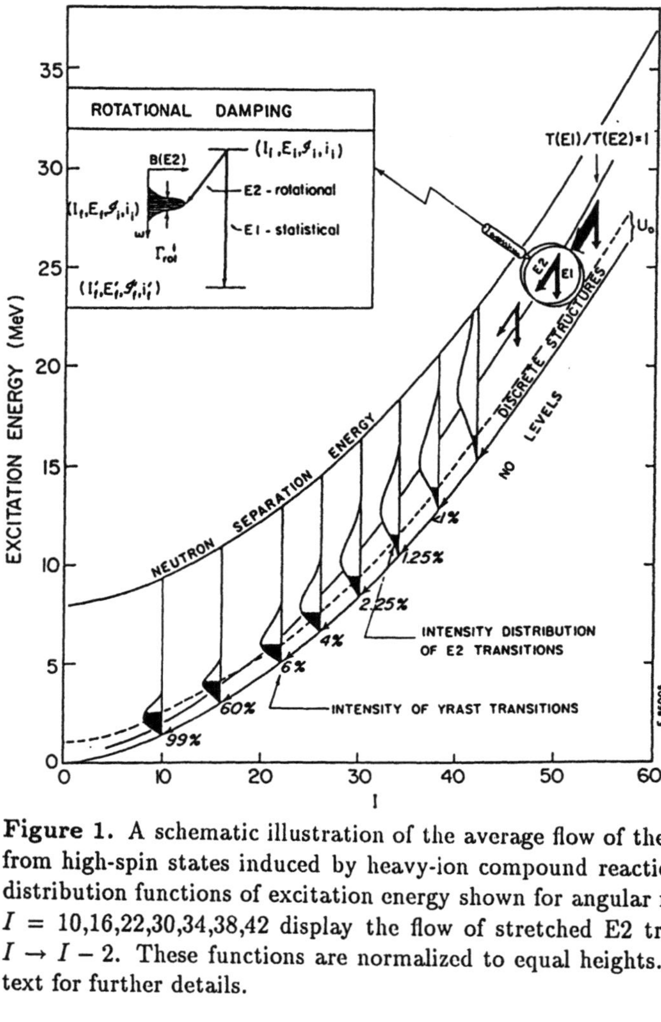

Figure 1. A schematic illustration of the average flow of the γ decay from high-spin states induced by heavy-ion compound reactions. The distribution functions of excitation energy shown for angular momenta I = 10,16,22,30,34,38,42 display the flow of stretched E2 transitions $I \rightarrow I-2$. These functions are normalized to equal heights. See the text for further details.

Two γ decay modes compete, statistical E1 transitions and rotational E2 transitions. The total E1 decay width is roughly proportional to the heat energy, that is the excitation energy minus the yrast energy, to the $\frac{5}{2}$th power. The rotational E2 decay width scales with the transition energy to the 5th power, that is roughly with the angular momentum to the 5th power. Thus there will be a line of equal competition between E1 and E2 transitions, labelled $T(E1)/T(E2) = 1$ on fig. 1. The γ cascades down in energy and angular momentum will have a tendency to follow this line. Starting above the line, E1 transitions will bring the decay down below the line. Starting below the line, E2 transitions will win, causing the decay to move parallel to the yrast line, bringing it up towards the equal competition line. In this way, the distribution function of excitation energy at a given value of the angular momentum will have its maximum close to the equal competition line. This is also the outcome of the more involved computer simulations of the decay, as displayed in fig. 1.

Along the yrast line, and somewhat up in heat energy, the rotational decay follows rotational bands. Above some heat energy, the bands lie so closely in energy that they will be mixed into more complicated states by even a small residual interaction, which was not included into the description of the pure rotational bands. This mixing of bands may cause many possible final states, as depicted in the insert of fig. 1. Such a wide distribution of rotational decay energies for each quantum state implies that the rotational motion is damped. The transition from undamped to damped motion occurs around a specific heat energy, and can be viewed as a phase transition. The place for this phase transition is denoted by the dashed line in the figure. The relative intensity of undamped transitions at each angular momentum is illustrated by shading the part of the distribution functions below the dashed line. With the present estimate of the phase transition energy, the bulk of rotational transitions are predicted to be damped. The width of the distribution of the rotational E2 strength is *the rotational damping width* Γ_{rot}. The purpose of the present paper is to discuss at which heat energy the rotational damping sets in, and how the damping width is estimated within the cranking model. Before continuing to this, we shall shortly review the theoretical descriptions of rotating nuclei.

Collective rotation is fully described within the nuclear shell model for light nuclei, such as sd-shell nuclei. The shell model provides a complete wave function for the states, including also all transition matrix elements. For heavier nuclei, the very large dimension of the valence nucleon state space prohibits the application of the full shell model, and truncation schemes must be applied. Although schemes have been invented for selecting a good shell model basis, rapidly rotating states such as for example superdeformed states are well out of reach of such descriptions for the moment.

The standard description of rotating nuclei is the cranked shell model, in which one has renounced the full description of the wave function. Likewise, instead of requiring an exact eigenvalue for the angular momentum, the average angular momentum projection on one axis is required to be fixed. In its simplest form, the cranking model is just a rotating independent particle model, but it can be extended to include a pairing or other residual interactions, treated with BCS transformations or the RPA approximation. In this way, one can calculate the moment of inertia of the ground band, and treat the crossing and the coupling between rotational bands, determined by the pairing interaction [2]. For the highest angular momenta, where there is no BCS solution, pairing can be treated in RPA [3]. At high excitation energy, RPA with the cranked wave functions and a residual dipole-dipole interaction is used to study the giant dipole vibration built on rotating states [4]. For specific few-nucleon configurations, the cranking model can be related to the particle-rotor model [5]. The particle-rotor model states are good eigenstates of angular momentum, built out of complete single particle states for the active particles, coupled to a quantal rotor which represents the inactive nucleons.

Due to its technical simplicity, the cranked mean field model can readily be applied to the highest angular momenta. This has extensively been done, first by using coarse grids in angular momenta, in order to display the general expectations of the shapes of rotating nuclei over the whole periodic table. These early investigations [6, 7] revealed profound differences among the different rotating nuclei, depending on the shell structure. Indeed, angular momentum has proven to be a third coordinate, besides neutron and proton number, for characterizing the shell structure of nuclei. Actually, the stability of superdeformed rotating states for masses around 150 was predicted six years before the superdeformed bands were discovered experimentally. More recently, cranked independent particle calculations have been carried to a higher perfection, now following the minute changes of deformation of each configuration step by step in angular momentum, (actually in frequency, governed by the average value of the angular momentum). Such investigations have been quite successful in explaining the various properties of the superdeformed states, such as moments of inertia, which display characteristic changes when one nucleon is added or subtracted [8].

These studies all aim at describing the cold rotating nucleus, that is, states close to the yrast line. In the present paper the extension of the cranking model to the rotational damping in warm rotating nuclei is discussed.

2 General description of band mixing

The formal development of rotational damping was given in reference [9], and shall not be repeated here. Instead, a more qualitative discussion will be given.

2.1 Mixing of two bands

Figure 2 gives a schematic illustration of the mixing of only two rotational bands at two consecutive angular momenta. The two bands are taken as basis bands, which diagonalize the collective transition quadrupole operator connecting states at angular momenta I and $I-2$.

The branching of the decay is seen to follow from the change in the mixing coefficient between the angular momenta I and $I-2$, and not just from the fact that the bands are mixed. Since the change in mixing comes from the change in the energy of the bands and their coupling, one can say in general that the damping of the rotational motion is caused by the change in the Hamiltonian between I and $I-2$, when expressed in terms of basis bands.

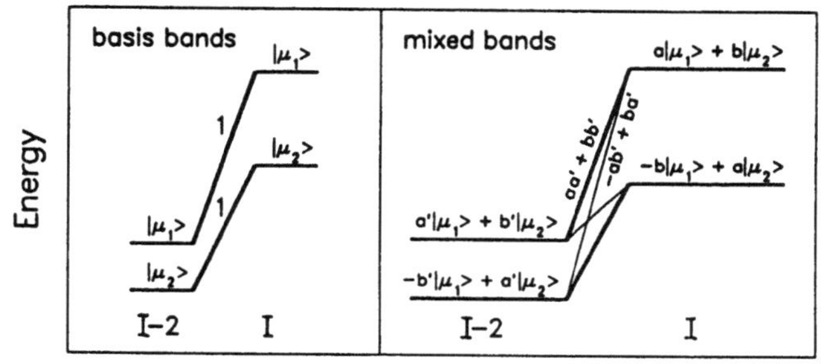

Figure 2. Schematic illustration of the mixing of two basis bands $|\mu_1\rangle$ and $|\mu_2\rangle$ into energy eigenstates described by the mixing coefficients a, b and a', b' at two consecutive values of the angular momentum, I and $I-2$. The thickness of the lines indicate the relative strength of the collective E2 transitions deexciting the states at angular momentum I.

2.2 Basis bands and their interaction

The basis bands, such as $|\mu_1\rangle$ and $|\mu_2\rangle$ displayed in figure 2, can be selected in many ways. The formal description of rotational damping can be given independently of the basis. An important ingredient will be the assumption of random signs of the matrix elements of the residual interaction, which mixes the bands.

Later, we shall take the bands as cranked mean field states. In addition to the assumption of random signs, the matrix elements may have strong selection rules, and we shall later consider a two body interaction acting between cranked mean field states.

With a random type residual interaction, the expansion coefficients X_μ^α of the energy eigenstates $|\alpha(I)\rangle$ at angular momentum I

$$|\alpha(I)\rangle = \sum_\mu X_\mu^\alpha(I)|\mu(I)\rangle \tag{1}$$

are distributed on average according to a Breit-Wigner distribution

$$\rho\overline{(X_\mu^\alpha)^2} = \frac{1}{2\pi}\frac{\Gamma_\mu}{(E_\alpha - E_\mu)^2 + \frac{\Gamma_\mu^2}{4}} \tag{2}$$

where ρ denotes the total level density and Γ_μ is a measure of the interval in energy eigenstates over which the basis band is mixed, that is Γ_μ is the spreading width of the basis band state.

The change in the Hamiltonian between I and $I-2$ in terms of the basis bands contains a diagonal part and a non-diagonal part.

$$\Delta H(I-2) = \sum_\mu [\epsilon_\mu(I-2) - \epsilon_\mu(I)]\,|\mu(I-2)\rangle\langle\mu(I-2)| + \sum_{\mu\mu'} \Delta H_{\mu\mu'}|\mu(I-2)\rangle\langle\mu'(I-2)| \tag{3}$$

The diagonal matrix element is minus twice the rotational frequency. Subtracting out the average frequency, one sees that the fluctuation in rotational frequency among the basis bands is one of the causes for rotational damping. The relative importance of the diagonal and the non-diagonal part of $\Delta H(I-2)$ depends upon the choice of basis bands.

2.3 Rotational damping

The states of the same structure as the mixed states $\alpha(I)$, but taken at $I-2$, form a basis of *rotational partner states.*

$$|\alpha_{rot}(I-2)\rangle = \sum_{\mu} X_{\mu}^{\alpha}(I)|\mu(I-2)\rangle \tag{4}$$

Except for the dispersion of E2 transition matrix elements among the basis bands μ, the set of states $\alpha(I)$ and their rotational partners at $I-2$ forms an alternative set of basis bands. The transition to the partner state carries the main part of the rotation strength

$$\langle\alpha_{rot}(I-2)||\mathcal{M}||\alpha(I)\rangle = (2I+1)^{\frac{1}{2}} \sum_{\mu\mu'} M_{E2}(\mu) X_{\mu}^{\alpha}(I) X_{\mu'}^{\alpha} \delta_{\mu\mu'} = (2I+1)^{\frac{1}{2}} \overline{M_{E2}(\mu)_{\alpha}} \tag{5}$$

where the average is taken over the basis states, weighted with their contribution to the energy eigenstate α. In contrast, only weak transitions go to other states, since the reduced decay probability to all final states is

$$\sum_{\alpha'} B(E2; \alpha(I) \rightarrow \alpha'_{rot}(I-2)) = \overline{M_{E2}(\mu)^2_{\alpha}} = \overline{M_{E2}(\mu)_{\alpha}}^2 + \sigma^2(M_{E2}(\mu)_{\alpha}) \tag{6}$$

These weak transitions go to states within a quite wide energy interval of width Γ_{μ}, and their total strength is given by the variance of the collective transition matrix elements M_{E2}. For high excitation energy, or for nuclei with a soft potential energy surface as function of deformation, these transitions may become important. However, for the cases considered here, well deformed nuclei with stiff potential energy surfaces, the main part of the strength is going to the rotational partner state, and it is the subsequent coupling between rotational partners at $I-2$ which causes the rotational damping.

In evaluating the rotational damping, we first restrict ourselves to the diagonal part of the Hamiltonian $\Delta H(I-2)$ in the basis bands. This causes the following matrix elements mixing the rotational partners:

$$\langle\alpha_{rot}|\Delta H|\alpha'_{rot}\rangle = -2\sum_{\mu} X_{\mu}^{\alpha} X_{\mu}^{\alpha'} \Delta\omega_{\mu} \tag{7}$$

Here $\Delta\omega_{\mu}$ denotes the deviation of the frequency in the μ basis band from the average.

With this form of the coupling, the technical evaluation [9] of the strength function of the rotational partner at $I-2$ proceeds via simplifying assumptions to yield analytic, yet realistic results: (i) random signs of the mixing coefficients are assumed besides the Breit-Wigner form. (ii) the individual values of $(\Delta\omega_\mu)^2$ are replaced by the average variance, denoted $(\Delta\omega_0)^2$.

With these simplifications, specific limiting behaviors of the strength function are found.

For $\Delta\omega_0 \gg \Gamma_\mu$, which is supposed to describe the situation at low heat energy, the rotational damping width is found to depend only upon $\Delta\omega_0$:

$$\Gamma_{rot} = 4\Delta\omega_0 \quad \text{for} \quad \Delta\omega_0 \gg \Gamma_\mu \tag{8}$$

In the opposite limit, which will describe the situation at high heat energy, the following expression is found

$$\Gamma_{rot} = 2\frac{(2\Delta\omega_0)^2}{\Gamma_\mu} \quad \text{for} \quad \Delta\omega_0 \ll \Gamma_\mu \tag{9}$$

Compared to the low energy expression, there is an extra factor, which, when interpreting Γ_μ^{-1} as a time scale, can be viewed as an effect of *motional narrowing* of the collective warm rotation [9]. Equivalent considerations of motional narrowing have recently been carried out within diffusion equation approaches for describing the giant dipole vibration. [10, 11]

At $\Delta\omega_0 = \Gamma_\mu$, the two expressions yield the same result, and they can be glued together at the energy where this equation is fulfilled. Actual numerical calculations of the strength function [12] roughly justify this way of piecing the solutions (8) and (9) together.

The non-diagonal part gives an additional width $\Gamma_{rot,\Delta V}$, which should be added quadratically to (8) for $\Delta\omega_0 \gg \Gamma_\mu$, and simply be added to (9) for $\Delta\omega_0 \ll \Gamma_\mu$.

3 Cranking model evaluation of damping width

Here, we shall specify the basis band states as cranked mean field states. This has the character of an ansatz, which should be examined in the future. In this connection it may be quite important that the mean field only changes little between I and $I-2$.

With the cranked mean field, one finds that the average residual interaction matrix element is rather small, and that the non-diagonal contribution to the damping width is very small [9].

3.1 Onset of damping

The residual interaction which mixes the basis bands is here taken as a two body random type interaction with no specific selection rules except parity and r-symmetry. The strength of this interaction is chosen to yield also the spreading width of single particle states.

The single particle width is estimated by use of the golden rule

$$\Gamma_{1p} = 2\pi\overline{|\langle 1p|V|2p1h\rangle|^2}\rho_{2p1h} \quad (10)$$

where the bar over the square matrix element denotes averaging. A two body interaction couples a one particle state to two particle-one hole states, the density of which is denoted by ρ_{2p1h}.

In an average basis band state, the neutron and proton orbitals are thermally occupied, and the density of levels, which can be reached by a two body interaction is estimated to be [9]

$$\rho_2 = \pi^2\frac{T^3}{d^4} \quad (11)$$

where T is the nuclear temperature, being proportional to the square root of the heat energy U. The spreading width of a basis band becomes

$$\Gamma_\mu = 2\pi\overline{V^2}\rho_2 \quad (12)$$

When the spacing ρ_2^{-1} between basis band states coupled by the residual interaction is larger than the estimated spreading width, the bands will not mix. Since the ratio between Γ_μ and ρ_2^{-1} is estimated to be proportional to the heat energy U to the 3rd power, we expect a rather sudden onset of damping as function of U.

For a typical rare earth nucleus, the onset of damping was estimated in ref [9] to occur around the heat energy

$$U_0 \simeq 700 \text{ keV} \quad (13)$$

Thus, a specific number of bands from the yrast line and up will remain undamped.

3.2 Number of unmixed bands

The number of unmixed bands of both parities and signatures can be evaluated from the standard form of the level density appropriate to the cranking model [13]

$$\rho(U,I) = \frac{\sqrt{\pi}}{12}a^{-\frac{1}{4}}U^{-\frac{5}{4}}\exp\left(2\sqrt{aU}\right) \quad (14)$$

With such a dramatically increasing level density, the number of unmixed bands below the energy U_0 increases rapidly with U_0.

Recently, the experimental method of fluctuation analysis of the two dimensional γ - γ ray spectra has been developed and applied to estimate the number of unmixed bands [14]. The decay along unmixed bands gives rise to the so called *ridges* in the two dimensional spectrum.In the fluctuation method,the effective number of paths $N_{path}^{(n)}$ of order n for the decay is obtained from the nth statistical moment of the spectrum. Thus, for example, $N_{path}^{(2)}$ is obtained by estimating the variance of the spectrum relative to a smooth spectrum.

The following figure shows the most recent extraction of the effective number of paths of order up to 5 [14], together with an estimate of this quantity from simulations of the decay flow, such as the one displayed on figure 1. Known discrete transitions have been removed. Also shown is the "true" number of paths, that is the number of unmixed bands below the energy U_0, calculated for two values of U_0. One sees that the simulated number of paths corresponds quite closely to the true number of paths, for all angular momenta. Furthermore, the figure displays the above mentioned strong sensitivity to U_0. The data, although with large error bars, supports the value $U_0 = 1$ MeV for the transition energy between undamped and damped motion for warm rare earth nuclei.

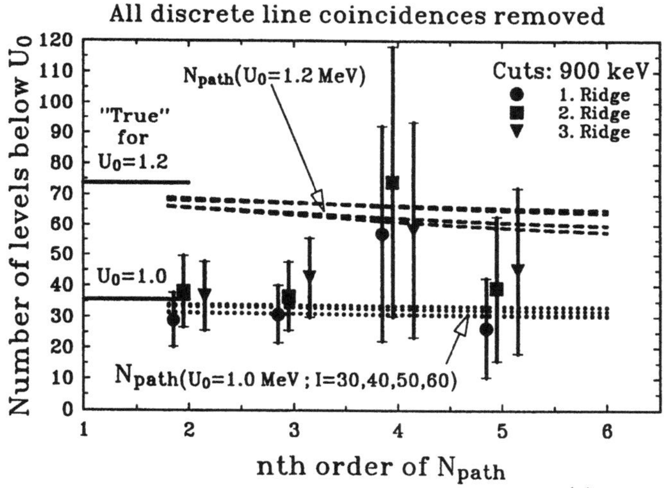

Figure 3. The effective number of decay paths $N_{path}^{(n)}$ of order n extracted from the fluctuation analysis of γ - γ energy correlation spectra for the nucleus ^{168}Yb. The average γ ray energy 900 keV corresponds to an angular momentum around 35. The numbers extracted from data are shown together with simulations of $N_{path}^{(n)}$, assuming two values of the energy for the onset of damping, $U_0 = 1$ and 1.2 MeV, dotted and dashed lines, respectively. The figure is taken from reference [14]

3.3 Dispersion in cranking rotational frequencies

We shall now move up in energy, into the damped region. Transitions in the damped region can best be studied in the so-called *central valley* region of two dimensional γ - γ spectra. The fluctuation analysis applied to the valley

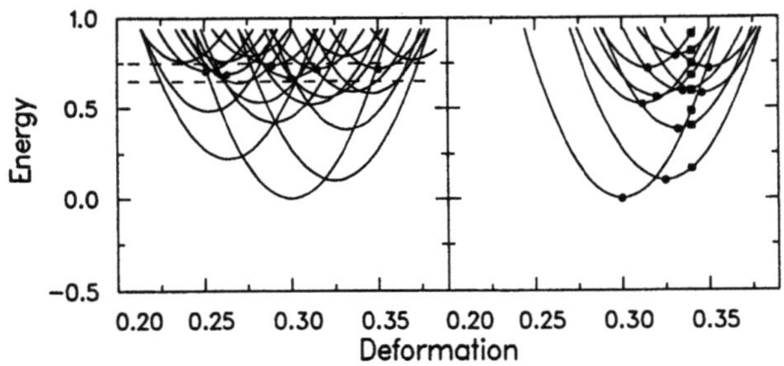

Figure 4 Schematic illustration of the energy of many cranked mean field states as a function of one deformation coordinate. On the left hand side the microcanonical sampling of the states for calculating the variance in rotational frequency is illustrated. On the right hand side, the technically more feasible sampling with a grid in deformation and a grand canonical sampling at each point is illustrated.

yields thousands of paths [14]. This is in accordance with the expected strong branching out of the decay expected in the damped region.

The key quantity describing the rotational motion in the damped region is the rotational damping width Γ_{rot}, determined by the dispersion in frequency $\Delta\omega_0$ and the spreading width of basis band states Γ_μ. The value of the latter quantity is here taken over from ref [9], whereas $\Delta\omega_0$ is calculated from a realistic cranked mean field, the Nilsson potential.

Figure 4 illustrates an important property of the present evaluation, namely that the rotational frequency of the mean field states should be evaluated at the equilibrium deformation. The left hand side illustrates the microcanonical averaging over frequencies for states within a certain energy interval. The right hand side illustrates the actual procedure followed, namely to sample the states over a grid of deformations. At each grid point, the states are given the probability of the canonical ensemble

$$P(E'_\mu, \beta) \propto \exp\left(-\frac{E'_\mu(\beta)}{T}\right) \tag{15}$$

Here E' denotes the energy in the rotating frame, and β denotes a specific deformation coordinate on the grid. The mean field states $|\mu\rangle$ may be specified by the set of occupied single particle levels:

$$g_n^\mu = \begin{cases} 1 & if\ n \in \mu \\ 0 & \text{otherwise.} \end{cases} \tag{16}$$

The canonical ensemble (15) implies that the average occupation of a single particle state n with energy e'_n in the rotating potential is given by the Fermi function

$$f_n = \left[1 + \exp\left(\frac{e'_n - \lambda}{T}\right)\right]^{-1} \tag{17}$$

The rotational frequency ω_f for the Fermi occupation, as well as the frequencies ω_μ of each of the mean field states μ are adjusted to yield the given angular momentum I:

$$I \;=\; I_f = \Sigma_n\, f_n i_n(\omega_f) \;=\; I_\mu = \Sigma_n\, g_n^\mu i_n(\omega_\mu) \tag{18}$$

where i_n is the alignment of the angular momentum in the orbit n. The alignment is defined as the contribution from the single particle state to the angular momentum on the cranking axis, $i_n = \langle n|j_x|n\rangle$. To first order in the difference $\omega_\mu - \omega_f$, we obtain

$$\frac{\partial I_\mu}{\partial \omega}(\omega_\mu - \omega_f) = -\Sigma_n i_n (g_n^\mu - f_n) \tag{19}$$

Averaging over the occupations (16), one obtains for the variance

$$(\Delta\omega_0)^2 = \frac{\Sigma_n f_n(1-f_n) i_n^2}{\left(\frac{\partial I_f}{\partial \omega}\right)^2} \tag{20}$$

This evaluation takes the rotational frequency at the grid point for all states μ. Extrapolating in stead the energy and angular momentum of each state from the grid point to the equilibrium position, by expanding to second order in deformation and frequency, one obtains after some algebra the variance

$$(\Delta\omega_0)^2 = \frac{\Sigma_n f_n(1-f_n)\left[i_n - \frac{\partial I_f}{\partial\beta} C_\beta^{-1} \frac{\partial e'_\mu}{\partial\beta} - \frac{\partial i_\mu}{\partial\beta} C_\beta^{-1} \frac{\partial E'_f}{\partial\beta} + \frac{\partial I_f}{\partial\beta} C_\beta^{-1} \frac{\partial^2 e'_\mu}{\partial\beta^2} C_\beta^{-1} \frac{\partial E'_f}{\partial\beta}\right]^2}{\left(\frac{\partial I}{\partial\omega}\right)^2 + \frac{\partial I_f}{\partial\beta} C_\beta^{-1} \frac{\partial I_f}{\partial\beta}} \tag{21}$$

where the stiffness matrix in the deformation coordinates is defined as

$$C_\beta = \left(\frac{\partial^2 E'_f}{\partial\beta^2}\right) \tag{22}$$

Illustrative results from these expressions are obtained for the cranked harmonic oscillator potential. In this case, the single particle states n are described by the quantum numbers $n_1\, n_2\, n_3$. Choosing the two deformation parameters ω_1 and $\omega_- = \omega_2 - \omega_3$, applying volume conservation, and using the expressions given in ref. [15], one obtains (after a considerable amount of algebra) the following expressions for large quantum numbers

$$(\Delta\omega_0)^2 \approx 8 \frac{T}{Ae_F} \frac{1}{\delta^2} \omega_{rot}^2 \quad \text{(fixed deformation } \delta\text{)} \tag{23}$$

$$(\Delta\omega_0)^2 \approx 8 \frac{T}{Ae_F} \omega_{rot}^2 \quad \text{(minimization with respect to deformation)} \tag{24}$$

where the expression (23) is evaluated at fixed deformation for all mean field states using (20), and expression (24) is evaluated by proper minimization of

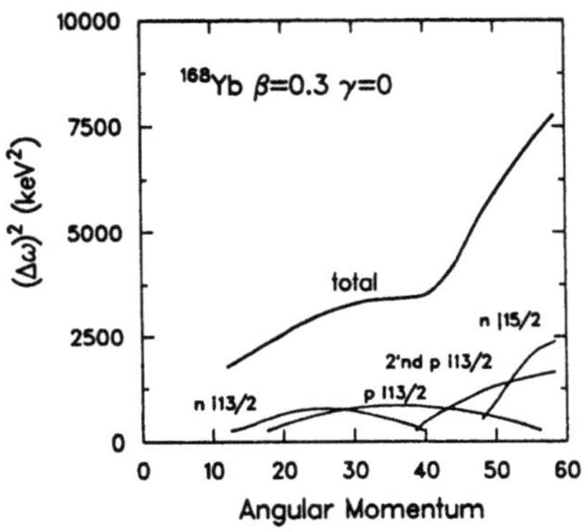

Figure 5. The fluctuation in rotational frequency and its major contributions in the sum over single particle orbitals, shown for the heat energy 4 MeV, and as function of angular momentum. The figure is calculated for the nucleus ^{168}Yb, at the fixed deformation $\beta = 0.3$, $\gamma = 0$.

the energy of the states using (21). In these formulas A is the number of particles in the rotating potential, e_F is the Fermi energy measured from the bottom of the potential, T is the temperature, ω_{rot} is the rotational frequency, and $\delta = 2(\omega_2 - \omega_3)/(\omega_2 + \omega_3)$ is a measure of the deformation in the plane perpendicular to the rotation axis. Since δ typically is of the order 0.3, one can see that the requirement of deformation equilibrium for each mean field state has a significant effect for the cranked harmonic oscillator. One can relate these expressions to the classical moment of inertia fluctuations of a hot Fermi gas, which yield the same result as equation (24).

The variance in frequency (21) together with the main contributions from the single particle states n is shown in figure 5. The nucleus is ^{168}Yb, and the deformation is the axially symmetric deformation with $\beta = 0.3$. The weighting function $f_n(1 - f_n)$ in (21) emphasizes the contributions to the variance from single particle levels close to the Fermi surface. For low rotational frequencies, the neutron Fermi surface is situated in the upper part of the $i\frac{13}{2}$ levels, and one such level with a considerable negative alignment is the most significant. At higher frequencies, levels from the shell above, proton $i\frac{13}{2}$ and neutron $j\frac{15}{2}$ orbits with large positive alignment, dive down in energy e'_n, and contribute significantly. Thus, one sees that the single particle shell structure reflects itself in the damping properties of the warm rotating nucleus. The excitation energy 4 MeV corresponds to a temperature of about $\frac{1}{2}$ MeV. Comparing the result on figure 5 to expressions (23), (24), one sees that the spin orbit potential included in the Nilsson potential greatly increases the fluctuations in alignment among the single particle states, and thereby also the rotational damping width.

3.4 Rotational damping width

Finally, the dispersion in frequency needs to be combined with the spreading width (12), using the equations (8), (9) to obtain the rotational damping width Γ_{rot}. The damping width calculated in this way is displayed in figure 6, and compared to experimental damping widths.

The kink on the calculated curves corresponds to the transition between the expressions (8) and (9), that is to the onset of motional narrowing. The full drawn parts of the curves denotes the region of heat energy where the decay flow occurs for the various values of the angular momentum. As discussed in connection with figure 1, the flow tends to follow the line of equal competition between E1 and E2 decay.

The data included in fig. 6 for the highest angular momentum value $I = 50$ only provides an upper limit to Γ_{rot}. This data point is obtained from subtraction of multiplicity gated singles spectra. The other two data points are obtained from analyzing the central valley in the two dimensional γ - γ spectra, together with the great valley in the three dimensional γ - γ - γ spectra [16, 17]. This analysis is still in a preliminary stage, and the data points should at present be taken with some care. Keeping this reservation in mind, the figure displays a quite nice agreement between data and calculated values.

On the figure one can imagine the calculated curves continued from low energies smoothly beyond the kink. These will be the curves obtained without motional narrowing. Only above angular momentum 40 will the decay flow be able to experience the motional narrowing. Thus, the data are certainly not conclusive on motional narrowing yet.

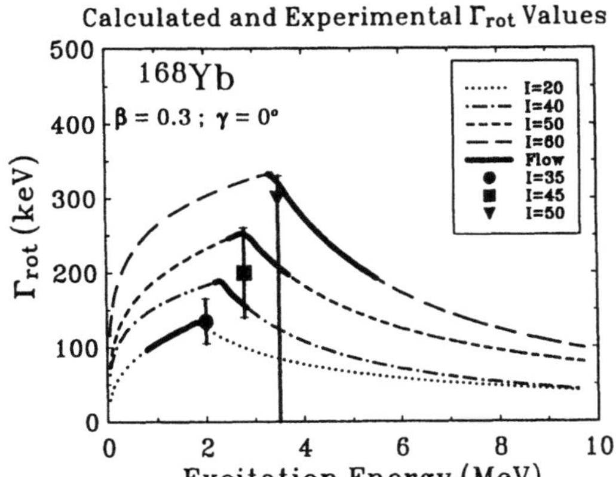

Figure 6. The rotational damping width calculated for the nucleus ^{168}Yb, at the deformation $\beta = 0.3$, $\gamma = 0$ is shown together with three recent experimental values. See text for details.

4 Conclusion

At this stage of comparison between the cranking model for warm nuclei and data, both theory and experimental procedures need to be consolidated. On the theoretical side, the cranked mean field description and the residual interactions await further investigations, and calculations including deformation fluctuations need to be carried out. On the experimental side, both the fluctuation method and the triple versus double analysis need to be more firmly established, and nuclei other than ^{168}Yb need to be studied.

The mixing between thermal and rotational motion in warm nuclei is a quite new subject, and it may have rich perspectives, such as seeing new kinds of phase transitions and chaotic motion.

Acknowledgement: This work was supported by the Commemorative Association for the Japan World Exposition Foundation.

References

[1] P.J. Twin, B.M. Nyakó, A.H. Nelson, J. Simpson, M.A Bentley, H.W. Cranmer-Gordon, P.D. Forsyth, D. Howe, A. R. Mokhtar, J.D. Morrison, J.F. Sharpey-Schafer and G. Sletten, Phys. Rev. Lett. **57** (1986) 811.

[2] R. Bengtsson and S. Frauendorf, Nucl. Phys. **A314** (1979) 27; **A327** (1979) 139.

[3] Y.R. Shimizu, E. Vigezzi and R. A. Broglia, Nucl. Phys. **A509** (1990) 80.

[4] M. Gallardo, M. Diebel, T. Døssing and R. A. Broglia, Nucl. Phys. **A443** (1985) 415.

[5] I. Hamamoto, Phys. Lett **193B** (1987) 399.

[6] K. Neergaard, V. V. Pashkevich and S. Frauendorf, Nucl. phys. **A262** (1976) 61.

[7] G. Andersson, S.E. Larsson, G. Leander, P. Möller, S.G. Nilsson, I. Ragnarsson, S. Åberg, R. Bengtsson, J. Dudek, B. Nerlo-Pomorska, K. Pomorski and Z. Szymanski, Nucl. Phys. **A 268** (1976) 205.

[8] T. Bengtsson, I. Ragnarsson and S. Åberg, Phys. Lett. **B208** (1988) 39.

[9] B. Lauritzen, T. Døssing and R.A. Broglia, Nucl. Phys. **A457** (1986) 61.

[10] Y. Alhassid and B. Bush, Phys. Rev. Lett **63** (1989) 2452.

[11] W.E. Ormand, P.F. Bortignon, R.A. Broglia, T. Døssing and B. Lauritzen, Phys. Rev. Lett **64** (1990) 2254.

[12] B. Lauritzen, R.A. Broglia, W.E. Ormand and T. Døssing, Phys. Lett. **B207** (1988) 238.

[13] S. Åberg, Nucl. Phys. **A477** (1988) 18

[14] B. Herskind, T. Døssing, S. Leoni, E. Vigezzi and A. Ikeda, in *Proc. Int. Conf. on High Spin Physics and Gamma-Soft Nuclei*, Pittsburgh, September 1990 (to be published in World Scientific)

[15] A. Bohr and B. R. Mottelson, "Nuclear Structure", vol. 2 (Benjamin, New York 1975), pp 84

[16] B. Herskind, J.J. Gårdhøje and K. Schiffer, in *Proc. Int. Conf. on High-Spin Nuclear Structure and Novel Nuclear Shapes*, Argonne National Laboratory, April 1988;

[17] B. Herskind, in *Slide Report of Workshop on Nuclear Structure in the Era of New Spectroscopy, Part B*, page 257; and to be published.

Collectivity and Chaoticity in Nuclear Dynamics

V.G. Zelevinsky

Institute of Nuclear Physics, SU-630090 Novosibirsk, USSR

Abstract. Collective and chaotic features of nuclear dynamics are discussed using simple criteria of complexity of wave functions and their coherence with respect to specific operators. Various physical phenomena are considered in this connection:

- coherent interaction of collective modes;
- fragmentation and spreading widths;
- mixing of compound states and dynamical enhancement;
- mean field as a smooth component of complicated dynamics;
- coupling through continuum and collectivization of widths;
- structure of giant resonances;
- statistical properties of unstable states as generalization of canonical random matrix ensembles.

1. Introduction

Collectivity and chaoticity are viewed commonly as extreme opposite cases of dynamics. In application to quantal many-body systems, there is no rigorous definition of such concepts. The paper will be an attempt to associate a physical pattern with those qualitative ideas. We will discuss the degree of complexity of corresponding wave functions and the observable manifestations of this complexity. It turns out that an intimate interrelationship exists between collective (regular, coherent) and chaotic (irregular, incoherent) types of motion in the system: order and chaos are tightly intertwined. The collective modes decay into complicated configurations which can be treated as chaotic. Less trivially, dynamical laws may create the synchronized self-organized behavior out of chaos.

In what follows, I use the simplest models still containing important physical aspects. The major part of the results is of very general nature and can be applied to various quantal systems.

We start with the states belonging to the discrete spectrum, or, strictly speaking, having small decay widths (radiation or weak interaction widths for low-lying states and narrow particle escape widths for well isolated resonances). In Sect.2 we discuss the characteristics of complexity of wave functions in terms of their components in the appropriate "natural" basis. It

Springer Proceedings in Physics, Vol. 58 New Trends in Nuclear Collective Dynamics
Eds: Y. Abe, H. Horiuchi, K. Matsuyanagi

allows us to distinguish roughly simple, coherent and incoherent (compound) states and to estimate the transition amplitudes between those classes. The remarkable example of parity nonconservation in fission will be given to illustrate the dynamical enhancement of weak interaction effects due to the chaoticity of compound states.

In Sect.3 we construct the phenomenological model of soft spherical nuclei by selecting the most coherent processes of the particle-phonon and phonon-phonon interactions. The selection is based on the same criteria of complexity of wave functions.

As a generic example of coupling of a collective mode to the complicated fine structure background, the problem of spreading widths is discussed in Sect.4. Their small variations for the isobaric analog resonances (IAR) throughout the periodic table are shown to be implied by the scaling of mixing matrix elements in accordance with the complexity of compound wave functions.

The self-averaging properties of strongly mixed levels are used in Sect.5 to show how the stochastic dynamics gives rise to the mean-field-type behavior. This example of secondary collectivity generated out of chaos sheds some light on the problem of the above mentioned "natural" basis.

The last chapters are devoted to the collective and chaotic aspects of nuclear dynamics in the continuum. In Sect.6 we show that the new coherence is generated by the coupling of overlapping unstable states via the common decay channels. Similar to the Dicke coherent state in optics, the states are formed synchronizing their decays and collectivizing the widths. In the language of the reaction theory, it corresponds to the separation of the direct processes from those proceeding through the compound nucleus. The structure of a giant resonance is analyzed from this viewpoint in Sect.7 as a result of the counterplay of the internal collectivization of multipole strengths and the external collectivization of decay widths.

Finally, in Sect.8 we generalize the Gaussian random matrix ensembles to study the statistical properties of unstable states. Analytical as well as numerical calculations show up new generic features like the cubic repulsion of complex energies, the disappearance of the level repulsion along the real energy axis at spacings less than widths and persistence of the Dicke width collectivization even for chaotic internal dynamics.

Sect.9 contains an incomplete list of open problems selected according to the personal interests of the author.

2. Structure of Wave Functions: Coherence and Randomness

At present, there exists a gap between the conventional description of the complicated compound states originated by the famous Niels Bohr's paper[1] and that of the collective modes[2]. For collective dynamics, the microscopic equations of motion seem to be an adequate tool provided with an approximate scheme for the practical solution. Compound states and reactions proceeding through such a stage are analyzed in statistical terms (global statistics using thermodynamical concepts and local statistics for correlations and fluctuations of energies and transition amplitudes utilizing usually random matrices).

From the studies of the classical chaos[3] it is known that, even for a small number of degrees of freedom, the dynamical solution can reveal the stochastic features emerging from instabilities of the phase space trajectories with respect to the variation of initial conditions. The quantal counterparts for the systems chaotic in the classical limit keep the fingerprints of chaoticity in the generic properties of energy spectra (level repulsion, Δ_3-statistics and so on)[4-6]. Certainly, the stochastic behavior is even more probable in a system with a large number of degrees of freedom, especially at high excitation energy where the level density grows exponentially. Here the dissipation and fluctuations become inherent features of collective motion so that the latter should be described by the Langevin-type random process. The collective mode is fragmented over complicated states covering the energy interval of the spreading width $\Gamma_\downarrow$. For a system under time reversal invariance, the Hamiltonian describing compound states can be treated, if the decay probability is negligible, as random with respect to its local properties which are similar to that of the Gaussian orthogonal ensemble (GOE) as it was discussed in detail in the review paper by Brody et al[7].

In typical many-body systems, as a rule, we have neither complete spectroscopic data nor the possibility to carry out the transition to the classical limit. In order to gain deeper insight into the structure of wave functions and to classify them at least in a schematic manner by their degree of complexity we expand the wave function $\Psi^{(\alpha)}$ of a state $|\alpha>$ over basis functions ψ_k from a complete set $\{|k>\}$,

$$\Psi^{(\alpha)} = \sum_{k=1}^{N} C_k^{(\alpha)} \psi_k \quad . \tag{2.1}$$

Here N stands for a number of components contributing significantly to the sum (2.1). This number can be very large and the $1/N$ expansion plays a guiding role in our estimates. Both sets $|\alpha>$ and $|k>$ are orthonormalized,

$$\sum_k C_k^{\alpha *} C_k^{\beta} = \delta^{\alpha\beta} \ , \quad \sum_\alpha C_k^{\alpha *} C_l^{\alpha} = \delta_{kl} \ , \tag{2.2}$$

and if we assume that there is no hierarchical scale of ranks among the amplitudes C_k^α, their typical order of magnitude is

$$|C_k^\alpha| \simeq 1/\sqrt{N} \ . \tag{2.3}$$

In the GOE quantities C_k^α are distributed uniformly over the unit sphere which may be attributed to the maximum chaoticity. In a sense, it is similar to the classical ergodicity; the wave function $\Psi^{(\alpha)}$ covers the whole accessible volume in the Hilbert space. In the limit of $N \to \infty$, each amplitude C_k^α is distributed normally with the zero mean value (isotropy in the N-dimensional space) and the variance $1/N$. This implies [2] the Porter-Thomas distribution for the probabilities $\mid C_k^\alpha \mid^2$ which are identified, for the isolated narrow resonances $\mid \alpha >$, with the decay widths to the channel matched to the simple state $\mid k >$ (the decay process accomplishes the analysis of the compound superposition $\mid \alpha >$). We recall this in Sect.8.

Here the key question arises: which basis set $\{\mid k >\}$ is to be used for the expansion (2.1)? In the limiting chaotic case of the GOE there is no such a problem since the statistical properties of matrix elements of Hamiltonians belonging to the GOE are invariant with respect to orthogonal transformations. If the dynamics is not fully chaotic this invariance is broken so one should look for the preferential or "natural" basis. The relative measure of complexity can then be defined for the family of states depending on certain parameters.

Supposedly, at least for a stable many-body system, the basis sought for should be related to the mean field (MF) although not necessarily the Hartree-Fock one. The reason is not that the MF or, in the simplest case, the shell model is a basis for all microscopic calculations. This fact itself is rather a consequence of deep properties of underlying dynamics. In Sect.5 we suggest the schematic scenario of the MF emerging from a chaotic motion as a result of self-averaging. On the other hand, the concept of coherence or collectivity acquires a definite meaning only after specific operators or external fields are demonstrated which induce the corresponding coherent response. For the MF basis those operators are of one- or two-body nature. It goes along with the instruments we have at our disposal to get meaningful experimental information.

To clarify the idea of coherence with respect to the specific agent, take as an example the harmonic oscillator coherent state

$$\psi^{(\alpha)} = \sum_k \exp\left(- \mid \alpha \mid^2 /2\right) \alpha^k (k!)^{-1/2} \psi_k \tag{2.4}$$

where ψ_k is the state with the definite quanta number equal to k. The coherent state (2.4) is labelled by the complex number $\alpha = \mid \alpha \mid \exp(i\phi)$ and the adjacent components of the superposition are synchronized since the phase difference ϕ is the same for all pairs. The average number of quanta ν is equal to $\mid \alpha \mid^2$ and the Poissonian distribution function of k,

$$\mid C_k^\alpha \mid^2 = \exp(-\nu)\nu^k/(k!) \ , \tag{2.5}$$

is reduced to the Gaussian one for $\nu \gg 1$ and $\mid k - \nu \mid \ll 1$, with the width of $\nu^{-1/2}$. The only difference as compared to the general expressions (2.1) and (2.3) is the exact phase tuning in the coherent state (2.5). But such a synchronization is, in some sense, fictitious. The main feature of the state (2.4) is that it is an eigenstate of the annihilation operator a, $a\psi^{(\alpha)} = 0$. One can make a phase transformation $\psi_k \rightarrow \exp(i\chi_k)\psi_k$ with the real k-dependent χ_k and transform the operators correspondingly, $a_{k'k} = k^{1/2}\delta_{k',k+1} \rightarrow k^{1/2}\delta_{k',k+1}\exp\{i(\chi_{k'} - \chi_k)\}$. We see that the coherence of a wave function is to be defined with respect to the specific operator, a in the case (2.4). In order to separate the coherence property from the unphysical phases, we can always have in mind the averaging procedure over the group of phase transformations. Then the random features of the basis are eliminated and the remaining picture has a closer resemblance to the classical phase space pattern with averaging over initial conditions. For the coherent state (2.4), there is the one-to-one correspondence between those pictures. We utilize the idea of the phase averaging to evoke the MF from chaotic dynamics.

It is tempting to introduce the entropy of the state $\mid \alpha >$ with respect to the given basis $\mid k >$,

$$S^{(\alpha)} = -\sum_k \mid C_k^\alpha \mid^2 \log\{\mid C_k^\alpha \mid^2\} \ . \tag{2.6}$$

It was shown[9] that this function might be a reasonable candidate for a characteristic of the localization of the state. The localized states have an ordered structure and the entropy is small whereas for the typical GOE states it reaches its maximum value $S \sim \log N$.

For the following, it will be useful to recall how collective modes (giant resonances (GR) or low-lying vibrations) can be derived from the residual interaction of quasiparticles. Instead of the powerful machinery like the random phase approximation (RPA) we use the ancestor model[10] which will be generalized for unstable states in Sect.7. The model space consists of N states, for example of 1 particle -1 hole $(1p - 1h)$ nature, with bare energies ε_k, $k = 1, \ldots, N$. The operator d (for example, that of the multipole moment) is supposed to have large matrix elements $d_k =< k \mid d \mid 0 >$ between the state $\mid k >$ and the ground state $\mid 0 >$. The model Hamiltonian H has a factorizable form

$$H_{kk'} = \varepsilon_k\delta_{kk'} + \lambda d_k d_{k'} \tag{2.7}$$

where λ is the appropriate coupling constant. The secular equation

$$\lambda P(E) \equiv \lambda \sum_k d_k^2/(E - \varepsilon_k) = 1 \tag{2.8}$$

determines the eigenvalues of H (2.7). The collective root of eq.(2.8) is shifted at $\lambda > 0$ $(\lambda < 0)$ up(down) in energy and the corresponding wave function

$\Psi^{(coll)}$ has the coefficients (2.1)

$$C_k^{(coll)} = \frac{d_k}{E_{coll} - \varepsilon_k} \left\{ \sum_k \frac{d_k^2}{(E_{coll} - \varepsilon_k)^2} \right\}^{-1/2} \quad . \tag{2.9}$$

In the case of strong collectivity, the energy shift $\Delta E = E_{coll} - \varepsilon_k$ exceeds the spread of bare energies ε_k so that

$$E_{coll} \simeq \varepsilon + \lambda \sum_k d_k^2 \equiv \varepsilon + \lambda D^2 \tag{2.10}$$

where the N-dimensional amplitude vector $D = \{d_k\}$ and the mean energy $\varepsilon = \sum_k d_k^2 \varepsilon_k / D^2$ are introduced. Then $C_k^{(coll)}$ coincide with the normalized components of D,

$$C_k^{(coll)} = d_k / (D^2)^{1/2} \quad , \tag{2.11}$$

in accordance with (2.3).

The deexcitation probability $| \Psi^{(coll)} > \rightarrow | 0 >$ induced by an operator Q with the matrix elements $q_k = < 0 | Q | k >$ depends on the angle ϑ between the N-dimensional vectors D and $Q = \{q_k\}$:

$$< 0 | Q | \Psi^{(coll)} > = \sum_k C_k^{(coll)} q_k \simeq (QD)/(D^2)^{1/2} = (Q^2)^{1/2} \cos\vartheta \quad . \tag{2.12}$$

Thus we have the enhancement factor $N^{1/2}$ for the amplitude and N for the probability of the collective transition if the vectors Q and D are not orthogonal. For low-lying quadrupole vibrations or GR, see Sections 3 and 7 respectively, $N \simeq A^{2/3}$, the capacity of the valence major shell. Coherence properties of the specific operator are accumulated in the angle θ being maximum for $Q//D$. The energies E^α of $N - 1$ noncollective roots of eq.(2.7) are located between ε_k, the coefficients C_k^α do not vanish for a small number of neighboring k only and the transition amplitudes $< 0 | Q | \alpha >$ are of the same order of magnitude as q_k.

In the case of the compound state $| \alpha >$, the number N of components can be very large, up to 10^6 for neutron resonances in heavy nuclei. The transition amplitude $Q_{k\alpha}$ between the simple $| k >$ and the complicated $| \alpha >$ states can be estimated as

$$Q_{k\alpha} = < k | Q | \alpha > = \sum_{k'} C_{k'}^\alpha < k | Q | k' > \simeq N^{-1/2} q \tag{2.13}$$

where q is the typical matrix element of Q between simple states; here we took into account that observables Q have definite selection rules so that, for the given $| k >$, only a small number of selected $| k' >$ do contribute. The suppression of matrix elements $Q_{k\alpha}$ by the factor $N^{-1/2}$ will be important in the discussion of the spreading width in Sect.4.

In the same way one can get the crude estimate for the transitions $Q_{\alpha'\alpha}$ between two complicated states,

$$Q_{\alpha'\alpha} = \sum_{k'k} C_{k'}^{\alpha'*} C_k^{\alpha} Q_{k'k} \simeq N^{-1/2} N^{-1/2} q N^{1/2} \simeq q N^{-1/2} \quad . \tag{2.14}$$

Here again q is the typical value of the matrix element $Q_{k'k}$ between two simple states. Selection rules reduce the two-fold summation to the onefold one $\sum C^* C q$ and the lack of coherence leads to the random walk factor $N^{1/2}$ instead of N. As a result we have the same attenuation parameter $N^{-1/2}$ as for transitions between simple and compound states, eq.(2.13). The attenuation (2.14) is not strong enough to compensate the effect of the high level density $\rho \sim N$ of compound states. It gives rise to the so called dynamical enhancement[11] of perturbations in the region of isolated compound resonances. It is necessary to stress that the estimate (2.14) holds only if, for each component $| k >$ of the initial state $| \alpha >$, the final state $| \alpha' >$ contains the component $| k' >$ required for the transition. It means that the states $| \alpha >$ and $| \alpha' >$ cover the whole available Hilbert space which corresponds to the fully chaotic case.

A perfect example of the dynamic enhancement is given by the parity nonconservation (PNC) in nuclear fission[12]. The parity violating asymmetry sp is observed in the experiment (n, f) where the momentum p of the fission fragment (for example, the heavy one) turns out to be correlated with the spin s of the captured polarized neutron. Certainly, the PNC occurs on the level of elementary weak interaction. Remarkably enough, the fundamental asymmetry is transferred to the macroscopic collective motion of hundreds of particles which looks like a violation of the second law of thermodynamics. Moreover, the observed asymmetry is enhanced by several orders of magnitude. Indeed, one can assume that the mixing of nuclear states of opposite parities is caused mainly by the weak interaction component of the MF,

$$V_v \sim G(sp)\rho(r)/m \tag{2.15}$$

where $G \simeq 10^{-5}/m^2$ is the Fermi coupling constant, m is the nucleon mass and $\rho(r)$ is the local nuclear density. The perturbation Hamiltonian (2.15) induces the mixing of single-particle orbitals with the same angular momentum and opposite parities. The mixing amplitude is $F_{s-p} \sim < + | V_v | - > /\Delta(\pm)$ where $\Delta(\pm)$ stands for the typical energy spacing of the mixed orbitals. As a result, $F \sim 10^{-7}$.

The estimate changes drastically if the mixing occurs between two compound states of chaotic nature, see eq.(2.14). In the region of neutron resonances, the level densities for both parities are about the same so one can identify the localization lengths for two classes of states. The typical mixing matrix element is now, according to (2.15), $< + \mid V_v \mid - > /N^{1/2}$ so that for the mixing amplitude we get in terms of the mean level spacing D

$$F_{comp} \sim < + \mid V_v \mid - > /N^{1/2} D \sim F_{s-p} N^{-1/2} \Delta(\pm)/D \sim F_{s-p} N^{1/2} \quad . \quad (2.16)$$

Thus, we have derived the dynamical enhancement factor $N^{1/2}$ which increases the PNC amplitude by three order of magnitude in heavy nuclei. These qualitative arguments are confirmed by the data which are compiled in ref.[13]. In all cases (^{233}U,^{235}U and ^{239}Pu) the asymmetry coefficient α in the fragment angular distribution,

$$W(\theta) = 1 + \alpha sp \quad , \quad (2.17)$$

is $\mid \alpha \mid \geq 10^{-4}$. The experiment reveals independence of the asymmetry of the properties of the exit channels like the specific mass distribution or the total kinetic energy of fragments. It should be so if the mixing occurs at the hot stage of the process, i.e. in the compound nucleus.

The scenario of the transfer of the PNC to the macroscopic observables was suggested by Sushkov and Flambaum[14]. The point is that the process can be described by the stationary wave function since the energy spread of neutrons is very narrow, about 0.03 eV. The wave function of a system of $(A+1)$ nucleons can be expanded in terms of two alternative "natural" bases: either the basis of the MF of the initial nucleus after the neutron capture or the basis of the cold elongated nucleus near the top of the external fission barrier. The first expansion explains the dynamical enhancement of the PNC in the hot compound nucleus. The second expansion isolates very few terms coupled to fission channels. Here the parity violation is transformed into the mixing of rotational states split in parity doublets due to the pear-shaped deformation. That is the stage where the observed angular distribution is finally formed. Details of the mass and kinetic energy distributions are defined after the saddle point when the parity is conserved. Similar arguments can be applied to various nuclear reactions (the so called $1/N$ classification). The expansion over two supplementary sets of states, external and internal, will be used extensively below in the discussion of dynamics in the continuum.

3. Coherent Phonon Coupling in Soft Spherical Nuclei

To illustrate the pure collective aspects of nuclear dynamics, we consider the low-lying spectra of non-magic nuclei which do not show up the distinct rotational band pattern. These nuclei are transitional between spherical and deformed ones so that here we have to deal with the finite Fermi-system in the vicinity of the phase transition.

The problem of the description of transitional nuclei is even more complex than that of phase transitions in macroscopic systems since the corrections which are negligible in the thermodynamic limit of large systems, in the finite system can become important and eventually dominant. At the present moment, we cannot formulate a regular microscopic theory. Using the general arguments of collectivity and symmetry we separate the most coherent contributions and construct the simple phenomenological model which fits experimental data surprisingly well. The model generalizes the ideology (unfortunately, not the formalism!) of the boson expansion method (BEM)[15] for the large amplitude motion. Details of the approach can be found in the original papers by Vorov and the author[16,17] followed by the reviews[18-20].

Let us summarize the main features of low-lying spectra of even-even transitional nuclei as revealed by the enormous amount of nuclear data.
(i) *Collectivity.* The properties vary regularly along the isotope chains; the quadrupole transition probabilities are enhanced up to several dozens of Weisskopf units.
(ii) *Adiabaticity.* The frequencies of strong transitions are decreasing along with the filling of the valence shell; the adiabatic factor $\tau = \omega/\varepsilon$ is typically small, $\tau \approx (1/3) \sim (1/5)$, where $\varepsilon \geq 2\Delta$ is the single-particle excitation energy associated with the Cooper pair breaking.
(iii) *Quadrupole symmetry* is obvious from the quantum numbers of levels, see below.

From these facts, we conclude that the first excited 2_1^+ state has a wave function of collective type ("quadrupole phonon"). In the popular model[15] with pairing and quadrupole forces the collective RPA mode is shifted, similar to (2.7) for $\lambda < 0$, down from the threshold of two-quasiparticle excitations. It absorbs the essential fraction of the quadrupole strength the corresponding collectivization factor is $N \simeq A^{2/3}$. The amplitude of this motion is proportional to $\omega^{-1/2}$ being of the same order of magnitude as the static quadrupole deformation of well deformed nuclei. This justifies the term "soft" we apply to nuclei. When the decreasing frequency reaches some limiting value the spherical shape becomes unstable. At this limit the approximate relationship between the parameters of collectivity and adiabaticity exists,

$$N\tau^3 \simeq 1 \ . \tag{3.1}$$

As was suggested by V.M. Galitsky, eq.(3.1) means the estimate of the maximum wavelength corresponding to the standing wave in the "volume" N; actually, eq.(3.1) holds throughout almost the whole transitional region. Note that the RPA predicts, as a rule, too early transition to the deformed shape. Evidently, higher order stabilizing factors are not taken into account properly.

According to (2.3), each single-particle mode transfers only a minor part of the strength to the vibrational excitation. Therefore we expect that the operator producing the phonon state 2_1^+ can be applied repeatedly with no significant change of its structure. Thus, we can introduce the boson image $d_\mu^\dagger(\mu = 0, \pm1, \pm2)$ of the operator creating a quadrupole phonon with the angular momentum $J = 2$ and its projection μ. By means of the operators $d_\mu^\dagger$ we generate the boson Hilbert space with the basis states labelled by the total phonon number $\nu = \sum_\mu d_\mu^\dagger d_\mu$, the numbers $n = (\nu - v)/2$ and n' of phonon pairs and triples, respectively, coupled to $J = 0$, and the total angular momentum quantum numbers J and M. The quantum number v (boson seniority, or the number of unpaired bosons) plays the most essential role in what follows. It gives the magnitude of the 5-dimensional angular momentum (the generator of the orthogonal group $O(5)$) in the same way as J does for the rotation group $O(3)$: the corresponding Casimir operators are equal to

$$C[O(5)] = v(v+3); \quad C[O(3)] = J(J+1) \ . \tag{3.2}$$

For almost all known low-lying levels linked by the strong quadrupole transitions one can uniquely assign the quantum numbers $\{\nu, v, n', J, M\}$ in such a way that the enhanced transitions obey the selection rules $\Delta\nu = \pm1, \Delta v = \pm1$ (the n' quantum number is redundant at low energies) and the level energies are growing, on average with increasing quantum numbers. Intruder states which could not find their place in such a scheme appear very seldom at $E \leq 2$ MeV whereas the collective transitions survive at least up to $E \simeq 4 \sim 5$ MeV.

The close correspondence between the experimental level scheme and the phonon one shows that the pure phonon effective Hamiltonian H can be constructed for the description of the level set under study. The simplest choice corresponds to the harmonic approximation of independent phonons,

$$H_0 = (1/2)(\pi^2)_{00} + (1/2)\omega_0^2(\alpha^2)_{00} \ , \tag{3.3}$$

where the collective coordinated α_μ and momenta π_μ are defined,

$$\alpha_\mu = (2\omega_0)^{-1/2}[d_\mu + d_{-\mu}(-)^\mu], \pi_\mu = -i(\omega_0/2)^{1/2}[d_\mu - d_{-\mu}(-)^\mu] \ . \tag{3.4}$$

The basis states $| \nu v n' J M >$ are eigenvectors of H_0 with degenerate eigenvalues $E_0(\nu) = \omega_0(\nu + 5/2)$.

The harmonic approximation is equivalent to the $SU(5)$ limit of the interacting boson model (IBM)[21]. The underlying physical picture is however different since we do not identify phonons with the images of nucleon pairs coupled to $J = 2$. The number ν of our phonons has nothing to do with the valence nucleon number and should not be conserved; by the same reason, we have no artificial cut-off of the boson space.

Certainly, the predictions of the harmonic approximation are badly violated. The energy multiplets are split and transition probabilities do not obey the simple rules following from (3.4) for the transition operator Q_μ being simply proportional to the collective coordinate α_μ. However the weaker $O(5)$ symmetry does apparently persist and v is still a good quantum number. For example, the states 0_2^+ ($v = 0$ corresponding to the number $n = 1$ of pairs with $J = 0$) usually have decay properties different to the other two-phonon triplet members, 4_1^+ and 2_2^+. The latter states have very similar wave functions which differ by the resulting angular momentum only. Therefore the branching ratio $B(E2; 3_1^+ \rightarrow 4_1^+)/B(E2; 3_1^+ \rightarrow 2_2^+)$, as well as others similar to that, can be predicted in the model-independent way in good agreement with data.

Why are the level classification and the $O(5)$ symmetry still valid in spite of the $SU(5)$ breakdown? The answer is: due to the specific character of the most coherent anharmonic terms. The phonon anharmonicity arises[15] as an outcome of several effects:
(i) the RPA phonons are not the ideal bosons; this should be improved by higher order terms of the BEM;
(ii) the phonons are not dynamically independent being made of the same interacting quasiparticles; these effects are contained in the corrections to the RPA;
(iii) the new type of collective motion ("virtual rotation"[17]) is induced by the slowly changing quadrupole deformation which defines the body-fixed frame and corresponding angular coordinates.

To rate various contributions on coherence, we need the dictionary to translate from the language of particles to that of phonons[22]. The phonon exchange generates the non-static particle interaction amplitude $I(\omega) \sim \gamma \boldsymbol{I}(\omega)\gamma$ where γ is the vertex for the phonon decay into quasiparticles and the harmonic phonon propagator $\boldsymbol{I}(\omega) = 2\omega_0/(\omega^2 - \omega_0^2)$ is introduced; ω stands for the running energy variable. On the other hand, the same process is described in the RPA as a sum of the ring diagrams leading to the pole term of the form

$$I_{kk'} = \lambda q_k \{\lambda P(\omega)/[1 - \lambda P(\omega)]\} q_{k'} \tag{3.5}$$

where $P(\omega)$ is the RPA analog of eq.(2.8). For the factorized interaction of type (2.7),

$$P(\omega) = \sum_k q_k^2 \varepsilon_k/(\omega^2 - \varepsilon_k^2); \quad \lambda P(\omega_0) = 1 \ . \tag{3.6}$$

Extracting the propagator $\boldsymbol{I}$ from the denominator (3.5) we obtain the estimate of the phonon-quasiparticle vertex

$$\gamma^2 = (q^2/2\omega_0)\lambda P(\omega) \left\{ \sum_k q_k^2 \varepsilon_k / [(\omega_0^2 - \varepsilon_k^2)(\omega^2 - \varepsilon_k^2)] \right\}^{-1} \ . \tag{3.7}$$

We are interested in the case of negative λ when the low-lying mode $\omega_0 \ll \varepsilon$ may exist. At $\omega \sim \omega_0$ eqs.(3.6) and (3.7) give the effective coupling constant expressed in terms of the average two quasiparticle excitation energy ε and the "volume" N,

$$\gamma^2 \simeq (\varepsilon^3/2\omega_0)/N \simeq \omega_0^2/N\tau^3 \quad . \tag{3.8}$$

Recalling the relationship (3.1) we see that for soft nuclei $\gamma \sim \omega$, i.e. the phonon coupling to quasiparticles is strong enough to produce significant anharmonic effects. Note that if one used the Tamm-Dancoff form (2.8) of the secular equation instead of the RPA form (3.6) the result, $\gamma^2 \sim \varepsilon^2/N \sim \omega_0^2/N\tau^2$, would be wrong since the difference of the two approximations is dramatic at low ω (the RPA restores the p-h symmetry). A crucial role is played by pairing correlations which repell quasiparticle energies beyond the pair continuum threshold 2Δ and create the low-lying mode described by the secular equation of the form (3.6) where the Bogolyubov coherence factors are hidden in q_k.

Proceeding further, we can estimate the many-phonon interaction blocks. The closed fermion loop with n phonon loops and n quasiparticle propagators G is

$$H_n \sim \int d\varepsilon Tr\{(G\gamma)^n\} \sim \gamma^n N \varepsilon^{1-n} \quad , \tag{3.9}$$

or, using again (3.1) and (3.8),

$$H_n \sim \omega_0 \tau^{n-4} \quad . \tag{3.10}$$

On the scale of the phonon frequencies ω, only cubic ($n = 3$) and quartic ($n = 4$) anharmonicities are strong whereas higher order processes are suppressed by adiabaticity; one can say that the typical time for the many-phonon collision is too short. The phenomenological model including, in addition to H_0 (3.3), cubic and quartic terms treated as small corrections, was developed very long ago[23]. But the perturbation theory is not valid and the anharmonic effects are proved to dominate in dynamics.

The cubic terms are overestimated by eq.(3.10). There is a significant suppression of odd-order loops due to the approximate p-h symmetry about the Fermi surface. In quantum electrodynamics the charge conjugation symmetry makes all n-photon loops vanish (the Furry theorem). In our problem, $H_3 \neq 0$ but supposedly it will be strongly reduced with respect to eq.(3.10). It is confirmed by fitting the data.

Among the quartic structures, $\{(\pi^4)_{00}, ((\pi^2)_L(\alpha^2)_L)$ for $L = 0$, 2 and 4, and $(\alpha^4)_{00}\}$, the most important are corrections $\sim (\alpha^4)_{00}$ to the potential energy which are large in the adiabatic limit, see (3.4). Thus, we come to the phenomenological model

$$H = H_0 + \lambda \left\{ \sum_\mu \alpha_\mu^\dagger \alpha_\mu \right\}^2 + \ldots = H^{(4)} + \ldots \tag{3.11}$$

where the dots include cubic and higher, $n \geq 5$, corrections. The quartic term in (3.11) is mere $\lambda\beta^4, \beta$ being the standard 5-dimensional radial coordinate. The Hamiltonian $H^{(4)}$ is obviously $O(5)$ symmetric which is necessary to serve as a good candidate for describing the data.

Here we are not able to go into details of calculations, fitting procedures and comparison with experiment. In what follows I just summarize the main results.

A. The strong quartic anharmonicity changes the structure of the ground state. The condensate of phonon pairs coupled to $J = 0$ is created and new boson-like excitations above this condensate should be compared to the excited nuclear levels. The level classification coincides with that in the harmonic scheme: quantum numbers v and n describe unpaired quadrupole bosons and excited pairs with $J = 0$ respectively. The seniority v is conserved due to the $O(5)$ invariance of $H^{(4)}$. Here the excited pairs with $J = 0$ play a role quite similar to that of s-bosons in the IBM.

B. The energy levels can be computed with the numerical diagonalization in the big phonon space. But it is instructive to start with the optimum construction of the phonon condensate in each sector with the fixed seniority separately. It can be done by the v-dependent canonical transformation. The method was suggested in ref.[24] and rediscovered and extended by many authors, for example [25,26], being the simplest version of the $1/N$ expansion in the quantum field theory. For the lowest states ($n = 0$ or 1) this approximate method gives results very close to those of numerical computations and the precision is higher for large v. The renormalized coupling of new excitations above the correct condensate is weak so that corrections to the first approximation can be found regularly.

C. In the limit of $\omega_0 \to 0$ we get from $H^{(4)}$ the primary standard of the 5-dimensional isotropic oscillator with pure quartic anharmonicity. Neglecting the small corrections mentioned in the previous paragraph we predict[16] the parameter-free asymptotic ($\lambda \to \infty$) values of the energy ratios in the yrast band ($n = 0, v = J/2$),

$$R_J = \frac{E_J - E_0}{E_2 - E_0} = \frac{(J+5)(J/7+1)^{1/3} - 5}{7^{2/3}9^{1/3} - 5} \quad . \tag{3.12}$$

Quite remarkably, eq.(3.12) describes precisely the yrast band of ^{100}Pd[17] up to 14^+ presenting a beautiful example of the new dynamical symmetry. The same is approximately valid for many other nuclides, especially for those with four valence particles or holes.

D. In most cases where the deviations from (3.12) exist, they can be described by the one-parameter correction

$$\delta R_J = \sigma J(J+1) \tag{3.13}$$

which we interpret as the virtual rotation[27]. In ref.[17] the systematics of the parameter σ was discussed. With the exception of near-magic nuclei, σ is almost constant along the isotope chains and does not change significantly at the transition to the stable deformation. The corresponding rotational energy (3.13) is close to that of the rigid rotor prompting the idea that the main effects of superfluidity reducing nuclear moments of inertia are already accumulated by the angular momentum dependence implied by the quartic anharmonicity.

E. Two-parameter calculations taking into account $H^{(4)}$ along with the virtual rotation give [17-20], as a rule, very good agreement with experimental spectra. The introduction of the third parameter connected with the cubic anharmonicity $H^{(3)} = \chi\beta^3 \cos 3\gamma$ slightly improves the fit. The χ value turns out to be small to preserve the $O(5)$ symmetry in accordance with assumptions. Fitted parameters change rather smoothly from one nucleus to another. The quality of description is usually superior to that of the IBM.

F. The transition probabilities are known to be more sensitive to the model assumptions than the level energies. The crucial point of the model is the choice of the effective quadrupole operator Q_μ. In the standard approach this is the collective coordinate α_μ. However, operators undergo renormalization due to the phonon interaction. Various corrections can be evaluated similar to the anharmonic terms in the Hamiltonian. The substitution of the phonon tail in the fermion loop for the external electromagnetic field of the same multipolarity does not change the graph structure. Therefore the main corrections to the quadrupole operator can be summarized as

$$Q'_\mu \simeq \delta H / \delta \alpha^\dagger_\mu \ . \tag{3.14}$$

In particular, for the quartic anharmonicity we have

$$Q_\mu = \alpha_\mu + æ\alpha_\mu(\alpha^\dagger \alpha)_{00} + \dots \tag{3.15}$$

Microscopic estimates show that the new parameter æ should be typically negative, $\mid æ \mid \simeq 0.1 \sim 0.3$. The operator (3.15) allows us to reproduce available data for enhanced transitions without artificial assumptions on the saturation of the collective space made in the IBM. Weak ("forbidden") transitions and quadrupole moments of excited states require for their description the additional two-phonon terms in eq.(3.15) related to the cubic anharmonicity as in (3.14). In the same way, one can construct effective magnetic operators and explain the data on regular trends in M1 transition probabilities between collective states.

G. In many cases the quality of description becomes better if the parameter ω_0^2 in (3.3) is taken to be negative. It means that the nucleus is already beyond the RPA point of the phase transition and the quartic potential is responsible for the stability. Such a situation resembles closely the γ-unstable case ($O(6)$ limit of the IBM). Then the wave functions of states with $v \neq 0$ could be localized near the deformed minimum of the γ-unstable potential due to the 5-dimensional centrifugal barrier. Therefore levels with $v \neq 0$ go down in energy against levels with $v = 0$. In particular, the 0^+ states ("two-phonon", $v = 0$, $n = 1$, and "three-phonon", $v = 3, n = 0$) can interchange their positions. This is known to be the case in the Pt isotopes [28]. But the same situation occurs for many other nuclides.

H. The soft phonon mode determines the low energy properties of neighbouring odd-A nuclei. As we have shown the coupling of phonons to quasiparticles is strong on the scale of phonon frequencies [22]. An unpaired quasiparticle is dressed by the phonon cloud. The structure of this cloud depends [29] on the scheme of the single-particle orbitals. If the typical angular momenta j are large, $j \gg 1$, the special perturbation theory with respect to j^{-1} can be developed [30] starting from monopole phonons where the exact solution, similar to the well known fixed source problem in the meson theory, is available. In this approximation, the phonon field corresponds to the coherent state (2.4). Then the regular corrections for the quasiparticle recoil in the annihilation and creation of phonons can be calculated. In the opposite case of $j \simeq 1$, the quasiparticle angular momentum vector changes its direction easily in the processes of interaction with phonons. Here a geometric chaoticity arises which kills many higher order diagrams. Again one has to deal with a problem of the summation of the most coherent contributions. The solution gives the resulting picture of the spectrum. In both cases, the qualitative pattern of split phonon-quasiparticle multiplets is similar. Up to now, the comparison with data taking into account the anharmonic effects in even-even nuclei has not been done yet.

The same phonon-quasiparticle coupling is important for particles in the continuum. It contributes to the width of GR in soft nuclei where there is no splitting of GR due to the static anisotropy. The influence of low-lying phonons which cannot be reduced to weak perturbation, significantly changes the single-particle strength functions [31]. The latter are calculated usually from the optical potential. It means that incoherent excitations are included in the average nuclear field. The coherent low-lying vibrations should be added separately which can change the shape of the strength function essentially.

I. The challenging problem of the self-consistent microscopic derivation of the collective Hamiltonian[32] remains unsolved. The higher order insertions into many-phonon loops should be taken into account in addition to the main terms we have estimated in (3.9). This corresponds to the response of incoherent degrees of freedom to the slow collective motion of large amplitude. The proximity of the phase transition and multiple level crossings accompanying the large scale evolution of the nuclear field make this problem extremely difficult, especially in what concerns the virtual rotation [27].

Apart from that, one must bear in mind that the actual shape and the phase transition point are governed by small terms, like the cubic and sextic anharmonic members of the full Hamiltonian (3.11) which fix the non-axiality of a nucleus.

4. Compound Mixing of Collective Modes

In Sect.2 we have sketched the mechanism of the formation of the collective state accumulating the major part of the strength. If such a state is located at the high level density which is the case for GR, the coupling to the fine structure background becomes important, being not included in the model Hamiltonian (2.7). The RPA Hamiltonian used for the low-lying mode in Sect.3 contains the MF, with or without pairing, and the interaction within the 1p-1h class imitated in eq.(2.7) by factorizable forces; residual interactions between different exciton classes are not included at this stage. At higher energies the number of classes involved in the chain of interactions is rather big. For simplicity, we discuss the hierarchy consisting of two groups only: the collective state $|\, 0 >$ and background states $|\, 1 >, \ldots |\, N >$.

We assume that the fine structure states are described by the truncated $(N \times N)$ part h of the full nuclear Hamiltonian. When expanded in terms of the simple MF states, the eigenvectors of h are given by the complicated superpositions (2.1). The dimensionality of this subspace characterizes the complexity of a typical eigenfunction, i.e. its degree of localization in the MF basis. We know that the matrix elements h_{12} between basis states are more or less random being of order of magnitude $a/N^{1/2}$ where a stands for a typical matrix element of h between single- or two-particle states. Let D be an average level spacing in the region under study (remember that all states have the same values of exact integrals of motion like the total angular momentum and parity). We are interested in the strong coupling limit when $a \gg D$. Since the localization range $\Delta E \simeq < h^2 > /D \simeq a^2/ND$ should coincide with ND, we have $a \simeq ND$ so that at $N \gg 1$ our estimates are self-consistent and the simple states are fragmented over the energy interval $\Delta E \simeq a$.

Without loss of generality we assume that the major part of the fine structure states is concentrated within the finite energy interval $(-a, a)$ where the origin of the energy scale is put in the middle. The convenient quantities to work with are the Green function

$$G(E) = (E - h)^{-1} \tag{4.1}$$

and its average trace in this subspace

$$g(E) = N^{-1}\mathrm{Tr} < (E - h)^{-1} > \quad . \tag{4.2}$$

The imaginary part of the analytic continuation $g(E + i0)$ gives the average level density

$$\rho(E) = -(N/\pi)\text{Im } g(E) \tag{4.3}$$

normalized according to $\int dE\rho(E) = N$. In eq.(4.2) we do not specify precisely the averaging procedure. Our concept of complexity implies the ergodicity, so the results have to be insensitive to the details of this procedure. In actual applications, one can average either over energy E or over random parameters of the Hamiltonian h.

The average trace (4.2) has no singularities near the origin $E = 0$ where it can be presented in the general form

$$g(E) = \zeta(E) - i\pi N^{-1}\rho(E) \tag{4.4}$$

with the real part $\zeta(E)$ originating from the principal value of $< (E-h)^{-1} >$ and vanishing at the origin, $\zeta(E) \sim E/a^2$, whereas the level density (4.3) is constant, $\rho(E) \simeq \rho(0) + O(E^2/a^3)$. The central level density $\rho(0) \simeq N/2a \simeq D^{-1}$ does not strongly depend on the details of distribution so that $\text{Im } g \simeq -\pi/2a \simeq (ND)^{-1}$.

These general properties can be illustrated using as an example the GOE of random real symmetric $N \times N$ matrices. According to our scaling rules, we write down the statistics of the GOE matrix elements as

$$< h_{12} >= 0, \; < h_{12}h_{34} >= (\delta_{13}\delta_{24} + \delta_{14}\delta_{23})a^2/4N \quad . \tag{4.5}$$

At $N \to \infty$, the average GOE Green function within the fragmentation interval is known to be

$$g(E) = 2a^{-2}(E - i\sqrt{\alpha^2 - E^2}) \; , \; | E |< a \quad . \tag{4.6}$$

It leads to the semicircle Wigner law for the level density

$$\rho(E) = (2N/\pi a^2)\sqrt{a^2 - E^2}\theta(a^2 - E^2) \quad . \tag{4.7}$$

In conformity with our expectations, we get from (4.7) and (4.4) $\zeta(E) = 2E/a^2$, $\rho(E) \simeq 2N/\pi a$. The specific GOE features like the sharp cut-off of the level density at $E = \pm a$ do not matter.

Now we switch on the coupling of the background states to the isolated resonance $| \; 0 >$ located at $E = E_0$. The problem is equivalent to that of the strength function of a doorway state [2,Vol.1]. Denoting $V_{0n} = V_{n0}$ the coupling matrix elements, the secular equation for the energies of resulting stationary states can be written down as

$$\Omega(E) \equiv E - E_0 - \sum_{n=1}^{N} V_{0n}^2/(E - \varepsilon_n) = 0 \tag{4.8}$$

where ε_n, $n = 1, \ldots, N$, are eigenvalues of h for the fine structure states. We assume the matrix elements V_{0n} to be random variables with the same scaling properties,

$$< V_{0n} >= 0, \ < V_{0n}V_{0m} >= \delta_{nm}v^2/N \ , \tag{4.9}$$

v being the typical matrix element coupling the resonance to the simple MF states. At such a scaling, the mean squared matrix element $< (V^2)_{00} >= v^2$ remains finite at $N \to \infty$. Suppose that compound states are mixed completely by the Hamiltonian h. As in the PNC discussion, it implies that each of the stationary compound states contains a component suitable for mixing with the collective mode. Therefore we can identify factors N in eqs.(4.5)and (4.9) so that ranges of fragmentation and delocalization coincide.

Experiments with insufficient resolution cannot reveal the fine structure of a resonance. In the average description, we reduce the influence of the effective continuous background to the centroid shift of the resonance, its broadening and distortion of the shape. By conventional definition, the spreading width $\Gamma_\downarrow$ is the energy uncertainty of the resonance state $| \ 0 >$ due to the damping into complicated configurations. After averaging over intrinsic states $| \ n >$, one gets from (4.2),(4.4),(4.8) and (4.9)

$$\Omega(E) = E - E_0 - \Delta(E) + (i/2)\Gamma_\downarrow(E) \ , \tag{4.10}$$

$$\Delta(E) = v^2\zeta(E) \ , \ \Gamma_\downarrow(E) = 2\pi v^2\rho(E)/N \ . \tag{4.11}$$

In practical applications, one should have $\alpha \equiv v^2/a^2 \leq 1$. Otherwise, the starting derivation of the collective resonance by isolating simple states (see eq.(2.7)) from the background would be unjustified. On the other hand, there is no reason to expect the essential asymmetry of the collective resonance ($E = E_0$) with respect to the fragmentation region $(-a, a)$. Therefore the estimates are valid made for the vicinity of the origin where $\Delta(E)$ is a linear function and $\Gamma_\downarrow$ is approximately constant,

$$\Delta(E) \sim (v^2/\alpha^2)E \sim \alpha E \ , \tag{4.12}$$

$$\Gamma_\downarrow(E) \simeq 2\pi v^2/ND \sim \pi v^2/a \sim \alpha a \ , \tag{4.13}$$

for example, in the GOE model $\alpha = 2v^2/a^2, \Gamma_\downarrow = 4v^2/a$.

In eqs.(4.11) and (4.13) we have two estimates for $\Gamma_\downarrow$: either in terms of coupling to compound states, $\Gamma_\downarrow \sim< V_{0n}^2 > /D$ (the Fermi golden rule), or through the original residual interaction, $\Gamma_\downarrow \sim v^2/a$. Equivalence of these estimates implies that the quantities growing exponentially with the level density ρ could be eliminated so that the observables as $\Gamma_\downarrow$ are expressed with the aid of smooth functions of the mass number and excitation energy.

Many authors [33-35] noted that the spreading widths of IAR have rather small variations throughout the periodical table. The reason for that can be understood qualitatively from the preceding arguments [36]. In this problem the state $| 0 >$ (the IAR) and compound states $| n >$ belong to the isospins $T_>$ and $T_< = T_> - 1$ respectively. The interaction h randomizing the background states conserves the isospin whereas the mixing V is caused by the isospin-violating part of the residual interaction; the physical sources of the isospin nonconservation are discussed in [37]. As we have shown, the spreading width $\Gamma_\downarrow$ does not depend drastically on N being determined by the degree

$$\alpha \sim v^2/a^2 \sim < V^2 > / < h^2 > \tag{4.14}$$

of the isospin violation in the simple states. Apart from fluctuations and possible regular shell effects, α should be on average a smooth function of the mass number growing with A not faster than proportional to the nuclear volume. The experimental information on isolated IAR confirms these arguments and, consequently, our scaling rules for complex wave functions and their couplings. A similar reasoning was used for the description of IAR in compound nuclear reactions[35].

Extending the approach to the many-classes hierarchy one is able to develop the kinetic description of multistep nuclear reactions and equilibration processes.

5. Mean Field out of Chaos

In Sect.2 we have mentioned the problem of the "natural" basis suitable for defining the degree of complexity of a wave function. Here we show schematically how the self-averaging of contributions of complicated intermediate states smoothes out the dynamics and gives rise to the MF-like picture of simple modes developing on the average background of incoherent excitations.

To make the derivation adapted to the standard language of the many-body problem, we start with generalizing the single-particle density matrix (SPDM) which is the c-number matrix

$$\rho_{12} = < \Psi_0 \mid a_2^\dagger a_1 \mid \Psi_0 > \tag{5.1}$$

in the space spanned by single-particle orbitals $| 1)$. Here a_1 and $a_2^\dagger$ are the annihilation and creation operators of particles and the expectation value (5.1) is taken for the ground state $| \Psi_0 >$ or, in the time-dependent MF approach, for the appropriate wave packet $| \Psi(t) >$.

For a Fermi system with the effective antisymmetrized interaction V, the conventional stationary MF approximation [38] determines the SPDM ρ^0 via the matrix equation

$$[\rho^0, h\{\rho^0\}] = 0 \tag{5.2}$$

where the self-consistent Hamiltonian matrix

$$h_{12}\{\rho\} \equiv \varepsilon^0_{12} + \mathcal{W}_{12}\{\rho\} = \varepsilon^0_{12} + \sum_{34} V_{13;42}\ \rho_{43} \tag{5.3}$$

is obtained from the original many-body Hamiltonian

$$H = \sum_{12} \varepsilon^0_{21} a^\dagger_2 a_1 + \sum_{1234} V_{12;34} a^\dagger_1 a^\dagger_2 a_3 a_4 \ . \tag{5.4}$$

According to (5.2), the matrices ρ^0 and $h\{\rho^0\}$ are to be diagonalized simultaneously providing us with their eigenvalues n_1 and ε_1 respectively. The occupation numbers n_1 being equal to 1 or 0 (which can be expressed as the operator condition$\rho^0\rho^0 = \rho^0$) define the Slater determinant $| k >$; ε_1 are the energies of the optimized single-particle orbitals. The total energy of $| \Psi_0 >$ is

$$E = \mathrm{Tr}(\varepsilon^0\rho^0) + (1/2)\mathrm{Tr}(\rho^0 V \rho^0) \ . \tag{5.5}$$

Here and below Tr stands for the trace in the single-particle space. Various Slater determinants give, in general, different self-consistent fields $\mathcal{W}$ so that the corresponding orbitals are not orthogonal. Practically, one considers usually the ground state only and expresses the excitation properties in terms of the MF defined for the ground state.

In problems of collective dynamics, the ground state is connected with selected excited states by strong one-body matrix elements. We dealt with such a case in Sect.3 for vibrational excitations. Similarly, according to the Alaga rules [2,vol.2], the multipole matrix elements within a rotational band are of the same order of magnitude for all allowed transitions. Then it is necessary to consider on equal footing all matrix elements

$$R^{cc'}_{12} =< c \mid a^\dagger_2 a_1 \mid c' > \tag{5.6}$$

between the collective states which form a band in the extended sense of this word. Such a generalization of the SPDM was suggested by Belyaev and the author [39,40]; for the review see [41] and references therein. This approach proved to be very fruitful, in particular for the collective motion of large amplitude [32] and for the high-spin 3-dimensional cranking [42].

Here we are interested in the average influence of many incoherent degrees of freedom. The adequate quantity will be the generalized density matrix (GDM) defined as a complete set of matrix elements, compare with (5.1) and (5.6),

$$R^{\alpha\beta}_{12} \equiv< \alpha \mid R_{12} \mid \beta >=< \alpha \mid a^\dagger_2 a_1 \mid \beta >\equiv< \alpha; 1 \mid \boldsymbol{R} \mid \beta; 2 > \tag{5.7}$$

where α and β run over the total Hilbert space truncated reasonably for actual calculations, and $\boldsymbol{R} = \boldsymbol{R}^\dagger$ is the formally defined density operator in the combined space $\mid \alpha; 1 >$. One-body operators are expressed by traces over single-particle variables,

$$Q = \sum_{12} q_{12} a_1^\dagger a_2 = \sum_{12} q_{12} R_{21} = \text{Tr}(q\boldsymbol{R}) \quad , \tag{5.8}$$

still remaining operators with respect to many-body states $\mid \alpha >$.

The kinematics of the GDM can be written down in terms of the commutation relations

$$[R_{12}, R_{34}] = \delta_{14} R_{32} - \delta_{23} R_{14} \tag{5.9}$$

which correspond, in the truncated Ω-dimensional single-particle space, to the Lie algebra $SU(\Omega)$. For any one-body operators like (5.8) two important identities,

$$[Q + q, \boldsymbol{R}] = 0 \quad , \tag{5.10}$$

and

$$[Q^{(1)}, Q^{(2)}] = \text{Tr}\{[q^{(1)}, q^{(2)}]\boldsymbol{R}\} \quad , \tag{5.11}$$

follow from (5.4); here the operators $Q, Q^{(1)}$ and $Q^{(2)}$ act upon the global ("external") variables whereas the corresponding $q, q^{(1)}$ and $q^{(2)}$, see eq.(5.8), act upon the single-particle ("internal") ones. These identities are violated by the truncation of the global space $\mid \alpha >$ up to the collective band (5.6) and even more in the standard SPDM (5.1). The saturation principle [32] suggested in order to minimize this violation is important for the self-consistent determination of the collective path.

The definitions (5.7) and (5.8) as well as the kinematic properties (5.9-11) do not depend on the type of particle statistics. With the aid of doubling the single-particle space [39], they can be extended to the systems with the anomalous Green functions $\sim < a_2 a_1 >$ as in the BCS-type description of the pairing correlations. The additional condition for the GDM emerges from the permutational symmetry,

$$R_{12} R_{34} \pm R_{32} R_{14} = \delta_{14} R_{32} \pm \delta_{34} R_{12} \tag{5.12}$$

where the upper sign stands for fermions and the lower one for bosons. It is easy to check all the equations using the basis of the states $\mid k >$ of non-interacting particles. A state $\mid k >$ from this set is characterized by the occupation numbers $n_1(k)$; for any single-particle basis $\mid k >$ these numbers should be integers and correspond to the type of statistics. In the conventional definition of the secondly quantized operators, the matrix elements $R_{21}^{k'k} = < k' \mid R_{21} \mid k >$ of the GDM are equal to

$$R_{21}^{k'k} = \delta_{12}\delta_{k'k}n_1(k) + (1-\delta_{12})\delta_{k',k-2+1}\sqrt{n_2(k)[1 \mp n_1(k)]}\;\eta_{12}^{k} \qquad (5.13)$$

where the same sign convention as in (5.12) was used. In the second term of eq.(5.13) the orbitals $\mid 1)$ and $\mid 2)$ should be different and the state $\mid k' >$ differs from $\mid k >$ by the p-h pair which is indicated by the Kronecker symbol $\delta_{k',k-2+1}$; the phase $\eta_{12}^{k} = 1$ for bosons and $\eta_{12}^{k} = (-1)^{\sum n_3(k)}$ for fermions where the occupation numbers $n_3(k)$ are summed for all orbitals $\mid 3)$ arranged between $\mid 1)$ and $\mid 2)$ in the ordered sequence of the single-particle states.

Now we use the non-interacting states $\mid k >$ for the expansion (2.1) of the stationary states $\mid \alpha >$ for a system with the arbitrarily strong interaction. The GDM matrix elements $R_{12}^{\alpha\beta} =< \alpha \mid R_{12} \mid \beta >$ between the physical states are

$$\begin{aligned} R_{12}^{\alpha\beta} =& \delta_{12}\sum_k n_1(k)C_k^{\alpha *}C_k^{\beta} + \\ & (1-\delta_{12})\sum_k \sqrt{n_1(k)[1 \mp n_2(k)]}\;\eta_{21}^{k}C_{k-1+2}^{\alpha *}C_k^{\beta} \;. \end{aligned} \qquad (5.14)$$

Assume that the states $\mid \alpha >$ and $\mid \beta >$ belong to the class of complicated states with a large number of significant components $\mid k >$. The smooth component of $R_{12}^{\alpha\beta}$ sought for reflects average properties of complex stationary states. At high level density this component can be discerned by averaging over the small interval containing many states of similar degree of complexity. If the states within the interval are uncorrelated the smooth component can be produced only by the summation of the positive combinations $\mid C_k^{\alpha} \mid^2$ in the first item of (5.14). Then the coherent sum over k cancels the smallness $\sim N^{-1}$ of each term. If the states $\mid \alpha >$ and $\mid \beta >$ differ by a coherent excitation which can be created by an one-body operator, the second item of (5.14) also could contain a nonfluctuating component for the corresponding orbital pair $\mid 1)$ and $\mid 2)$. This is obviously the case for a collective band. Here we consider the fully incoherent dynamics neglecting the mean value of the second item.

Denoting the averaging procedure by the bar we get

$$\overline{C_k^{\alpha *}C_l^{\beta}} = \delta^{\alpha\beta}\delta_{kl}\overline{\mid C_k^{\alpha} \mid^2} \equiv \delta^{\alpha\beta}\delta_{kl}f_k^{\alpha} \qquad (5.15)$$

where f_k^{α} is the (unknown) distribution function for the simple states $\mid k >$ fragmented over the region of physical states $\mid \alpha >$. Matrix elements (5.14) become now random variables with mean values which are reduced to the average occupation numbers $\overline{n_1^{\alpha}}$ for the single-particle orbitals $\mid 1)$ in complicated states near $\mid \alpha >$,

$$\overline{R_{12}^{\alpha\beta}} = \delta_{12}\sum_k n_1(k)\delta^{\alpha\beta}f_k^{\alpha} = \delta_{12}\delta^{\alpha\beta}\overline{n_1^{\alpha}} \qquad (5.16)$$

Binary products of the GDM elements contain the quartic combinations of the random coefficients C_k^α. As a rule, in similar cases [7] the pair-wise contractions of type (5.15) dominate due to the maximum number of independent summations. In particular, it gives the simplest way to derive the Green functions for the canonical random matrix ensembles, eq.(4.6). However, in the two-body operators, the summation over many intermediate states brings in comparable contributions from the quartic contractions which are irreducible with respect to splitting into pairs.

The recipe which exhausts all contributions non-vanishing after averaging over random phases of C_k^α can be formulated as follows: one has to make all possible pair-wise identifications among lower and upper indices independently, each pair being formed from one index belonging to one of C_k^α and another one belonging to one of conjugate $C_l^{\beta *}$. As it should be, the correlation functions will be invariant under the arbitrary phase transformation $C_k^\alpha \to \exp[i(\phi_\alpha + \phi_k)]C_k^\alpha$ of the basis vectors $|\ \alpha>$ or $|\ k>$. For quadruplets of amplitudes such a procedure leads to

$$\overline{C_k^{\alpha *} C_l^\beta C_m^{\gamma *} C_n^\lambda} = (\mathcal{P}_{\alpha\beta\gamma\lambda}\mathcal{P}_{klmn} + \tilde{\mathcal{P}}_{\alpha\beta\gamma\lambda}\tilde{\mathcal{P}}_{klmn})\Phi_{k;m}^{\alpha;\gamma} + (\mathcal{P}_{\alpha\beta\gamma\lambda}\tilde{\mathcal{P}}_{klmn} + \tilde{\mathcal{P}}_{\alpha\beta\gamma\lambda}\mathcal{P}_{klmn})\Psi_{k;m}^{\alpha;\gamma} \tag{5.17}$$

where the contraction symbols

$$\mathcal{P}_{\alpha\beta\gamma\lambda} = \delta_{\alpha\beta}\delta_{\gamma\lambda},\ \tilde{\mathcal{P}}_{\alpha\beta\gamma\lambda} = (1 - \delta_{\alpha\beta})\delta_{\alpha\lambda}\delta_{\beta\gamma} \tag{5.18}$$

and the irreducible quartic correlation functions

$$\Phi_{k;m}^{\alpha;\gamma} = \overline{|\ C_k^\alpha\ |^2|\ C_m^\gamma\ |^2};\ \Psi_{k;m}^{\alpha;\gamma} = \overline{C_k^{\alpha *} C_m^\alpha C_m^{\gamma *} C_k^\gamma} \tag{5.19}$$

are introduced. By virtue of the unitarity conditions (2.2), the summation over one of four indices reduces the functions (5.19) to the pair distribution function (5.15), for example,

$$\sum_k \Phi_{k;m}^{\alpha;\gamma} = f_m^\gamma\ ,\ \sum_k \Psi_{k;m}^{\alpha;\gamma} = \delta^{\alpha\gamma} f_m^\alpha\ . \tag{5.20}$$

We can use the averaging procedure to derive from (5.14) and (5.17) the pair correlation function of the GDM elements:

$$\overline{R_{12}^{\alpha\beta} R_{34}^{\gamma\lambda}} = \mathcal{P}_{1234}\sum_{km}\left\{\mathcal{P}_{\alpha\beta\gamma\lambda}\Phi_{k;m}^{\alpha;\gamma} + \tilde{\mathcal{P}}_{\alpha\beta\gamma\lambda}\Psi_{k;m}^{\alpha;\gamma}\right\} n_1(k)n_3(m)$$
$$+\tilde{\mathcal{P}}_{1234}\sum_k\left\{\mathcal{P}_{\alpha\beta\gamma\lambda}\Psi_{k-1+3;k}^{\alpha;\gamma} + \tilde{\mathcal{P}}_{\alpha\beta\gamma\lambda}\Phi_{k-1+3;k}^{\alpha;\gamma}\right\} n_1(k)[1 \mp n_3(k)]\ . \tag{5.21}$$

Formally, the one-fold sum in the second line of eq.(5.21) is small $\sim N^{-1}$ as compared with the two-fold sum in the first line. Nevertheless, the second sum should be kept since the operator product $R_{12}R_{34}$ includes the additional summation over intermediate states compensating this smallness:

$$\begin{aligned}\overline{(R_{12}R_{34})^{\alpha\lambda}} &= \sum_{\beta} \overline{R_{12}^{\alpha\beta} R_{34}^{\beta\lambda}} \\ &= \delta^{\alpha\lambda} \sum_{k} f_k^{\alpha} \left\{ \mathcal{P}_{1234} n_1(k) n_3(k) + \widetilde{\mathcal{P}}_{1234} [1 \mp n_1(k)] n_3(k) \right\} .\end{aligned} \tag{5.22}$$

Eq.(5.22) contains, owing to the relationships (5.20), only the pair distribution f and presents an analog of the Wick theorem. The term with the operator $\widetilde{\mathcal{P}}$ in eq.(5.22) distinguishes two kinds of statistics and describes the effect of fluctuations due to the off-diagonal components of the GDM. For the one-body operators $Q^{(1,2)} = \mathrm{Tr}(q^{(1,2)} R)$ we obtain from (5.22)

$$\begin{aligned}\overline{(Q^{(1)}Q^{(2)})^{\alpha\lambda}} = \delta^{\alpha\lambda} \sum_{k;12} f_k^{\alpha} \{ & q_{11}^{(1)} q_{22}^{(2)} n_1(k) n_2(k) \\ & + (1 - \delta_{12}) q_{21}^{(1)} q_{12}^{(2)} [1 \mp n_1(k)] n_2(k) \} .\end{aligned} \tag{5.23}$$

The point is that the averaging procedure, defined almost uniquely via invariance with respect to phase transformations, generates the statistical ensemble which maintains, for the mean quantities, all exact relations (5.9-12). Thus, our coarse-grained dynamics satisfies all kinematical and quantum-statistical restrictions as well as conservation laws.

The mean energy $E^{\alpha} = \overline{H^{\alpha\alpha}}$ of a state $| \alpha >$ is, according to our procedure applied to the Hamiltonian (5.4),

$$E^{\alpha} = \sum_{k} f_k^{\alpha} \left\{ \sum_{1} \epsilon_{11}^{0} n_1(k) + (1/2) \sum_{12} V_{12;21} n_1(k) n_2(k) \right\} \tag{5.24}$$

which resembles closely the MF result (5.5) averaged out with the distribution function f.

Finally, consider the exact operator equations of motion

$$[R_{12}, H] = [\varepsilon^0 + (1/2)\mathcal{W}\{R\}, R]_{12} . \tag{5.25}$$

As distinct from the MF Hamiltonian (5.3), the self-consistent field ω enters eq.(5.24) with the factor 1/2. However, the averaging (5.22) restores the exchange contributions resulting in

$$\overline{[R, H]_{12}^{\alpha\beta}} = \delta^{\alpha\beta} \sum_{k} f_k^{\alpha} [h_k, \rho_k] \tag{5.26}$$

where ρ_k and $h_k = h\{\rho_k\}$ are the SPDM for the Slater determinant $\mid k >$ and the corresponding MF Hamiltonian (5.3), respectively. Hence, we have, instead of (5.2), the set of SPDM's ρ_k, each producing its own MF h_k, and being weighted in the dynamics (time derivative of the total GDM R) with the distribution function f^α. In some sense, the MF is resurrected from the complete chaos due to the self-averaging. Certainly, for actual applications, collective modes should have been included in addition.

The central problem of the method, namely that of the correct distribution function f_k^α, is still to be solved. The simplest way out is using the information approach. Restricting the information available by the mean energy, which fixes the region where the complicated states $\mid \alpha >$ are located, we come to the maximization of the entropy (2.6), $S^\alpha = -\sum_k f_k^\alpha \log f_k^\alpha$. This results in the thermal distribution function

$$f_k^\alpha = \left\{ \sum_l \exp(-E_l/T^\alpha) \right\}^{-1} \exp(-E_k/T^\alpha) \quad . \tag{5.27}$$

This recipe is by no means unique. The supplementary information can be extracted from equations of motion for higher order products of the GDM elements. It is interesting that it can lead to the non-zero temperature of the ground state or low-lying states. This is related to the longstanding problem of partial occupancies of single-particle orbitals near the Fermi surface [43]. This effective temperature will serve as a measure of the smearing out the single-particle motion by the background of the incoherent degrees of freedom. The average coupling to this background determines the residues of the single-particle Green function as well as regular parts of this function.

The generalization of the method sketched in this section hopefully will make it possible to derive for the simple modes new equations of the Langevin type with random forces generated by the complicated intermediate states and responsible for the dissipation and fluctuations of simple modes.

6. Coupling Through Continuum and Collectivization of Widths

Any excited state of a quantum system is, strictly speaking, quasistationary rather than stationary. Excitation and deexcitation processes supply, through the response function, the information on the intrinsic properties of the system. In the previous sections we disengaged ourselves from this fact and carried out the analysis as if the states under study were genuine discrete ones. In nuclei, such an approximation can be adequate not only for low-lying states with small radiation widths but also for the states in the particle continuum. Thus, the neutron resonances in the slow neutron scattering correspond to the compound states which are well isolated due to the smallness of their widths Γ, æ$\sim \Gamma/D \ll 1$. By this reasoning one is able to treat the neutron resonances [2,5,7] as eigenvectors of a random matrix belonging to the GOE.

The situation changes when, with the excitation energy and level density increasing, the resonance widths become wider and the states overlap. The lifetime of states shortens and the coupling to the decay channels dominate the whole dynamics [44]. We show that the new type of collective behavior is possible in this domain which segregates direct and compound processes and suppresses the Ericson fluctuations [45], at least for the small number of open channels [46,47].

To describe the internal dynamics in the presence of decay channels, we use the effective Hamiltonian for the resonances embedded into continuum [48] which can be derived either from the nuclear reaction theory [49] or with the use of the projection operator formalism [37]. Let the hermitian $N \times N$ matrix H (real and symmetric for a T-invariant system) describes the intrinsic dynamics. The external dynamics is associated with the decay into the open channels $c = 1, \ldots, k$; the matrix elements A_n^c for the decay of internal states $| n >$ into channels c are real amplitudes which are supposed to be energy independent in the region $\Delta E \approx a$ where the N intrinsic levels coupled strongly to those channels are located. For simplicity we neglect here the channel coupling and potential scattering and focus on processes going through the intermediate unstable states.

The scattering matrix for the $c' \to c$ process at the energy E can be written down as

$$S^{cc'}(E) = \delta^{cc'} - iT^{cc'}(E) = \delta^{cc'} - i \sum_{mn} A_m^c \mathcal{G}_{mn}(E) A_n^{c'} \tag{6.1}$$

where the effective propagator in the internal space

$$\mathcal{G}(E) = (E - \mathcal{H})^{-1} \tag{6.2}$$

has the poles at the complex points

$$\mathcal{E}_n = E_n - (i/2)\Gamma_n \tag{6.3}$$

of the eigenvalues of the effective nonhermitian Hamiltonian

$$\mathcal{H} = H - (i/2)W \ . \tag{6.4}$$

The Hamiltonian (6.4) has acquired the antihermitian part $(i/2)W$ as a price for the elimination of the channel variables. But for the S-matrix (6.1) the hermitian matrix cannot be arbitrary being defined completely by the unitarity condition. Indeed, from $S^\dagger S = 1$ it follows that W is expressed as a sum over the open channels,

$$W_{mn} = \sum_c A_m^c A_n^c \tag{6.5}$$

of products of the same amplitudes which label the entrance and exit states in eq.(6.1). The complex eigenvalues (6.3) give energies E_n and widths Γ_n of the unstable states with the pure exponential decay, $\sim \exp(-i\mathcal{E}_n t)$. Representation (6.1) is used extensively in the statistical reaction theory[50,51].

The origin of the antihermitian Hamiltonian (6.5) can be understood from the consideration of propagation (6.2) through the medium coupled to the decay channels,

$$W_{mn} \simeq \sum_c \int \Pi(dp)_c(m \leftarrow c)(c \leftarrow n)^*/[E - E(p) + i0]^{-1} \qquad (6.6)$$

where the sum enumerates all decay channels, closed as well as open, and the integral runs over the momentum variables for particles in the continuum for the given channel. The principal value of (6.6) being given by the real part of (6.6) corresponds to the virtual (reversible) processes. It covers all channels and gives the additional hermitian contribution to be included in the internal part H in (6.4). The δ-function imaginary part of (6.6) corresponds to irreversible processes of allowed decays into the open channels. For the amplitudes with the smooth energy dependence (in particular, far away from the reaction thresholds) this part leads to eq.(6.5).

For well separated narrow levels, W in eq.(6.4) is a weak perturbation. The off-diagonal elements $W_{m\neq n}$ are, therefore, of minor importance whereas the diagonal ones, $W_{nn} = \sum_c | A_n^c |^2$, coincide with the widths of the resonances. The same approximation is frequently used outside the range of its applicability. For instance, in the treatment of the inelastic photon scattering[52] in the GR region, the authors stress the sensitivity of the (γ, γ') cross section to the degree of overlap among intermediate dipole states. However they used for the fitting the expression

$$(d\sigma/do)_{0\to f} \sim \left| \sum_n d_{fn}d_{n0}\{(E_n+E'+i\Gamma_n/2)^{-1}+(E_n-E-i\Gamma_n/2)^{-1}\} \right|^2 \quad (6.7)$$

where E and E' are the photon energies and d is the dipole operator. This form borrowed from atomic spectroscopy is not valid in the most interesting case of overlapping levels: just there one cannot neglect the off-diagonal elements of W responsible for the coupling of internal states via the common decay channels. Eq.(6.7) corresponds to the perturbation theory with respect to W. For overlapping levels, the exact diagonalization of W brings the exact complex energies to the denominator of the T-matrix (6.1) and substitutes the bare dipole amplitudes in eq.(6.7) by the exact ones which are complex in order to ensure the unitarity.

Let us consider the structure of the effective Hamiltonian $\mathcal{H}$, eq.(6.4). The separability of the antihermitian part W implies, in conformity with the similar properties of the collective Hamiltonian(2.7), that the rank of the matrix W is equal to the number of the factorizable terms in the sum (6.5), i.e. to the number k of the open channels. In practical applications, $k < N$ and, usually, $k \ll N$. It means that $N \times N$ matrix W can have not more than k nonzero eigenvalues γ_n which are easily shown to coincide with those of the $k \times k$ matrix in the channel space

$$X^{ab} = \sum_n A_n^a A_n^b \quad . \tag{6.8}$$

In the simplest one-channel case the only nonzero eigenvalue of W is its trace, i.e. the summed width

$$\gamma_1 = \sum_n A_n^2 = \mathrm{Tr} W \equiv \mathcal{W} \tag{6.9}$$

whereas the remaining $N-1$ eigenvalues $\gamma_2, \ldots, \gamma_N$ vanish. Thus, if the antihermitean part W of the total effective Hamiltonian $\mathcal{H}$ plays the main role in dynamics, the collectivization of widths occurs so that the total width is absorbed by the specific collective state which becomes short-lived in contrast to other eigenstates which are stationary.

The collectivization of widths is well known in optics. Let N two-level atoms (or even classical radiators) be confined to the volume with a size less than the wavelength of radiation. The common radiation field synchronizes the radiators so that the coherent state, so called Dicke state [53], is formed with a width $\gamma(N) \sim N\gamma(1)$, i.e. N times more than the typical width $\gamma(1)$ of a single atom. In our problem the intrinsic levels ("radiators") are coupled through the common decay channel; the physics is the same. For the case of k open channels k states share the total width and $N-k$ are stable.

The total non-hermitian Hamiltonian (6.4) cannot be diagonalized by an orthogonal transformation since the matrices H and W do not commute in general. On the basis of the stationary states of the internal hermitian Hamiltonian H the total matrix takes a form (for simplicity, consider the one-channel case)

$$\mathcal{H}_{12} = \varepsilon_1 \delta_{12} - (i/2) B_1 B_2 \tag{6.10}$$

where ε_1 are energies of intrinsic states and B_1 are their decay amplitudes obtained from A_n by the orthogonal transformation. The secular equation for the complex energies $\mathcal{E}_n$ (6.3) is [47]

$$1 + (i/2)K(E) \equiv 1 + (i/2) \sum_1 B_1^2/(\mathcal{E} - \varepsilon_1) = 0 \quad . \tag{6.11}$$

The scattering matrix has poles in the points ε_n,

$$T(E) = K(E)\{1 + (i/2)K(E)\}^{-1} \ . \tag{6.12}$$

Eq.(6.11) is equivalent to the set of two coupled equations,

$$(\Gamma/4)\sum_{1} B_1^2/\{(E-\varepsilon_1)^2 + \Gamma^2/4\} = 1 \ , \tag{6.13}$$

and

$$(\Gamma/4)\sum_{1} B_1^2\varepsilon_1/\{(E-\varepsilon_1)^2 + \Gamma^2/4\} = E \ , \tag{6.14}$$

for the energies $\mathcal{E} = E - i\Gamma/2$. In the degenerate case, $\varepsilon_1 = \varepsilon$, we again get the Dicke state (6.9) with $\Gamma_D = \sum_1 B_1^2 = \mathcal{W}$ and $N-1$ stable states. It is easy to obtain the solution of eqs.(6.13-14) for the strong coupling to the continuum. Assume that the levels ε_1 are distributed more or less uniformly in the interval $(-\alpha, \alpha)$ around $\varepsilon = 0$ so that the mean level spacing is $D = 2\alpha/N$, see Sect.4. The strong coupling means that the original levels are overlapped,

$$æ \sim < A^2 > /D \sim < B^2 > /D \sim \mathcal{W}/\alpha > 1 \ . \tag{6.15}$$

Then one gets easily the Dicke solution with

$$\Gamma_D = \mathcal{W}\{1 - O(\alpha^2/\mathcal{W}^2)\} \ , \ E_D = \sum_{1} B_1^2\varepsilon_1/\mathcal{W} \approx 0 \tag{6.16}$$

and narrow $N-1$ states with the energies, located intermittently in between ε_1, and the widths $\Gamma \sim æ^{-2}\mathcal{W}/N$ which are reduced by the factor $1/æ^2$ against the typical values $< A^2 > \sim < B^2 > \sim \mathcal{W}/N$ they had before the coupling via continuum was included. Often the parameterization (6.1) of the S-matrix is used with the arbitrary complex parameters $\mathcal{E}_n$. We see that in the case of overlapped levels the arbitrary choice of widths is forbidden, being able to violate unitarity [54].

The collectivization of widths leads to the segregation of direct and compound processes where the word "direct" relates actually to any fast process independently of its specific mechanism. In the strong overlap limit, $æ \gg 1$, there is a distinct separation of the time scales:

$$\tau_d \ll \tau_f \ll \tau_r \ll \tau_c \tag{6.17}$$

or, for the corresponding widths,

$$\Gamma^{\uparrow} \sim \mathcal{W} \gg \Gamma_{\downarrow} \sim a \gg D \gg \Gamma_c \tag{6.18}$$

where $\tau_d \sim \hbar/\omega$ is the lifetime of the broad resonance enveloping the very narrow compound resonances their widths being $\Gamma_c \sim \mathcal{W}/N æ^2$; $\tau_f \sim \hbar/a$ corresponds to the fragmentation time, see the discussion of the spreading width $\Gamma_\downarrow$ in Sect.4; $\tau_r \sim \hbar/D$ is similar to the recurrence time, and the lifetime $\tau_c \sim \hbar/\Gamma_c$ of compound states exceeds the recurrence time, $\tau_c \sim \hbar æ^2 N/\mathcal{W} \sim \hbar æ N/a \sim æ\tau_r$, so that these long-lived states certainly are completely equilibrated. The specific shape of the cross section $\sigma \sim\mid T \mid^2$, with the broad wings from the Dicke resonance and the flattened interference pattern in the interior, was discussed in [55].

It is interesting to note that there is no room for the Ericson fluctuations [45] in the picture obtained above for the one channel case. In the weak coupling limit resonances do not overlap whereas at the strong coupling the collectivization of widths again makes the remaining narrow resonances to separate. The situation is not so simple in the many channel case. As was shown by Sokolov, the collectivization of widths takes place when the average partial width $\gamma^c \sim< (A^c)^2 >$ becomes comparable with the mean level spacing D. Therefore three regimes are possible: the weak coupling with $\gamma = \sum_c \gamma^c \ll D$; the intermediate coupling when, at the number of open channels k big enough, $1 \ll k \ll N$, the levels are overlapped, the average total width $\gamma \simeq k < \gamma^c >$ is already larger than D but the collectivization condition is not met, $< \gamma^c > \ll D$; and, finally, the strong coupling at $< \gamma^c > > D$. One can expect the conventional picture of Ericson fluctuations to be valid in the intermediate case but the observable pattern should change after the transition to the strong coupling limit where k broad levels are emerged. The whole subject is important for applications and deserves to be elaborated in detail.

The realistic shell model calculations [56], taking into account the continuum properly, were carried out independently of this general theory. The results confirm the predictions of the width redistribution depending on the channel number in the expected way, at least for $k = 1, 2$ and 4. The phase transition to the strong coupling regime turns out to be rather sharp as a function of the overlap parameter æ.

In eq.(6.10) we have used the basis of eigenvectors of the hermitian part H of the total Hamiltonian $\mathcal{H}$ (6.4). It can be useful to diagonalize firstly the antihermitian part W. In the one channel case W has only one nonzero eigenvalue $\mathcal{W}$ (6.9). Then one can diagonalize the hermitian part H in the $(N-1)\times(N-1)$ subspace of the zero eigenvectors of W. The resulting matrix has a very simple structure. It contains the complex energy $\mathcal{H}_{11}$ with the imaginary part $(-\mathcal{W}/2)$, real energies $\mathcal{H}_{\nu\nu}$ of states $\nu = 2, \dots, N$ and the real coupling elements $\mathcal{H}_{1\nu} = \mathcal{H}_{\nu 1}$. Hence, on this basis we have the characteristic picture of the doorway state coupled with the decay channel whereas all other states have access to the continuum through the doorway state only. Remarkably, two apparently opposite pictures, the uniform coupling of the individual states to the continuum and the coupling through the doorway state are strictly equivalent being just different mathematical representations of the same physics.

The second method of diagonalization results in the scattering amplitude

$$T(E) = \mathcal{W}\{E - \mathcal{H}_{11} - \Lambda(E)\}^{-1} \tag{6.19}$$

where the roots of the denominator determine the quasistationary states and

$$\Lambda(E) = \sum_{\nu=1}^{N} \mathcal{H}_{1\nu}^2/(E - \mathcal{H}_{\nu\nu}) \ . \tag{6.20}$$

In the case of the dense spectrum of states ν, we can go to the continuous limit and reduce, similar to (4.10), the effect of $\Lambda(E)$ in the denominator of (6.19) to the energy shift and the spreading width of the doorway resonance, $\Lambda(E) \to \Delta(E) - i\Gamma_{\downarrow}(E)/2$. In this approximation the widths $\Gamma_{\downarrow}$ and $\Gamma^{\uparrow}$ are added to give the total width of the resonance.

Evidently, the first (second) method of diagonalization is more convenient in the case of the weak (strong) coupling since the form of the matrix is already fitted to the real conditions.

7. Giant Resonances and Two Types of Collectivity

The formation of the collective state due to the coherent interaction in the discrete spectrum was discussed in Sect.2 and developed further in Sect.3. Such a collectivity can be called internal one. On the other hand, in Sect.6 we introduced the external collectivity associated with the coupling of overlapping levels through their common decay channels. For the GR and other collective states embedded into continuum, the coexistence and the interrelation of those two kinds of collective behavior become vital for giving birth to the final observed picture [57].

To illustrate the interplay of the two couplings, we consider the solvable toy-model including, as in the prototype paper [10], the factorizable internal forces (2.7) together with the additional antihermitian interaction (6.4),

$$\mathcal{H}_{12} = H_{12} - iW_{12}/2 = \varepsilon_1\delta_{12} + \lambda d_1 d_2 - (i/2)A_1 A_2 \ , \tag{7.1}$$

where again we show only the one channel case. The simple generalization of the formulae of Sect.2 gives the scattering amplitude (6.1)

$$T(E) = \Omega^{-1}\{[1 - \lambda P(E)]R(E) + \lambda Q^2(E)\} \tag{7.2}$$

where the response functions for two types of interaction are

$$\{P(E); Q(E); R(E)\} = \sum_{1}(E - \varepsilon_1)^{-1}\{d_1^2; A_1 d_1; A_1^2\} \tag{7.3}$$

and the secular equation for complex energies $\mathcal{E}$ is

$$\Omega(\mathcal{E}) \equiv [1 - \lambda P(\mathcal{E})][1 + iR(\mathcal{E})/2] + (i/2)\lambda Q^2(\mathcal{E}) = 0 \quad . \tag{7.4}$$

For a closed system A_1 vanish and eq.(7.4) coincides with (2.8). The collective root shifted outside the region $\Delta E \sim a$ accumulates the major part of the strength but its energy is real so that the collective state is stable. On the other hand, in the open system with no internal interaction, $\lambda \rightarrow 0$ and the secular equation coincides with (6.11); thus, we obtain the Dicke state displaced along the imaginary energy axis precisely in the same manner as the stable collective state is displaced along the real axis. The Dicke state carries the total width but no multipole strength.

For the general case of an open system with the internal and external collectivities, various intermediate situations are possible. Generally, two collective resonances will coexist separated distinctly in the strong interaction limit. One of them is situated in the range of discrete levels ε_1 whereas another is shifted outside this range to the energy $E = E_{coll}$, see (2.5). The two resonances share the total multipole strength as well as the total decay width. In addition, the unshifted resonance has a characteristic structure due to the interference of the main Dicke peak with the uncollective solutions of eq.(7.4) located within the same range [57].

The gross features of the two-resonance pattern depend, roughly speaking, on the angle ϑ between the two N-dimensional vectors $\boldsymbol{A} = \{A_1\}$ and $\boldsymbol{d} = \{d_1\}$, cp.(2.12). In the degenerate case $\varepsilon_1 = \varepsilon$, which presents a qualitatively true picture if both the internal and external couplings are strong as compared with the energy splitting of the intrinsic levels, the energies E_{coll} of the GR and $\mathcal{E}_D$ of the Dicke resonance are equal, respectively, to

$$R_{coll} = \varepsilon + \lambda \boldsymbol{d}^2 \ , \ \mathcal{E}_D = \varepsilon - i\boldsymbol{A}^2/2 = \varepsilon - i\mathcal{W}/2 \quad . \tag{7.5}$$

The one channel scattering amplitude (7.2) can be expressed as

$$T(E) = \frac{(E - E_{coll})\mathcal{W} + \lambda(\boldsymbol{A}\boldsymbol{d})^2}{(E - E_{coll})(E - \varepsilon_D) + i\lambda(\boldsymbol{A}\boldsymbol{d})^2} \quad . \tag{7.6}$$

If the transitions from the internal states to the ground state and to the decay channel are detuned entirely, the vectors $\boldsymbol{A}$ and $\boldsymbol{d}$ will be almost orthogonal, $(\boldsymbol{A}\boldsymbol{d}) = 0$. Then the GR at $E = E_{coll}$ is decoupled from the continuum so that one can observe the unshifted Dicke resonance only,

$$T(E) \rightarrow T_{\perp}(E) = \mathcal{W}/(E - \mathcal{E}_D) = \mathcal{W}/(E - \varepsilon + i\mathcal{W}/2) \quad . \tag{7.7}$$

In the opposite, parallel, situation $\boldsymbol{A}//\boldsymbol{d}$ and the "Dicke GR" arises combining both types of collectivity, namely the shift λd^2 and the width $\mathcal{W}$,

$$T(E) \rightarrow T_{//}(E) = \mathcal{W}/(E - E_{coll} + i\mathcal{W}/2) \quad . \tag{7.8}$$

The rest of the states at $E = \varepsilon$ have neither shift nor width. At intermediate values of the angle ϑ between $\boldsymbol{A}$ and $\boldsymbol{d}$ the two resonance peaks share the total width approximately in the ratio of $\sin^2\vartheta/\cos^2\vartheta$ for the unshifted and shifted peaks respectively.

Certainly, in the realistic case of nondegenerate intrinsic levels and several decay channels, the observed pattern becomes more complicated. We have looked at it in the model with the two decay channels which were to imitate the nucleon and radiation channels of the GR[57]. The amplitude vector $\boldsymbol{A}^{\gamma}$ for the gamma decay was supposed to be parallel to the internal multipole excitation vector $\boldsymbol{d}$ whereas the nucleon emission vector $\boldsymbol{A}^{n}$ was characterized by the angle ϑ. The more or less universal energy behavior of cross sections in $(N, N), (\gamma, N)$ and (γ, γ) channels is predicted for nondegenerate levels and not very small values of $\cos\vartheta$. The displaced GR carries a significant fraction of a total width; the cross section maxima are located at a little different energies in different channels. The most essential qualitative prediction is that of the unshifted structure of the GR which is a manifestation of the Dicke resonance and its interference with the noncollective 1p-1h excitations. Such a low-lying component ("precursor") is well known for the dipole GR in light nuclei where a rather small number of open channels and increased mean spacing of intrinsic levels help to carry out the high resolution measurements. Thus, the structure, very similar to the predicted one, is observed at the vicinity of the unperturbed shell model dipole transitions of ^{16}O in (γ, p_0), (γ, n_0) and $(e, e'p_0)$[58,59] as well as in the fully realistic RPA calculations in the continuum [60].

The analysis of the structure of collective resonances embedded into the continuum can be extended to include the incoherent coupling to the background of compound states, see Sect.4, and the coupling to low-lying collective vibrations, see Sect.3. Such an approach was utilized in the consideration of the mixing phase of the IAR [36]. It would be interesting to study in a similar way the GR in hot nuclei since one can expect the redistribution of strengths and widths among the unshifted and shifted components with increasing temperature. Apart from that, it is known that the microscopic calculations of partial decay widths of GR disagree regularly with data (for example [61,62]). These calculations neglect, as a rule, coupling of internal states via the common decay channels. The validity of such approximations must be checked carefully.

8. Level and Width Statistics for Unstable States

The goal of this chapter is to generalize the canonical Gaussian random matrix ensembles [63] in order to take into account the finite lifetime of excited states. It is obvious that the level correlations will be modified by the level widths which bring in the energy uncertainties. There is no reason to expect the level repulsion, even at the chaotic dynamics, for level spacings less than level widths. Another important question is concerned with the collectivization of widths explained in Sect.6: is this effect preserved for complicated states involved in the chaotic dynamics? It is necessary to answer these questions to be able to apply the theory to real nuclear processes.

We try to construct the statistics for the unstable states as closely as possible to that of stationary states. The dynamics of a system will be described by the effective nonhermitian Hamiltonian $\mathcal{H}$ (6.4). Assuming a very complicated character of motion in the energy interval under study, both parts of $\mathcal{H}$ will be treated as random matrices. For the hermitian part H, we postulate the GOE statistical properties used earlier (4.5) in the description of compound states randomizing the collective mode. Using the same scaling rules for the matrix elements, the distribution function for the real eigenvalues $E_1, \ldots, E_N$ of the Gaussian canonical ensemble of random Hamiltonians is

$$dP_\beta(E_1, \ldots, E_N) \sim \prod_{m<n} \{ \mid E_m - E_n \mid^\beta \} \times$$
$$\exp\{ -\beta N \sum_n E_n^2/a^2 \} dE_1 \ldots dE_N \qquad (8.1)$$

where $\beta = 1$ corresponds to the GOE (below we restrict ourselves to this case), $\beta = 2$ and $\beta = 4$ correspond to the Gaussian unitary and simplectic ensembles respectively. The distinctive feature of these ensembles is the level repulsion [2,7],

$$dP(s) \sim s^\beta ds \ , \ s \to 0 \ , \qquad (8.2)$$

at small spacings between nearest neighbors, $s \ll D$. It is believed that the repulsion (8.2) reflects the stochastic character of motion and destruction of integrals of motion which could prevent the interaction of states possessing different values of these integrals. Here and below we do not write down the awkward normalization constants in the formulae like (8.1-2).

The linear repulsion (8.2) for a T-invariant system is actually of the very general nature being independent of details of statistical hypotheses and of the long range level correlations. For the completely integrable dynamics, one expects the Poisson formula for the level spacing distribution,

$$dP(s) \sim \exp(-s/D) ds \ , \qquad (8.3)$$

with no level repulsion. But a weak randomizing perturbation with small mixing matrix elements of typical value $\sigma \ll D$ result [64] in the appearance of linear repulsion in very narrow spacing range $s \leq \sigma$. A similar effect occurs in so called band matrices where the matrix elements strictly vanish outside the band of a definite size around the main diagonal. For example, in the problem [65] of 3×3 matrices H where $H_{23} = H_{32} \equiv 0$ while the other elements are distributed as in the GOE, the level spacing distribution has a logarithmic singularity,

$$dP(s) \sim s \log (D/s) \quad . \tag{8.4}$$

Again, the slightly smoothed analog of this problem, with the matrix element $H_{23} = H_{32}$ having the Gaussian distribution function with the zero mean value and very small variance $\sigma^2 \ll D^2$, shows up the level repulsion at tiny spacings, $s \leq \sigma$.

Considering the unstable states we have to deal with complex although symmetric matrices. In ref.[66] the Gaussian ensemble of complex matrices was investigated where real and imaginary parts of matrix elements are Gaussian distributed and uncorrelated. The simple generalization of the GOE gives for the distribution function of the complex eigenvalues $Z_1, \ldots, Z_N$

$$dP_c(Z_1, \ldots, Z_N) \sim \prod_{m<n} \{ \mid Z_m - Z_n \mid^2 \} \times \\ \exp \{ -2N \sum_n \mid Z_n \mid^2 /a^2 \} \prod_n d(\mathrm{Re}Z_n) d(\mathrm{Im}Z_n) \quad . \tag{8.5}$$

The relation between the GOE and the complex ensemble (8.5) is especially clear in the limit of $N \to \infty$. The GOE level density (4.7) obeys the Wigner semicircle law being confined within the energy interval $(-a, a)$. The complex eigenvalue density (8.5) is uniform over the circle of the radius a (or the ellipsis if the real and imaginary parts of matrix elements were of different mean square values). The Wigner law is restored if this uniformly occupied circle is projected onto the real axis which corresponds to integrating with respect to Π ImZ; a similar result is valid for the distribution of ImZ. Of course, we cannot use the complex ensemble (8.5) for our aims since the distribution of the physical widths Γ must be limited by the half-space $\Gamma_n > 0$. Note that the ensemble (8.5) predicts the repulsion of the eigenvalues in the complex plane to be quadratic in the spacing $r_{mn} = \mid Z_m - Z_n \mid$ expressed in Cartesian coordinates or, equivalently, cubic in polar coordinates,

$$dP_c(r) \sim r^2 dx dy \sim r^3 dr \ , \ r \to 0 \quad . \tag{8.6}$$

The effective nonhermitian Hamiltonian (6.4) contains new parameters, namely the amplitudes A_n^c so one needs their statistical properties. In the region of isolated resonances it is assumed usually[2,7] that the distribution of the decay amplitudes coincides with that of the components C_c^n of the wave function of the resonance $| n >$ which correspond to the simple basis state $| c >$ matched to the channel c. It implies that the decay implements the analysis of the intrinsic states. The isotropic distribution of C over the N-dimensional sphere (see Sect.2) determines for $| C |^2$, and, consequently, for the widths of isolated resonances (at $N \to \infty$ and for the one channel case),

$$dP_1(\gamma) \sim \gamma^{-1/2} \exp(-\gamma/2 < \gamma >) d\gamma \quad . \tag{8.7}$$

The Porter-Thomas distribution (8.7) agrees with the data for isolated neutron resonances. According to our scaling rules, the mean value $< \gamma > \sim$ $<| C |^2> \sim N^{-1}$.

In the most interesting case of overlapping levels we need to specify the statistical properties of the decay amplitudes A_n^c independently of *a priori* unknown properties of the coefficients of expansion of wave functions. However it is quite natural to expect that, in the domain of chaotic dynamics, the matrix elements of the complete nuclear Hamiltonian between compound states and any simple states, either bound of unbound (decay channels), have similar statistical features. Therefore, in conformity with the estimates of Sect.2, we assume the Gaussian distribution for the decay amplitudes,

$$< A_n^c > = 0 \ , \ < A_n^c A_{n'}^{c'} > = N^{-1} \delta^{cc'} \delta_{nn'} \eta^c \quad , \tag{8.8}$$

where we neglect, for simplicity, the correlations between the amplitudes describing the decays of the same intrinsic state into different channels; this restriction could be lifted easily with the transition to the eigenchannel representation. Similarly, we consider the matrix elements of the two parts of the effective Hamiltonian, H and W, to be uncorrelated.

For the Gaussian distribution of the decay amplitudes (8.8), the diagonal matrix elements $W_{nn} \equiv \gamma_n$, being equal to the widths of resonances in the case of no overlap, are statistically independent. In the simplest case of k equivalent channels, $\eta^c = \eta$, the additional orthogonal invariance in the channel space arises, and γ_n are distributed according to χ_k^2 law,

$$dP(\gamma_n) \sim \gamma_n^{k/2-1} \exp(-N\gamma_n/2\eta) d\gamma_n \quad , \tag{8.9}$$

which becomes the Porter-Thomas distribution (8,7) for the one channel case and, for the general k-channel case, leads to

$$< \gamma_n > = k\eta/N \ , \ < \gamma_n^2 > - < \gamma_n >^2 = 2k\eta^2/N^2 = (2/k) < \gamma_n >^2 \quad . \tag{8.10}$$

The second eq.(8.10) is used frequently as a definition of the effective number of open channels in the treatment of data, especially for the fission process. Evidently, it can be reasonable for separated resonances only when γ_n coincides with the width of a resonance, but not in the case of the strong overlap with the trend of the width collectivization. Note that the trace $\mathcal{W} = \sum \gamma_n$ of the matrix W fluctuates very weakly,

$$< \mathcal{W} >= k\eta \ , \ < \mathcal{W}^2 > - < \mathcal{W} >^2 = 2k\eta^2/N = (2/kN) < \mathcal{W} >^2 \quad . \quad (8.11)$$

For the many channel case, $k \gg 1$, the k-dependence is important for the manifestation of Ericson fluctuations, Sect.7.

Since the antihermitian part W has only k nonzero eigenvalues γ^c, $c = 1, \ldots, k$, it is instructive to derive the distribution function for these quantities which are the eigenvalues of the matrix (6.5) being the widths of the broad Dicke states in the strong coupling limit. Using, for example, the Dyson stochastic equation method[67,68], i.e. constructing a random process for the eigenvalues of the matrix with the given properties (8.8) of the elements and looking for the stationary solution of the Fokker-Planck equation, we get

$$dP(\gamma^1, \ldots, \gamma^k) \sim \prod_{a<b} \{ \mid \gamma^a - \gamma^b \mid \} (\prod_c \gamma^c)^{(N-k-1)/2} \times$$
$$\exp \{ -(N/2\eta) \sum_c \gamma^c \} d\gamma^1 \ldots d\gamma^k \quad . \quad (8.12)$$

Here we have the linear repulsion of the non-zero widths. The difference between (8.12) and (8.1) is caused by the specific structure of the matrix X (6.8).

Now we write down our main result, namely the joint distribution function for the complex eigenvalues $\mathcal{E}_n = E_n - i\Gamma_n/2$ of the nonhermitian Hamiltonian (6.4) under the statistical assumptions discussed above. Lengthy, although straightforward calculations [47] allow us, in the one channel case, to carry out the transformation from the original matrix elements to the eigenvalues $\mathcal{E}_n$ and to obtain

$$dP(E_1, \ldots, E_N; \ \Gamma_1, \ldots, \Gamma_N) \sim$$
$$\prod_{m<n} \{(E_m - E_n)^2 + (\Gamma_m - \Gamma_n)^2/4\} \{(E_m - E_n)^2 + (\Gamma_m + \Gamma_n)^2/4\}^{-1/2} \times$$
$$\exp\{ -N(\sum_n E_n^2/a^2 + \sum_{m<n} \Gamma_m\Gamma_n/2a^2 + \sum_n \Gamma_n/2\eta)\} \prod_n dE_n \prod_n d\Gamma_n/\sqrt{\Gamma_n}. \quad (8.13)$$

Let us stress the main features of this result.

(i) At small spacings in the complex plane the eigenvalues repel each other according to the cubic law (8.6). It can be supported also by the numerical results for the dissipative spin systems[69].
(ii) From the first line of r.h.s. of eq.(8.13), we conclude that energies and widths are correlated. Qualitatively, the picture is determined by the overlap parameter æ$= \eta/a$. For æ $\ll 1$, the typical widths are $\Gamma \sim \eta/N$ so that the ratio $\Gamma/D \sim \Gamma N/a \sim$ æ is small so that it corresponds to the weak overlap. Then the terms of the exponent (8.13) being bilinear in Γ can be omitted, and the distribution function at large real spacings $| E_m - E_n | \gg \Gamma \sim$ æD becomes decomposed in the product of independent distributions (8.1) for energies and (8.7) for widths.
(iii) At small spacings $| E_m - E_n | <$ æD, even at æ $\ll 1$, the linear repulsion of levels disappears so that instead of (8.3) $P(s)$ will be constant at $s \to 0$.
(iv) In the opposite case of æ $\gg 1$ the probability density (8.13) is small unless the term in the exponent proportional to $\sum \Gamma_m \Gamma_n$ vanishes. This happens if only one of the widths, let say Γ_1, is large. It means that the only configuration has the essential probability, namely that with the width collectivization. Hence, this phenomenon survives in the conditions of chaotic dynamics being merely a consequence of general quantum principles.

(v) In the regime of the width collectivization, the typical widths of the long-lived states turn out to be $\Gamma_{n>2} \sim (\eta/N)$æ$^{-2}$, see Sect.6. For these states the overlap parameter is $\Gamma/D \sim$æ$^{-1}$ so that here again we return to the weak overlap case and, at the spacings exceeding $D/$æ, to the GOE distribution function. The level density of the narrow states is again the Wigner one (4.7) and the widths distribution is the Porter-Thomas one with the small $< \Gamma > \sim \eta N^{-1}$æ$^{-2}$.

Approximately the same picture takes place also in the case of many channels where some preliminary numerical and analytical results are available which show up the phase transition to the regime of the strong coupling.

9. Conclusion

In what follows I would not like to repeat the main results of the previous chapters. Instead I just give a list (certainly, incomplete and restricted by personal interests) of the open problems of top priority:

1. The search for a quantitative measure of degree of complexity of wave functions.
2. The microscopic derivation of the collective Hamiltonian for soft spherical nuclei.
3. The application of the phenomenological approach of Sect.3 to odd-A nuclei.

4. New fields of application of the effective nonhermitian Hamiltonian (Sect.6), including various nuclear reactions as well as atomic, molecular and solid state phenomena[70].
5. The microscopic structure of GR and partial decay widths.
6. The non-zero temperature generalization of the theory.
7. Various problems connected with statistics of unstable levels, in particular, the distribution function for the residues of the scattering amplitude in the poles corresponding to the complex energies of resonant states.
8. Many channel statistical problems including those with a hierarchy of several classes of decay amplitudes; experimental manifestations of Ericson fluctuations and the corresponding theoretical description.
9. The analysis of the information available on the neutron resonances in the broad energy range in order to extract the realistic distribution function of complex energies.

The author highly appreciates the fruitful collaboration with S.T. Belyaev, P. von Brentano, O.K. Vorov and V.V. Sokolov.

References

[1] N. Bohr, *Nature* **137** (1936) 344.

[2] A. Bohr and B. Mottelson, *Nuclear Structure* **Vol.1** Benjamin, New York (1969); **Vol.2** Benjamin, New York (1974).

[3] B. V. Chirikov, *Phys. Rep.* **52** (1979) 263.

[4] M. V. Berry, *Ann. Phys.* **131** (1981) 163.

[5] O. Bohigas and M. J. Gianonni, *Lec. Not. Phys.* **209**,Springer, Berlin, Heidelberg 1984 1.

[6] P. Pechukas, *Phys. Rev. Lett* **52** (1984) 1.

[7] T. A. Brody, J. Flores, J. B. French, P. A. Mello, A. Pandey and S. S. M. Wong, *Rev. Mod. Phys.* **53** (1981) 385.

[8] F. Yonezawa, *Journ. Non-Cryt. Solids* **35-36** (1980) 29.

[9] J. Reichi, *Europhys. Lett.* **6** (1988) 669.

[10] G. Brown and M. Bolsterli, *Phys. Rev. Lett.* **3** (1959) 472.

[11] R. J. Blin-Stoile, *Phys. Rev.* **116** (1960) 181.

[12] G. V. Danilyan, B. D. Vodennikov, V. P. Dronyaev, V. V. Novitsky, V. S. Pavlov and S. P. Borovlyov, *JETP Lett.* **26** (1977) 197.

[13] A. K. Petukhov, G. A. Petrov, S. T. Stepanov, D. V. Nicolaev, T. K. Zvezdkina, V. I. Petrova, E. S. Markova, V. V. Ivanov and V. F. Morozov, *JETP Lett.* **32** (1980) 324.

[14] O. P. Sushkov and V. V. Flambaum, *Yad. Fiz.* **31** (1980) 55.

[15] S. T. Belyaev and V. G. Zelevinsky, *Nucl. Phy.* **39** (1962) 582.

[16] O. K. Vorov and V. G. Zelevinsky, *Yad. Phys.* **37** (1983) 1393.

[17] O. K. Vorov and V. G. Zelevinsky, *Nucl. Phys.* **A439** (1985) 207.

[18] V. G. Zelevinsky, *Nuclear Structure, Reactions and Symmetries* ed. by R. A. Meyer and V. Paar, World Scientific, 1986, **Vol.2**, 1125.

[19] V. G. Zelevinsky, *Nuclear Structure*, Dubna, 1985, 173.

[20] O. K. Vorov and V. G. Zelevinsky, *Modern Developments in Nuclear Physics* ed. by O. P. Sushkov, World Scientific, 1988, 281.

[21] A. Arima and F. Iachello, *Ann. Phys.* **99** (1976) 253; **111** (1978) 201; **123** (1979) 468.

[22] S. T. Belyaev, *Yad. Phys.* **1** (1965) 3.

[23] S. T. Belyaev and V. G. Zelevinsky, *Sov. Phys. Izvestia* **28** (1964) 127.

[24] V. G. Zelevinsky, *JETP* **46** (1964) 1853.

[25] L. D. Mlodinov and N. Papanicolaov, *Ann. Phys.* **128** (1980) 314.

[26] J. D. Feranchuk and L. J. Komarov, *Phys. Lett.* **88A** (1982) 211.

[27] V. G. Zelevinsky, *Sov. Phys. Izvestia* **48** (1984) 2054.

[28] R. F. Casten, *Nucl. Phys.* **A347** (1980) 173.

[29] S. T. Belyaev and V. G. Zelevinsky, *Yad. Phys.* **1** (1965) 13.

[30] S. T. Belyaev and V. G. Zelevinsky, *Yad. Phys.* **2** (1965) 615.

[31] S. E. Muravjev, B. A. Tulupov and M. G. Urin, *Yad. Phys.* **47** (1988) 62.

[32] V. G. Zelevinsky, *Nucl. Phys.* **A337** (1980) 40.

[33] R. Melze and P. von Brentano and H. Paetz gen. Schieck, *Nucl. Phys.* **A432** (1985) 363.

[34] E. Kuhlmann, *Phys. Rev.* **C20** (1985) 527.

[35] H. L. Harney, A. Richter and H. A. Weidenmüller, *Rev. Mod. Phys.* **58** (1986) 607.

[36] V. G. Zelevinsky and P. von Brentano, in press.

[37] N. Auerbach, J. Hüfner, A. K. Kerman and C. M. Shakin, *Rev. Mod. Phys.* **44** (1972) 48.

[38] D. J. Thouless and J. G. Valatin, *Nucl. Phys.* **31** (1962) 211.

[39] S. T. Belyaev and V. G. Zelevinsky, *Yad. Phys.* **16** (1972) 1195.

[40] S. T. Belyaev and V. G. Zelevinsky, *Yad. Phys.* **17** (1973) 525.

[41] V. G. Zelevinsky, *Prog. Theor. Phys. Suupl.* **74-75** (1983) 251.

[42] V. G. Zelevinsky, *Nucl. Phys.* **A344** (1980) 109.

[43] E. Quint, *Limitations of the Mean-Field Description for Nuclei in the Pb-region*, Observed with the (e, e'p) Reaction. Thesis, Amsterdam, 1988.

[44] N. Bohr, R. Peierls and G. Placzek, *Nature* **144** (1939) 200.

[45] T. E. O. Ericson, *Ann. Phys.* **23** (1963) 390.

[46] V. V. Sokolov and V. G. Zelevinsky, *Phys. Lett.* **B202** (1988) 10.

[47] V. V. Sokolov and V. G. Zelevinsky, *Nucl Phys.* **A504** (1989) 562.

[48] I. Yu. Kobzarev, N. N. Nicolaev and L. B. Okun, *Yad. Phys.* **10** (1969) 864.

[49] C. Mahaux and H. A. Weidenmüller, *Shell-model approach to nuclear reactions*, North Holland, Amsterdam, 1969.

[50] H. A. Weidenmüller, *Ann. Phys.* **5** (1984) 357.

[51] J. J. M. Verbaarschot, H. A. Weidernmüller and M. R. Zirnbauer, *Phys. Rep.* **129** (1985) 367.

[52] T. J. Bowles, R. J. Holt, R. M. Laszewski, R. D. McKeown, A. M. Nathan and J. R. Specht, *Phys. Rev.* **C24** (1981) 1940.

[53] R. H. Dicke, *Phys. Rev.* **93** (1954) 99.

[54] P. A. Moldauer, *Phys. Rev.* **135** (1964) 642.

[55] V. V. Sokolov and V. G. Zelevinsky, *Proceeding of the 24-th Winter School*, Leningrad, 1989, 3.

[56] P. Kleinwächter and I. Rotter, *Phys. Rev.* **C32** (1985) 1742.

[57] V. V. Sokolov and V. G. Zelevinsky, *G. Alaga Memorial Volume*, Fizika, **22** (1990).

[58] E. Fuller, *Phys. Rep.* **127** (1985) 185.

[59] V. F. Dmitriev, D. M. Nikolenko, S. G. Popov, I. A. Rachek, D. K. Toporkov, E. P. Tsentalovich, B. B. Woitsekhowski and V. G. Zelevinsky, *Nucl. Phys.* **A464** (1987) 237.

[60] V. F. Dmitriev and V.B. Telitsin, *Yad. Fiz.* **50** (1989) 357.

[61] A. Bracco, J. R. Beene, N. Van Giai, P. F. Bortignon, F. Zardi and R. A. Broglia, *Phys. Rev. Lett.* **60** (1988) 2603.

[62] N. Van Giai, P. F. Bortignon, A. Bracco and R. A. Broglia, *Phys. Lett.* **B233** (1989) 1.

[63] F. J. Dyson, *Journ. Math. Phys.* **3** (1962) 140, 157, 166.

[64] M. Robnik, *J. Phys.* **A20** (1987) L495.

[65] L. Molinari and V. V. Sokolov, *J. Phys.* **A22** (1989) L999.

[66] J. Ginibre, *Journ. Math. Phys.* **6** (1965) 3.

[67] F. J. Dyson, *Journ. Math. Phys.* **3** (1962) 1191.

[68] S. Tanaka and S. Sugano, *Phys. Rev.* **B34** (1986) 6880.

[69] R. Grobe and F. Haake, *Z. Phys.* **B68** (1987) 512.

[70] H. A. Weidenmüller, *Nucl. Phys.* **A518** (1990) 1.

Index of Contributors

Alhassid, Y. 41

Berry, M.V. 177, 183

Cassing, W. 143

Døssing, T. 239

Fröbrich, P. 93

Hannuschke, S. 143
Hashimoto, Y. 187
Herskind, B. 239

Itabashi, H. 187
Iwamoto, A. 125
Iwasawa, Y. 187

Marumori, T. 187
Matsuo, M. 219
Mosel, U. 143

Niita, K. 143

Sakata, F. 187

Terasaki, J. 187
Tsukuma, H. 187

Weidenmüller, H.A. 1, 11, 21, 31
Wolf, G. 143

Yamamoto, Y. 187

Zelevinsky, V.G. 255

MIX
Papier aus verantwortungsvollen Quellen
Paper from responsible sources
FSC® C105338

If you have any concerns about our products,
you can contact us on
ProductSafety@springernature.com

In case Publisher is established outside the EU,
the EU authorized representative is:
Springer Nature Customer Service Center GmbH
Europaplatz 3, 69115 Heidelberg, Germany

Printed by Libri Plureos GmbH
in Hamburg, Germany